Chemical History
Reviews of the Recent Literature

Chemical History
Reviews of the Recent Literature

Edited by

Colin A. Russell and Gerrylynn K. Roberts
The Open University, Milton Keynes, UK

RSC Publishing

Cover image Courtesy of David R Roberts

ISBN 0-85404-464-7

A catalogue record for this book is available from the British Library

Published by The Royal Society of Chemistry,
Thomas Graham House, Science Park, Milton Road,
Cambridge CB4 0WF, UK

Registered Charity Number 207890

For further information see our web site at www.rsc.org

Typeset by Macmillan India Ltd, Bangalore, India
Printed by Biddles Ltd, King's Lynn, Norfolk, UK

Dedication

Dedicated to the memory of

Dr W. A. ("Bill") Smeaton
1925–2001

much valued friend and
eminent co-worker in the history of chemistry

Preface

This volume is the natural successor to *Recent Developments in the History of Chemistry*,[1] edited by one of us, published in 1985. Its primary intention was to familiarize newcomers to the history of chemistry with some of the more important developments over the previous twenty years or so. We had specially in mind the large number of chemistry teachers who were at that time expecting to introduce at least a small amount of historical material into their curricula. While they might have been familiar with some of the older standard histories, or with historical introductions in school textbooks, they had no means of discovering the large amount of recent historical research or of sharing in its excitement. The original impetus came from a meeting of the Historical Group of the Royal Society of Chemistry though the actual work was given to the History of Chemistry Research Group at the Open University. Most of the writing, and all the editing, was undertaken by members of that Group, though we were glad to include several chapters from distinguished scholars beyond our ranks. The text was keyed in to what is now an ancient computer and was, we are told, the first book to be published by the Royal Society of Chemistry entirely from camera-ready copy.

We believe this publication is timely, not simply because it bestrides a convenient gap of twenty years since the previous volume. This is a moment when chemistry is being challenged as never before in modern times. Some of the reasons are undoubtedly located in history, and historical studies can offer highly relevant insights. In addition, they can be valuable in humanizing the task of chemistry teachers and, when judiciously used, can increase the appeal of the subject as a whole. The present President of the Royal Society of Chemistry has recently urged the need for a better perception of chemistry by the public and Government, and for better communication by chemists.[2] We hope this book may make a modest contribution to both those ends.

The original intention for the previous volume was that each chapter should both provide a literature update and at the same time tell a connected story. In the event that was possible for only a few chapters, largely because of the kind of material that emerged. Because we were writing chiefly for chemists, we had no compunction in using an unashamedly chemical framework for the book, as is also true for the present volume. It soon became evident that other needs were being incidentally met, and we were very glad that practising chemists not in teaching, as well as historians of science in general, found the book of some value. We hope this may continue to be the case.

Chemical History: Reviews of the Recent Literature continues many of these practices, and we have adopted a similar (though not identical) chapter plan. Once again, most authors are or have been full-time members of the same History of Chemistry

Research Group, though we are glad to welcome several others, including three new-comers to the project. The period covered is about the same, *i.e.* the twenty years since the appearance of *Recent Devlopments in the History of Chemistry*. We have kept to this limit fairly strictly but on some occasions have referred to publications before 1985 where there were special reasons to do so. This is especially true of Chapter 5 (on physical organic chemistry); since the previous volume had no chapter on that topic, the author had a more extended brief. Furthermore, owing to the unusually long period of gestation for this work, terminal dates vary slightly from chapter to chapter.

Another difference from the first volume is that we have had the use of a computerized literature search programme organized by the Royal Society of Chemistry Search Services and involving a world-wide scan *via Chemical Abstracts*. It has thrown up large numbers of papers and articles, and very occasionally books, which would never have come to our attention otherwise. We have to say that it has not been possible to check all the originals in Russian, Japanese and Chinese and one or two less common languages. In all other cases the originals have been thoroughly examined. English translations have been provided of all foreign language titles, where the text language is other than English, French or German. We have indicated that for the benefit for readers who might find the information helpful before ordering material by Inter Library Loan.

It has been our practice to use, as far as possible, Royal Society of Chemistry conventions in citing literature. Thus, initials rather than first names are employed and standard journal abbreviations are used. One exception to frequent scientific practice is that, in most cases, titles of papers and articles are added. We find that this custom from the historical literature has much to commend it in the present context.

Finally, as editors we wish to thank all our contributors for their work, for their patience in sometimes waiting interminably for editorial comment, and for their cooperation at all times. We specially wish to record our appreciation of the work of Alec Campbell (1917–1999); Chapter 3 reflects his deep engagement with the chemical literature to the time of his death. We are very grateful to the Open University for travel support, for a most generous use of Inter Library Loans and for funding for several years the considerable costs of the computerized literature survey. To individual members of the Library staff our gratitude is immense.

Colin A. Russell
Gerrylynn K. Roberts
May 2005

References

1. C. A. Russell, ed., *Recent Developments in the History of Chemistry*, Royal Society of Chemistry, London, 1985.
2. S. Campbell, *RSC News*, August, 2004.

Table of Contents

CHAPTER 1

Getting to Know History of Chemistry

COLIN A. RUSSELL AND GERRYLYNN K. ROBERTS

History of Chemistry Research Group, The Open University

1.1 Trends in History of Chemistry Literature

In the twenty or so years between the appearance of this volume and that of its predecessor,[1] the history of chemistry may be said to have undergone something of a sea-change. Partly, of course, this reflects changing fashions or 'trends', as it might be more polite to call them, within the history of science itself. A strong tendency to provide exclusively social explanations for scientific events was already under fire in the 1980s and the extreme form of sociological reductionism observable in the 1960s is now rarely seen. Instead, we have a much broader approach, with a widespread recognition that, while science undoubtedly is a social phenomenon, it is certainly much more than that. Hence respect is given to all kinds of interpretative schemes, including also those that take seriously the internal structures of scientific theory and practice.

Some of the most impressive work in the history of chemistry over the last twenty years has been performed by people who are, or have been, chemists by profession. This is a remarkable swing of the pendulum. Such efforts should not necessarily be consigned to the dustbin of 'Whiggishness' (where all history is read from the standpoint of the present day and seen as an almost inevitable progress to the current situation of triumph and success). Some contributions by workers active in chemical research demonstrate a quite sophisticated approach to history (even their own). These stand in the tradition of 19th-century historians of science who, contrary to popular belief, were not in their dotage and were very alert to events that proved unfavourable to their own approach to chemistry.[2] For a critique of the rather dated view that there is something inferior about history written by professional chemists, it would be hard to recommend anything more appropriate than a brief essay on 'Historians of chemistry' in a book by the eminent biochemist Joseph Fruton.[3] Another writer facing this issue is R. W. Cahn, author of a history of materials science,[4] a closely related field to chemistry. Examining the question as to whether 'it is acceptable for someone trained as a scientist' to write the history of a science, he comments on historical treatments of his own science, which have all been written by working scientists. His conclusion represents a substantial body of opinion: it is 'far better to be read by other scientists, who may on the whole be assumed to be in favour of science as a vocation'.[5] Today, controversy between self-styled 'historians of chemistry' and those they dubbed 'chemist-historians' has been subsumed in an alliance in which the latter hold a distinguished place. It is striking how frequently those who write as chemists use well-developed techniques of historical research, bringing fresh insights as they look back on their own subject. So let us see how in practice the history of chemistry has altered in character during the last couple of decades.

In the first place, *historical interest appears to have shifted from the classical to the modern period*, and especially to the very recent past. Thus a new historical interest in physical organic chemistry is so considerable that Chapter 5 is devoted to it alone. There must be several reasons for this stress on 20th-century developments and it is tempting to speculate. It could be argued, for example, that there is little new to learn about classical branches of chemistry and that the subject is virtually exhausted. However, some major incursions in the last few years have demonstrated the fallacy of such a view, for in fact the development of (say) organic chemistry has been far more complex than most of us have realized until quite recently. In its institutional aspects, there is much to be learned about the design and construction of laboratories and about the growth and decline of individual schools all over Europe. However, it takes time for these perceptions to take root.

Another explanation for the retreat to modernity may well lie in the need for contemporary chemists to gain public acceptance at a time when, partly for environmental reasons, the whole of chemistry is in some state of disrepute.[6] One way to win favourable public attention is to display recent research in the best light, and that may sometimes be accomplished by retailing a story of gradual progress until the present enlightenment has arrived. This would help to explain the large number of references in this volume to work by chemists on the history of their own branch of study. The remarkable recent emphasis on fullerenes may not be unconnected with the award of the 1996 Nobel Prize in chemistry to (and subsequent Knighthood of) Harold Kroto and (in Britain at least) with political mileage to be gained by such publicity.

Furthermore, the emergence of oral history as a systematic technique of historical investigation may have favoured interest in the history of contemporary chemistry. Properly handled interviews of practitioners of chemistry of the recent past can provide a rich repository of historical source material on the informal and affective aspects of doing science, which seldom enter the written record.[7,8] A study of the origins of the Krebs cycle is a powerful example of the potential of oral history.[9] The specialist history of chemistry centre, the Chemical Heritage Foundation in Philadelphia, has undertaken a systematic programme of chemical oral histories.[10]

A second feature of the current literature is a *strong emphasis on chemical biography*. Biographical writing in general has acquired a new vogue, even eclipsing the novel in some quarters. History of science is no exception,[11] though the number of major biographies of important chemical figures is relatively small. However, there are plenty of essays on a smaller scale that describe the lives of lesser mortals; and there are some that attempt two-page coverage of chemical giants. Such ambitious cameos are not necessarily to be derided. They may well convey to a newcomer far more vivid impressions than lengthy tomes replete with footnotes could hope to do. Indeed, the professional historian may have something to learn from them in terms of communication skills and even content.

A sub-set of chemical biography is of course that written by the subjects themselves. The current efflorescence of autobiographical writing by chemists may well be a further reflection of their desire to portray their science in a good light and so help in its general promotion. The most conspicuous array of autobiographies is the large number published in the American Chemical Society series "Profiles, Pathways and Dreams: Autobiographies of Eminent Chemists", a milestone in preserving our knowledge of many of the greatest figures in the subject. Edited by J. I. Seeman, whose portrait is curiously the first one encountered in every volume, they are rather variable in quality and interest, and most do not approach the level of discerning analysis that historians of science have come to expect. Yet they are most valuable as primary sources, and doubtless will be used as such in the future by new generations of grateful historians.

A third characteristic of many of the works cited in this book is the *strong emphasis on obviously useful chemistry.* This, of course, is what one might expect from a strategy of popularization through history. Many people believe that environmental attacks on the subject may best be met with positive statements as to the positive good imparted by chemical research and application. There is therefore much activity in the history of medical chemistry (Chapter 8) and also of industrial chemistry. However, utilitarian aspects also appear as the history of all kinds of other topics is recounted. Often this is entirely justified by the facts, but sometimes one may wonder how far the treatment is determined by considerations of public image or even public and private funding. In such cases it is wise to adopt an attitude of judicious scepticism. A concern for the image of chemistry is indeed one reason why chemists are interested in the history of their subject.[12] As chemistry becomes ever more specialized and in danger of losing its core identity, so history becomes a way of reasserting that identity.[13,14]

We are now in a position to examine a whole battery of scholarly approaches available to modern students of the history of chemistry. There is no shortage of material!

1.2 General Histories of Chemistry

For a general overview of the history of chemistry the obvious place to start is the numerous books that may be called general histories of the subject. The daunting nature of the task has not inhibited occasional attempts to chronicle the whole of history of chemistry for nearly two centuries. As the subject gets bigger, the technicalities become more complex and the history gets longer, authors of such works are faced with a project of exponentially increasing difficulty. Yet, in the last twenty years, no fewer than six major efforts have been made to supply up-to-date histories of chemistry. Before mentioning these, it may be helpful to glance back at the situation in the quarter century before 1985.

The 1960s saw seven brave efforts to 'tell the whole story', as one publisher put it. First there were two books in series dealing with scientific discovery in general. Neither offered footnotes or even bibliography, though one provided a simple but clear text[15] and the other compensated with lavish illustration.[16] On a different scale, and of immense academic value, was a trio of works that received attention in the first volume but must certainly be noticed here. From the USA came a magisterial work by Aaron Ihde that has recently been reprinted,[17] and another by Henry Leicester.[18] Even more ambitious was the four-volume *History of Chemistry* by J. R. Partington, published in London and probably the most valuable single tool ever provided for historians of chemistry.[19] Attacked by a few on the grounds that it offered facts rather than interpretations (which is certainly true), it nevertheless presents such a staggering wealth of material that it is hard to see how it can be eclipsed. Certainly none of the six more recent histories comes anywhere near it for detail of information, nor do they attempt or claim to do so. Unlike them, however, it may be hard to acquire copies (it is long out of print), it cannot be read as a consecutive narrative, and it offers no synthesis of the historical data.

Then, in the 1990s, came a veritable explosion of general histories of chemistry. Each has its merits and its limitations and the following account will inevitably reflect the predispositions of the present authors. For that reason, readers are specially referred to independent reviews, particularly those in *Ambix*, volume 40.

First, a book originally in French provides a healthy antidote to Anglocentricity and offers an excellent treatment of such figures as Berthollet, Gay-Lussac and (of course) Lavoisier.[20] It will appeal particularly to historians seeking a fresh approach to their subject. The same may also be true of a book by David Knight, whose title conveys a more accurate flavour of its contents than does its sub-title[21]. Various characteristics of chemistry are described in a very roughly chronological order (occult, mechanical, independent, fundamental and so on, to finish with a subject that is teachable, reduced and finally a service science). It has been criticized (by a French reviewer!) as being too British in emphasis but commended as a good introduction for non-chemists. It is clearly intended to be read within the context of the humanities.

An almost polar opposite to Knight's work is one whose title might have conveyed the impression of being rather similar. Entitled *The Historical Development of Chemical Concepts*, and written by a distinguished Polish chemist,[22] it conveys chemical insights with accuracy, but it sits as lightly to recent historical scholarship in the West as the previous author does to developments in 20th-century chemistry. Chemists will find it challenging, and many readers will learn much that is new about the development of chemistry in Poland.

Three other books, in different ways, will serve as good introductions to beginners. A text by John Hudson[23] is explicitly aimed at chemistry teachers and provides a straightforward internalistic account of the growth of chemistry with little regard to the social or even the biographical dimensions of the subject. Some will regret a tendency to old-fashioned positivism, though within its chosen limitations it is clearly successful. A more catholic treatment awaits the reader in Bill Brock's ambitious *Fontana History of Chemistry*, appearing at almost the same time.[24] Here is a rather longer account (744 pages), organized on a series of themes derived from important chemical texts, and displaying insights from the social history of science as well as a courageous effort to interpret historically quite modern developments. Graced with a lengthy bibliographical essay it must be one of the best introductions for a beginner who intends to prosecute the subject with due seriousness. Yet a book from America, which offers a general history of chemistry at a fairly popular level, may well turn out to be the best for an absolute beginner.[25]

It is thus apparent that the 'best' general history of chemistry depends largely on the intended readership. Without in any way limiting the scope of these six volumes it is possible to summarize the single most appropriate choice for a given *clientèle* as follows:

Newcomers	*Cobb and Goldschmidt*[25]
Chemistry teachers	*Hudson*[23]
Chemists	*Mierzecki*[22]
Humanities scholars	*Knight*[21]
Historians of chemistry	*Bensaude-Vincent / Stengers*[20]
General use	*Brock*[24]

Prices vary enormously, the volume by Mierzecki being approximately ten times as expensive as that by Brock.

Finally, several older histories of chemistry have recently appeared as reprints. One is a classic from 1866 and 1869, the famous *Histoire de la chimie* of Ferdinand Hoefer, reprinted in 1980.[26] Another is the Clows' classic of 1952, *The Chemical Revolution*.[27] It does not deal with the Lavoisierian transformation of chemistry (as might have been expected) but with the revolutionary changes in applied chemistry that made the Industrial Revolution possible.

1.3   Other General Works

In addition to complete histories of the science, several other types of general works may be noted. First, there are several volumes of *collected essays* that deal, mainly or exclusively, with the history of chemistry. An example of these is the published proceedings of the 'Symposium on Alchemy, Chemistry and Pharmacy', held during the International Congress of History of Science at Liège in 1997.[28] There is also a volume of essays on the history of chemistry in the multi-volume Italian publication, *Storia della scienza*.[29] The five-year European Science Foundation Programme, 'The Evolution of Chemistry in Europe, 1789–1939',[30] which was launched in 1993, has resulted in a number of collections of essays: on Lavoisier,[31] on aspects of the chemical industry,[32–35] on chemical textbooks,[36] and on chemical education and institutions.[37–39]

Furthermore, there is a series of reprints of original papers that in some cases originated well before 1985. They are usually unedited, often with the original pagination. Nevertheless, the Variorum series, in particular, provides a useful way to access and store such material. A collection by Allen Debus relates mainly to chemistry in its early phases[40] and is a highly authoritative introduction to such matters. Debus has also recently edited a collection of 'classic' articles on alchemy and early chemistry reprinted from *Ambix*, the Journal of the Society for the History of Alchemy and Chemistry.[41] Another volume in the Variorum series relates to a later period and reflects the author's well-known interest in the triple relationships between chemistry, ideas of nature, and society in the century or so following the French Revolution.[42] A third, by John Brooke, displays his concern for the history of ideas within chemistry, almost entirely within the 19th century.[43]

Secondly, there are *books that cover the history of chemistry in particular places.* The development of chemical science in modern China is the theme of one recent book that is probably the only source in the West of such information.[44] Setting a new trail in analysis of national trends was the multi-authored book *Chemistry in America 1876–1976.*[45] This presented a vast mass of empirical data about most conceivable aspects of American chemistry: vocational, professional, educational and industrial. Like the *History of Chemistry* by Partington, to which it bears a curious resemblance in its fact-collecting skills and deadpan presentation, it is unlikely to be replaced for a long time. As will be seen in subsequent chapters there are many smaller studies available of aspects of chemistry in individual countries or regions. One of general interest is a discussion on chemistry in 19th-century Baden,[46] while another localizes chemistry to Essex and East London.[47]

Thirdly, several books have appeared in the last two decades taking a *single chemical theme* and charting its development over, sometimes, many centuries. A recent example concerns itself with the role of language in chemical change and particularly focuses on the work of Lavoisier.[48] A broader treatment of the same theme comes in a discussion on the language of chemistry from the beginnings of alchemy to c.1800.[49] Just before 1985 came a book of similar genre engaging with the interpretation of form, notably in mineralogy, crystallography, and its wider relationships with chemistry.[50] On a smaller scale there is a paper on the aether in late 19th-century chemistry.[51] Other examples will be encountered in chapters dealing with the individual branches of chemistry, such as organic and inorganic.

A fourth kind of work is the *period study*: Chemistry in general is considered within a restricted time-span. A collection of essays dealing with the rather wide theme of law and order in 18th-century chemistry has appeared.[52] A book about chemistry in Britain from 1760 to 1820 includes an examination of chemistry in the Scottish Enlightenment, the rise of pneumatic chemistry, the work of Priestley and Davy, and a study of the London chemical community.[53] Despite possible unfamiliarity with the author's sociological emphasis, chemists will find useful material on the way their subject interacted with the culture of that critical period. A work of a very different kind traces the changing relationship between chemistry and physics and, largely avoiding the social history of science, presents an ingenious and well-researched synthesis of ideas in the two sciences.[54] The author, Mary Jo Nye, has also written another book that, by its intriguing sub-title, implies a triple concern with a period (1800–1850), a theme of chemical change (dynamics of matter) and a social dimension (dynamics of disciplines).[55] An unusual paper covers a specific period and place in a study of chauvinism in 19th-century European chemistry.[56]

A different kind of general treatment relates to *source materials.* It is very important, though often quite difficult, to know just what is available in the secondary literature. The best historical works have either full literary references in footnotes, or separate bibliographies or both. To some extent, it is hoped that the present book will meet that need, but other assistance is available. For example, a book published in

Germany, though written in English, offers bibliographical help for the 17th, 18th and 19th centuries.[57] Similar works in related fields have much to offer the historian of chemistry itself, as is the case with a bibliography of pharmacy.[58] Sometimes the bibliography may be narrowly focussed, but extremely useful none the less; a recent example is that relating to the Scottish chemist Joseph Black.[59] Ultimately, of course, the historian of chemistry or of anything else has to depend on those hidden primary sources generally termed archives. They need to be conserved and responsibly handled.[60] A handlist of the archives of the chemical industry discloses vast amounts of relevance to ordinary chemical history in deposits left or retained by industrial firms.[61]

1.4   Chemical Biographies

As noted above, chemical biography has acquired a new popularity and many examples will occur in the chapters that follow. Meanwhile it is convenient to note a few developments in what might be called general chemical biography. There are important studies of several men in the last 200 years or so who may justly be called 'universal chemists'. Their work has affected chemistry as a whole, none more obviously so than A. L. Lavoisier. His work is discussed in Chapter 2 and in a complete issue of *Ambix*.[62] A new biography has appeared in French,[63] and another in English should be welcomed by the general reader.[64] European reactions to Lavoisier, particularly in Italy, Poland, Belgium, and France itself are described in a recent collection of papers.[65]

An opponent of Lavoisier was Joseph Priestley, whose life has been ritually celebrated at the so-called 'Priestley Conferences' organized jointly by BOC and the Royal Society of Chemistry. The last occasion for a sustained analysis of his life and work was at the 250th anniversary of his birth, just before the period covered by this book.[66] There has also been a reprinting of two of Priestley's political works, though they say little about his science.[67] Another chemist who was reluctant to accept the new chemistry (at least at first) was Humphry Davy, and another two volumes have been added to the long list of his biographies.[68,69]

Influenced by Davy, and like him entitled to be seen as a founder of electrochemistry, the Swedish chemist J. J. Berzelius has also been the subject of much biographical attention. He was a man of enormous influence through his widely-translated textbooks, his network of former research students, an immense correspondence and his famous annual reports, the *Jahresberichte*. In addition to a substantial literature on Berzelius in the 1970s and early 1980s, there has been a book on his travels in France,[70] chiefly of interest to mineralogists, and another on his close collaboration with H. G. Trolle-Wachtmeister (1782–1821).[71] The latter, in Swedish with an English summary, is remarkable for its detailed account of laboratory practice, where again mineralogy looms large. A student and disciple of Berzelius has also received a biography: E. Mitscherlich, famous for his law of isomorphism.[72]

Another chemical 'giant' probably wins the prize for the greatest number of biographical studies in recent years. The founder of Europe's first major chemical research school, at Giessen, was Justus Liebig. A recent book describes how chemistry developed in the German states before 1840 and includes Liebig's provocative essay on the (poor) condition of chemistry in Prussia.[73] It includes forty documents from the Prussian state archives. Liebig played an immensely important part in the establishment of agricultural chemistry.[74] One recent paper stresses the 'factual' basis of history in archival deposits and differentiates such an account from the more conventional 'apocryphal' legends associated with this charismatic figure.[75] A further attempt at demythologization considers Liebig's claims concerning his *Agricultural Chemistry*.[76] There is an excellent account of Liebig as a social climber, greatly helped by the profits from his famous Extract of Meat,[77] and another on his relationship with the pharmacist P. L. Geiger.[78]

It is, however, in the correspondence of any individual that we see most clearly what they were really like, and in this respect also Liebig has received much attention. Since 1988, no fewer than six sets of his letters have been published. His correspondence with Berzelius was first published by Carrière, but now a facsimile is available of that 1898 edition.[79] Other correspondents are recalled in Liebig's exchanges with Schloßberger (a co-inventor of the word ester),[80] with the Brunswick pharmacist F. J. Otto,[81] with the Giessen University chancellor von Linde,[82] and with Hermann Kolbe.[83] Most important of all, perhaps, are the volume of correspondence between Liebig and Hofmann, edited by W. H. Brock,[84] and the same author's excellent full-length biography of the German chemist.[85]

Chiefly remembered for his reinstatement of Avogadro's hypothesis and his subsequent reform of atomic weights was the Italian chemist Stanislao Cannizzaro. His life and (especially) his work have been commemorated in a book by John Bradley, appearing in German[86] and English[87] editions. There have also been three volumes of Cannizzaro correspondence, edited by L. Paoloni[88–90]

A volume that comes halfway between a chemical biography and a thematic history is one on William Prout and the nature of matter, taking the subject to 1985, long after Prout's demise.[91] Another combines biography with national trends in chemistry. This is a life of the Melbourne chemist D. O. Masson, contriving at the same time to provide a helpful analysis of the growth, organization, and professionalization of the subject in Australia.[92]

A chemist whose achievements straddled the territories of organic, inorganic and physical chemistry was Edward Frankland. Also responsible for popularizing modern chemical notation, the laboratory instruction of students, the identification of pollutants in domestic water supplies and the doctrines of the 'chemical bond', Frankland was perhaps the leading chemist of Victorian England. His biography has been published,[93] following the discovery of several very large deposits of his letters, notebooks *etc.*, nearly all in private hands. The methodology of handling such archival material has also been described.[94]

Several scientists have been commemorated whose work has been of equal importance for physics and chemistry. Most notable, perhaps, has been a continuing interest in Michael Faraday. A complete issue of the *Bulletin for the History of Chemistry* is devoted to Faraday's chemistry.[95] His correspondence is currently being published in a far fuller form than before, the first volume[96] having appeared in 1991. A good deal of the material from this early period relates directly to chemistry. Faraday's work at the Royal Institution included chemistry and physics and is succinctly described by a former Director.[97] Chemical matters are less prominent in two other publications on the life of Faraday. Geoffrey Cantor's valuable book concentrates on Faraday's religion[98] and a collection of essays explores miscellaneous aspects of his life and work.[99] A large biography of Rutherford,[100] dating back to 1983, has been followed by an edited volume of essays on his colleague Frederick Soddy.[101] The inventor of the isotopic tracer technique, and a co-discoverer of hafnium, was George de Hevesy. A biography has appeared,[102] together with the reports of a Festschrift in his honour.[103]

Finally several general sources may be mentioned. Amongst recent collected biographies is one dealing with Nobel Laureates in chemistry,[104] and another with 154 chemists honoured by the American Institute of Chemists, from 1926 on.[105] A second volume of a series on American chemists and chemical engineers has appeared, eighteen years after the first.[106] Chemists continue to have a prominent part in a new edition of *A Biographical Dictionary of Scientists* (with a modified title).[107] A recent Supplement to the *Dictionary of National Biography* includes a number of chemists who had somehow escaped inclusion in the original volumes (including Crum Brown, Couper, Fenton, Lapworth, Newlands, Odling, Sugden, and Tilden, as well as people associated with industrial chemistry like Allhusen, Brunner, Castner, Deacon, Gamble, Gilchrist, Glover, Gossage and Griess).[108] The brand new edition of this famous multi-volume work, appeared in 2004 and contains many more chemists from the modern period. Biographical accounts of chemists recently deceased may of course be found in obituary notices, though only the most distinguished will find space in the crowded columns of *The Times* or in *Biographical Memoirs of Fellows of the Royal Society*. Others may be found in *Chemistry in Britain*, or (from 2004) in its successor *Chemistry World* or in *RSC News*. Usually though, these are so short as to be of little value; an index of such obituaries may be found in the Library of the Royal Society of Chemistry in London. The now defunct *Journal of the Royal Institute of Chemistry* did rather better, as does the *Journal of the Society of Chemical Industry*.

Finally, though the philosophy of science may or may not have historical elements, a new journal has notably appeared, *Hyle*, aiming to deal with philosophical matters relating to chemistry. It is a challenge to those holding a contrary if conventional view that 'chemistry doesn't have a philosophy'. Aspects of the philosophy of chemistry may be found in Brooke's volume of collected essays.[43] Several papers have emphasized chemistry in France. This is hardly surprising when we recall that here was the home of Comte, father of positivism, a theme that permeated the thought of G. Wyrouboff.[109] A more general study attempts a fresh look at the whole question,[110] and a further investigation into modern French chemistry comes to the astonishing conclusion that it was still being impeded by positivist scepticism until the 1950s, when in some places atomism was still regarded with hostility. Not surprisingly, it suggests that organic chemistry was seriously impeded.[111]

1.5   History of Chemistry and the Internet

One of the most significant changes in research and teaching of our subject since the appearance of *Recent Developments in the History of Chemistry* has been the emergence and explosive growth of the world wide web and the internet. For making scholars aware of what is available in traditional repositories such as libraries and archives, as a repository itself of source materials on the subject, and as a means of publication, the world wide web, together with the means of communication afforded by the internet, is transforming activity in the history of chemistry, which is of course also the case in every other subject, including chemistry itself. To be sure, for many historical topics, physical rather than virtual contact with resources will remain essential. There is also an affective dimension to physical contact with resources, which forms part of the enjoyment of historical research. Indeed, reading at length on screen is not necessarily a congenial activity and not necessarily the best way to make use of materials. However, it is undeniable that, used appropriately, the world wide web and the internet can greatly facilitate research. In what follows, addresses for accessing websites (URLs) are given sparingly, not least because it is a hazard of writing about the web that no sooner is an address printed than it changes. More importantly, history of chemistry websites are now so numerous that any attempt at a comprehensive printed listing would be unwieldy, the web itself is the best place for such information. Concentrating on resources that do not require library or institutional subscriptions for access, the aim here is to give a few lead references to various categories of resource to help readers get started.

The most immediate impact of the world wide web has been to make available catalogues of libraries and archives of major academic and other institutions. Thus precious, and still essential, research trips to facilities distant from a scholar can be well planned and efficiently executed. As national and international digitization projects[112] come to fruition over the next decade, the experience of engaging with collections is likely to change considerably as it will become possible for materials to 'travel' to scholars rather than essential for scholars to travel to all materials. Already, many journals in the field have online as well as paper versions, while some journals are now available only online.

Another useful category of web resource for the researcher is the bibliography. Indeed, paper-based bibliographies, such as this volume and its predecessor, may become obsolete over the next decade. Unlike library catalogues, which are produced and checked by institutional professionals and therefore present information to a high standard that will be up to date, bibliographies published on the web are of varying accuracy and may be dependent on the enthusiasm of dedicated individuals for updating. As with any source, it is important for the user to be aware of its provenance, purpose, and authority. The *Wellcome Bibliography for the History of Medicine* (formerly *Current Work in the History of Medicine: An International Bibliography*), produced by the Wellcome Institute for the History of Medicine has a much wider remit than its title indicates.[113] An important online bibliography is that produced under the auspices of *Hylé: International Journal for Philosophy of Chemistry*.[114] It is organized by topic, with individual topic bibliographies having been compiled by eminent researchers in the history of chemistry. There are also bibliographic components to web-projects on Boyle[115] and Lavoisier,[116] both of which provide additionally a range of resources relating to these individuals, including digitized primary sources. In addition to providing resources about Liebig, the Liebig Museum's website includes a virtual tour of Liebig's laboratory.[117]

An efficient way to locate internet resources is to use 'gateway' websites. In the early days of the world wide web, these tended to be compiled by enthusiastic individuals who saw the potential of the medium, were motivated to deploy it in their own teaching, and were willing to share their efforts outside their own institutions. Several sites containing considerable numbers of history of chemistry links have been compiled by members of university chemistry departments. One of the earliest and still most useful gateway sites is 'Classic Chemistry'.[118] Designed to complement an individual's own college-level teaching, it makes no claims to be comprehensive. The site consists largely of transcribed, annotated texts of classic papers (some full-text, some only excerpted) and links to a wide range of other history of chemistry websites. In addition to the links, among its very useful features is a glossary of archaic chemical terms and a history of chemistry calendar of anniversaries. Other resources designed for the support of secondary school teaching are provided by the Chem Team website,[119] which is organized around chemistry tutorial topics, but also includes a section on 'Classic Papers' and a chemists' 'Photo Gallery'.

A welcome trend now is the development of official, institutional sites, which are less dependent on the energies of enthusiastic individuals. An important international gateway site for the history of science is run by ECHO (exploring and collecting history online) at George Mason University's Center for History and New Media.[120] In the UK, the Resource Discovery Network (RDN) is 'dedicated to providing high quality Internet resources for the learning, teaching, and research community'.[121] A wide range of history of chemistry links is provided by the Physical Sciences Information Gateway (PSIgate),[122] which is managed by a consortium of academic libraries in Manchester under the auspices of the RDN. The Chemistry Biology Pharmacy Info Center[123] is a similar Swiss site. Most museums now also have useful and informative sites. For example, it is possible to study a number of topics in the history of chemistry *via* the National Museum of Science and Industry's 'Ingenious' project, which 'brings together images and viewpoints to create insights into science and culture', using some 30 000 images from the Museum's collections.[124] The Royal Society of Chemistry Library's website has links to a range of the RSC's historical resources.[125]

Furthermore, the organizations that deal with the history of chemistry all now have websites, which individuals interested in the subject can consult for information on activities and events as well as, in some cases, for access to materials. The Society for the History of Alchemy and Chemistry,[126] which publishes the journal *Ambix* is the 'senior' scholarly organization in the field, having been established in 1937. The Royal Society of Chemistry,[127] the American Chemical Society[128] and the Gesellschaft Deutscher Chemiker[129] all have specialist history of chemistry groups. The ACS publishes the *Bulletin for the History of Chemistry*, while the DCG Fachgruppe Geschichte der Chemie has published its *Mitteilungen* since 1988. The European Association for Chemistry and Molecular Sciences' (EuCheMS) Working Party on the History of Chemistry's website[130] reports on its projects, especially its Millennium Project. The website of the major specialist history of chemistry research and resource centre – the Chemical Heritage Foundation[131] in Philadelphia – is an important resource in itself.

Finally, the potential of the internet for encouraging discussion of the history of chemistry should be mentioned. Several of the groups listed above help to support an e-mail-discussion list that is open to anyone interested in the subject.

CHEMHIST[132] is an international electronic forum and news bulletin set up to carry information and discussion related to the history of chemistry and chemical industry. Founded in 1997, it soon became the 'official messenger' of the IUHPS/DHS Commission on the History of Modern Chemistry (CHMC).

References

1. C. A. Russell, ed., *Recent Developments in the History of Chemistry*, Royal Society of Chemistry, London, 1985.
2. C. A. Russell, '"Rude and disgraceful beginnings": a view of the history of chemistry from the nineteenth century', *Br. J. Hist. Sci.*, 1988, **21**, 273–294.
3. J. S. Fruton, *A Skeptical Biochemist*, Harvard University Press, Cambridge, MA, 1992, pp. 173–181.
4. R. W. Cahn, *The Coming of Materials Science*, Pergamon, Oxford, 2001.
5. R. W. Cahn, 'The history of physical metallurgy and of materials science', *Acta Metall. Sin.* 1997, **33**, 157–164.
6. C. A. Russell, 'Chemical industry and the environment: a view from history', *Chem. Australia*, 1996, **63**, 265–266.
7. I. Hargittai, *Candid Science [I] Conversations with Famous Chemists*, ed. M. Hargittai, Imperial College Press, London, 2000.
8. I. Hargittai, *Candid Science III: More Conversations with Famous Chemists*, ed. M. Hargittai, Imperial College Press, London, 2002.
9. F. L. Holmes, *Hans Krebs*, 2 vols, Oxford University Press, Oxford, 1994.
10. Chemical Heritage Foundation, "Oral History Collection", http://www.chemheritage.org/exhibits/ex-nav2.html [accessed 15 May 2005].
11. M. Shortland and R. Yeo, eds, *Telling Lives in Science: Essays on Scientific Biography*, Cambridge University Press, Cambridge, 1996.
12. C. A. Russell, 'Chemistry in society', in *The New Chemistry*, ed. N. Hall, Cambridge University Press, Cambridge, 2000, pp. 465–484.
13. W. H. Brock and G. K. Roberts, 'Chairman's and Editor's Remarks', *Ambix*, 1994, **41**, 1–3.
14. B. Bensaude-Vincent and I. Stengers, *Histoire de la chimie*, Éditions la Découverte, Paris, 1993; trans. by D. van Dam, Harvard Univiversity Press, Cambridge, MA, 1997.
15. B. S. Morgan, *Men and Discoveries in Chemistry*, John Murray, London, 1962.
16. C. -A. Reichen, *A History of Chemistry*, Leisure Arts Ltd., London, 1964.
17. A. J. Ihde, *The Development of Modern Chemistry*, Harper & Row, New York, 1964.
18. H. M. Leicester, *The Historical Background of Chemistry*, Dover, New York, 1971.
19. J. R. Partington, *A History of Chemistry*, 4 vols, Macmillan, London, 1961–1970.
20. B. Bensaude-Vincent and I. Stengers, *Histoire de la chimie*, Éditions la Découverte, Paris, 1993; trans. by D. van Dam, Harvard Univiversity Press, Cambridge, MA, 1997.
21. D. M. Knight, *Ideas in Chemistry: A History of the Science*, Athlone Press, London, 1992.

22. R. Mierzecki, *The Historical Development of Chemical Concepts*, Kluwer, Dordrecht, 1991.
23. J. Hudson, *The History of Chemistry*, Macmillan, Basingstoke, 1992.
24. W. H. Brock, *The Fontana History of Chemistry*, Fontana, London, 1992.
25. C. Cobb and H. Goldschmidt, *Creations of Fire*, Alumni, New York, 1995.
26. F. Hoefer, *Histoire de la chimie*, 2 vols, 2nd edn., reprint, Gutenberg, Paris, 1980.
27. A. Clow and N. L. Clow, *The Chemical Revolution*, Batchworth, London, 1952; reprinted by Gordon & Breach, Philadelphia, 1992.
28. M. Bougard, ed., *Alchemy, Chemistry and Pharmacy*, Proceedings of the XXth International Congress of History of Science, Liège, 1997, vol. XVIII, Brepols, Turnhout, Belgium, 2002.
29. D. M. Knight, ed., *Storia della scienza*, ed.-in-chief, S. Petruccioli, Istituto della Enciclopedia Italiana, Rome, 10 vols, 2001–2004, 2003, vol. VII.
30. C. Meinel, 'Modern chemistry in a historical perspective', *Reflections: Newsletter of the [ESF] Standing Committee for the Humanities*, 2000, **4**, 11-15.
31. F. Abbri and B. Bensaude-Vincent, eds, *Lavoisier in European Context: Negotiating a New Language for Chemistry*, Science History Publications, Canton, MA, 1995.
32. R. Halleux and E. Homburg, eds, 'Strategies of chemical industrialisation: from Lavoisier to Bessemer', *Arch. Int. d'Hist. Sci.*, Special Issue, 1996, **46**, 1–125.
33. R. Fox and A. Nieto-Galan, eds, *Natural Dyestuffs and Industrial Culture in Europe, 1750–1880*, Science History Publications, Canton, MA, 1999.
34. A. S. Travis, H. G. Schröter, E. Homburg, P. J. T. Morris, eds, *Determinants in the Evolution of the European Chemical Industry, 1900–1939: New Technologies, Political Frameworks, Markets and Companies*, Kluwer Academic, Dordrecht, 1998.
35. E. Homburg, A. S. Travis and H. G. Schröter, eds, *The Chemical Industry in Europe, 1850–1914: Industrial growth, Pollution and Professionalization*, Kluwer Academic, Dordrecht, 1998.
36. B. Bensaude-Vincent and A. Lundgren, eds, *Communiction in Chemistry Textbooks and their Audiences, 1789–1939*, Science History Publications, Canton, MA, 2000.
37. A. L. Janeira, M. E. Mara and P. Pereira, eds, *Demonstrar ou Manipular? O Laboratório de Química Mineral da Escola Politécnica de Lisboa na sua Época (1884-1894)*, Livraria Escolar Editoria, Lisboa, 1996.
38. D. M. Knight and H. Kragh, eds, 'Aspects of European Chemistry, 1900–1940', *Centaurus*, Special Issue, 1997, **39**, 291-381.
39. D. M. Knight and H. Kragh, eds, *The Making of the Chemist: The Social History of Chemistry in Europe, 1789–1914*, Cambridge University Press, Cambridge, 1998.
40. A. G. Debus, *Chemistry, Alchemy and the New Philosophy, 1550–1700: Studies in the History of Science and Medicine*, Variorum, Aldershot, 1987.
41. A. G. Debus, ed., *Alchemy and Early Modern Chemistry: Papers from Ambix*, The Society for the History of Alchemy and Chemistry, London, 2005.

42. T. H. Levere, *Chemists and Chemistry in Nature and Society, 1770–1878*, Variorum, Aldershot, 1994.

43. J. H. Brooke, *Thinking about Matter. Studies in the History of Chemical Philosophy*, Variorum, Aldershot, 1995.

44. J. Reardon-Anderson, *The Study of Change: Chemistry in China, 1840–1949*, Cambridge University Press, Cambridge, 1991.

45. A. Thackray, J. L. Sturchio, P. T. Carroll and R. Bud, *Chemistry in America 1876–1976: Historical Indicators*, Reidel, Dordrecht, 1985.

46. F. A. J. L. James, 'Science as a cultural ornament: Bunsen, Kirchhoff and Helmholtz in mid-nineteenth-century Baden', *Ambix*, 1995, **42**, 1–9.

47. M. James, *The History of Chemistry in Essex and East London*, Essex Section Trust of the Royal Society of Chemistry, London, 1991.

48. M. Beretta, *The Enlightenment of Matter: The Definition of Chemistry from Agricola to Lavoisier*, Science History Publications, Canton, MA, 1993.

49. M. P. Crosland, 'The language of chemistry from the beginnings of alchemy to c.1800', art. 259 in *Fachsprechen: Languages for Special Purposes*, ed. L. Hoffmann, H. Kalverkämper and H. E. Wiegand, de Gruyter, Berlin, 1999, vol. 2, 2477–2485.

50. N. Emerton, *The Scientific Reinterpretation of Form*, Cornell University Press, Ithaca, NY, 1984.

51. H. Kragh, 'The aether in late nineteenth century chemistry', *Ambix*, 1989, **36**, 49–65.

52. A. M. Duncan, ed., *Laws and Order in Eighteenth-Century Chemistry*, Oxford University Press, Oxford, 1995.

53. J. Golinski, *Science as Public Culture: Chemistry and Enlightenment in Britain, 1760–1820*, Cambridge University Press, Cambridge, 1992.

54. M. J. Nye, *Before Big Science: The Pursuit of Modern Chemistry and Physics 1800–1940*, Twayne Publishers, New York, 1996.

55. M. J. Nye, *From Chemical Philosophy to Theoretical Chemistry: Dynamics of Matter and Dynamics of Disciplines, 1800–1950*, University of Califonia Press, Berkeley, CA, 1993.

56. A. J. Rocke 'Pride and prejudice in chemistry: chauvinism and the pursuit of science', *Bull. Hist. Chem.*, 1992/3, **13/14**, 29–40.

57. V. Wehefritz and Z. Kobats, eds, *Bibliography on the History of Chemistry and Chemical Technology, Seventeenth to the Nineteenth Century*, 3 vols, Saur, München, 1994.

58. J. H. Gregory and E. C. Stroud, *The History of Pharmacy: An Annotated Bibliography*, Garland, New York, 1995.

59. J. G. Fyffe and R. G. W. Anderson, *Joseph Black: A Bibliography*, Science Museum, London, 1992.

60. C. A. Russell, 'Records of chemistry: combustion or conservation?', *Bull. Hist. Chem.*, 1991, **9**, 3–7.

61. P. J. T. Morris and C. A. Russell, *Archives of the British Chemical Industry 1750–1914*, British Society for the History of Science, Faringdon, Oxfordshire, 1988.

62. *Ambix*, 1989, **36** (no.1).

63. B. Bensaude-Vincent, *Lavoisier, mémoires d'un revolution*, Flammarion, Paris, 1993.

64. A. Donovan, *Antoine Lavoisier: Science, Administration, and Revolution*, Blackwell, Oxford, 1993; and Cambridge University Press, Cambridge, 1996.

65. B. Bensaude-Vincent and F. Abbri, eds, *Lavoisier in European Context: Negotiating a New Language for Chemistry*, Science History Publications, Canton, MA, 1995.

66. J. Needham *et al.*, numerous papers in *Oxygen and the Conversion of Future Feedstocks*, Royal Society of Chemistry, London, 1984.

67. P. N. Miller, ed., *Joseph Priestley: Political Writings*, Cambridge University Press, New York, 1993.

68. D. M. Knight, *Humphry Davy: Science and Power*, Blackwell, Oxford, 1992.

69. J. Z. Fullmer, *Young Humphry Davy: The Making of an Experimental Chemist*, American Philosophical Society, Philadelphia, PA, 2000.

70. C. G. Bernhardt, *Through France with Berzelius: Live Scholars and Dead Volcanoes*, Pergamon Press, Oxford, 1985.

71. J. Trofast, *Excellensen och Berzelius: Hans Gabriel Trolle-Wachtmeisters kemiska verksamhet*, Atlantis, Stockholm, 1988.

72. H. W. Schütt, *E. Mitscherlich: Baumeister am Fundament der Chemie*, Deutsches Museum, München, 1992.

73. R. Zott and E. Heuser, eds, *Die streitbaren Gelehrten. Justus Liebig und die Preußischen Universitäten. Kommentierte Edition eines historischen Disputes*, ERS Verlag, Berlin, 1992.

74. U. Schling-Brodersen, 'Liebig's role in the establishment of agricultural chemistry', *Ambix*, 1992, **39**, 21–31.

75. P. Munday, 'Liebig's metamorphosis: from organic chemistry to the chemistry of Agriculture', *Ambix*, 1991, **38**, 135–154.

76. M. R. Finlay, 'The rehabilitation of an agricultural chemist: Justus von Liebig and the seventh edition', *Ambix*, 1991, **38**, 155–167.

77. P. Munday, 'Social climbing through chemistry: Justus Liebig's rise from the *niederer Mittelstand* to the *Bildungsbürgertum*', *Ambix*, 1990, **37**, 1–19.

78. U. Thomas, 'Philipp Lorenz Geiger and Justus Liebig', *Ambix*, 1988, **35**, 77–90.

79. W. Lewicki, *Berzelius und Liebig: Ihre Briefe 1831–1845*, Jürgen Cromm Verlag, Göttingen, 1991.

80. F. Heße and E. Heuser, *Justus von Liebig und Julius Eugen Schloßberger in ihren Briefen von 1844–1860*, Bionomica Verlag, Mannheim, 1988.

81. E. Heuser, ed., *Justus von Liebig und Friedrich Julius Otto in ihren Briefen von 1838-1840 und 1856–1867*, Bionomica Verlag, Mannheim, 1988 and 1989.

82. E. -M. Felschow and E. Heuser, eds, *Universität und Ministerium im Vormärz. Justus Liebigs Briefwechsel mit Justin von Linde*, Verlag der Ferber'schen Universitäts-Buchhandlung, Giessen, 1992.

83. A. J. Rocke and E. Heuser, eds, *Justus von Liebig und Hermann Kolbe in ihren Briefen von 1846–1873*, Bionomica Verlag, Mannheim, 1994.

84. W. H. Brock, ed., *Justus von Liebig und August Wilhelm Hofmann in ihren Briefen (1841–1873)*, Verlag Chemie, Weinheim, 1984.

85. W. H. Brock, *Justus von Liebig: The Chemical Gatekeeper*, Cambridge University Press, Cambridge, 1997.

86. J. Bradley, *Cannizzaro's Methode: der Schlüssel zur modernen Chemie*, Franzbecker, Bad Salzdetfurth, 1992.

87. J. Bradley, *Before and after Cannizzaro*, Whittles, Latheronwheel, Caithness, 1992.

88. L. Paoloni, ed., *Lettere a Stanislao Cannizzaro. Scritti e carteggi 1857–1862*, Seminaro di Storia della Scienza, Luglio, 1992.

89. L. Paoloni, ed., *Lettere a Stanislao Cannizzaro 1863–1868*, Seminaro di Storia della Scienza, Luglio, 1993.

90. L. Paoloni, ed., *Lettere a Stanislao Cannizzaro 1868–1872*, Seminaro di Storia della Scienza, Luglio, 1994.

91. W. H. Brock, *From Protyle to Proton: William Prout and the Nature of Matter 1785–1985*, Hilger, Bristol, 1985.

92. L. W. Weickhardt, *Masson of Melbourne: The Life and Times of David Orme Masson*, Royal Australian Chemical Institute, Parkville, VIC, 1989.

93. C. A. Russell, *Edward Frankland: Chemistry, Controversy and Conspiracy in Victorian England*, Cambridge University Press, Cambridge, 1996.

94. C. A. Russell and S. P. Russell, 'The archives of Sir Edward Frankland: resources, problems and methods', *Br. J. Hist. Sci.*, 1990, **23**, 175–185.

95. *Bull. Hist. Chem.*, 1991, no. 11.

96. F. A. J. L. James, ed., *The Correspondence of Michael Faraday*, vol. 1, *1811 - Dec. 1831*, Institution of Electrical Engineers, London, 1991.

97. J. M. Thomas, *Michael Faraday and the Royal Institution: The Genius of Man and Place*, IOP Publishing, Bristol, 1991.

98. G. Cantor, *Michael Faraday: Sandemanian and Scientist*, Macmillan, Basingstoke, 1991.

99. D. Gooding and F. A. J. L. James, eds, *Faraday Discovered: Essays on the Life and Work of Michael Faraday, 1791–1867*, Macmillan, Basingstoke, 1985.

100. D. Wilson, *Rutherford, Simple Genius*, Hodder & Stoughton, London, 1983.

101. G. B. Kauffman, ed., *Frederick Soddy (1877–1956): Early Pioneer in Radiochemistry*, Reidel, Dordrecht, 1986.

102. H. Levi, *George de Hevesy: Life and Work*, Hilger, Bristol, 1985.

103. G. Marx, ed., *George de Hevesy 1885–1966, Festschrift*, Akadémia Kiadó, Budapest, 1988.

104. L. K. James, ed., *Nobel Laureates in Chemistry 1901-1992*, American Chemical Society, Washington, 1993.

105. R. B. Seymour and C. H. Fisher, *Profiles of Eminent American Chemists*, Litarvan Enterprises, Sydney, 1988.

106. W. D. Miles and R. F. Gould, *American Chemists and Chemical Engineers*, Gould Books, Guildford, CT, 1994, vol. 2.

107. T. I. Williams, ed., *Collins Biographical Dictionary of Scientists*, 4th edn., HarperCollins, Glasgow, 1994.

108. C. S. Nicholls, ed., *Dictionary of National Biography, Missing Persons*, Oxford University Press, 1993.

109. J. Jacques, 'Grégoire Wyrouboff (1843-1913) et la chimie positive', *C. R. Acad. Sci., Paris*; sér. 2c, 1999, **2**, 467–470.
110. B. Bensaude-Vincent, 'Atomism and positivism: a legend about French chemistry', *Ann. Sci.*, 1999, **56**, 81–94.
111. G. Bram and N. T. Anh, 'The difficult marriage of theory and French organic chemistry in the 20th century', *J. Mol. Struct. (Theochem)*, 1998, **424**, 201–206.
112. 'British Library national digital archive endorsed by MP committee', PublicTechnology.net: e-government and public sector IT news, http://www.publictechnology.net [22 July 2004, accessed 15 May 2005].
113. Wellcome Institute for the History of Medicine, 'Wellcome Bibliography for the History of Medicine', http://library.wellcome.ac.uk/ [accessed 15 May 2005].
114. Hyle, 'Collected Bibliography in History & Philosophy of Chemistry', http://www.hyle.org/service/bibilio.htm [accessed 15 May 2005].
115. Robert Boyle Project, http://www.bbk.ac.uk/boyle [accessed 15 May 2005].
116. Panopticon Lavoisier, http://moro.imss.fi.it/lavoisier/entrance/panotxt.html [accessed 15 May 2005].
117. Das Liebig-Museum in Giessen, http://www.liebig-museum.de/homepage.html [18 April 2004, accessed 15 May 2005].
118. Carmen Giunta, 'Classic Chemistry', http://web.lemoyne.edu/~giunta/ [accessed 15 May 2005].
119. The Chem Team, http://dbhs.wvusd.k12.ca.us/webdocs/ChemTeamIndex.html [accessed 15 May 2005].
120. ECHO (exploring and collecting history online – science, technology and industry), http://echo.gmu.edu/index.php [1996–2004, accessed 15 May 2005].
121. Resource Discovery Network, http://www.rdn.ac.uk/ [accessed 15 May 2005].
122. Physical Sciences Information Gateway, http://www.psigate.ac.uk/newsite/about.html [2001, accessed 15 May 2005].
123. Chemistry Biology Pharmacy Info Center, 'Chemistry, Biology and related Disciplines in the WWW', http://www.infochembio.ethz.ch/links/en/ history_chem.html [01 May 2005, accessed 15 May 2005].
124. National Museum of Science and Industry, 'Ingenious', http://www.ingenious.org.uk/ [2003, accessed 15 May 2005].
125. Royal Society of Chemistry, Library and Information Centre, http://www.rsc.org/Library/index.asp [2005, accessed 17 May 2005].
126. The Society for the History of Alchemy and Chemistry, http://www.ambix.org [accessed 15 May 2005].
127. Royal Society of Chemistry Historical Group, http://www.chem.qmw.ac.uk/rschg/ [consulted 15 May 2005].
128. American Chemical Society, 'HIST – Division of the History of Chemistry', http://www.scs.uiuc.edu/~mainzv/HIST/ [December 2002, accessed 15 May 2005].
129. Gesellschaft Deutscher Chemiker, 'Fachgruppe Geshcichte der Chemie', http://www.gdch.de/strukturen/fg/geschichte.htm [13 April 2005, accessed 15 May 2005].

130. European Association for Chemistry and Molecular Sciences, 'EuCheMS Working Party on the History of Chemistry', http://www.chemsoc.org/networks/enc/ fecs/fecshistory.htm [2001, accessed 15 May 2005].
131. Chemical Heritage Foundation, http:\\www.chemheritage.org [accessed 15 May 2005].
132. 'CHEM-HIST, History of Chemistry Electronic Discussion Group', http://www.uni-Regensburg.de/Fakultaeten/phil_Fak_I/Philosphie/ Wissenschaftsgeschichte [1997, accessed 15 May 2005].

CHAPTER 2

Chemistry Before 1800

NOEL G. COLEY

History of Chemistry Research Group, The Open University

2.1 Introduction

Many of the studies in pre-1800 chemistry discussed in *Recent Developments in the History of Chemistry*[1] are still important and useful to the historian. In part, my purpose then was to propose the history of chemistry as a subject for the sixth form and undergraduate chemistry syllabus. That remains an aim and the present chapter is concerned additionally with new trends in recent research on the history of alchemy and chemistry up to the end of the eighteenth century. In 1984, Allen Debus expressed his view of the value to be derived from studying the history of chemistry in association with intellectual, political and social history.[2,3] Since then he has continued his investigations of the Paracelsians and their importance to the early development of a philosophy of chemistry that suffused scientific and other studies, including medicine. New works on the seventeenth-century scientific revolution have appeared and there has been much discussion of the late eighteenth-century chemical revolution, with revised views on the significance of Lavoisier's work for the scientific, economic and cultural history of pre-revolutionary France. The question whether 'chemistry' actually began with Lavoisier, or whether the chemical revolution was only the reform of a pre-existing discipline has been debated with persuasive arguments on both sides.[4] A new bibliographical study of chemical publications has appeared[5] and the recent past has also seen attempts to improve both the public image of chemistry and recognition of the role of chemists and chemistry in society.[6]

2.2 Early Chemistry in General Histories of Chemistry

Historians of chemistry, along with some historians of science, have begun to take a fresh look at the nature and content of the subject and to reassess some long-held ideas.[7] The history of chemistry has recently undergone something of a revival after a period following the publication of Partington's major survey,[8] the first three volumes of which deal with chemistry before 1800. As discussed in Chapter 1, several new general histories of chemistry have appeared in recent years, each with a particular aim and readership in view. Each affords some space to early chemistry. W. H. Brock's *Fontana History of Chemistry*, valuable for its broad scope and for the fresh insights it brings to understanding how chemistry has developed since early times, is aimed accurately at a general readership.[9] It deals most successfully with early chemistry and with developments in the eighteenth and nineteenth centuries, placing chemistry in its historical and social context. By contrast, Fred Aftalion's book,[10] focussing on the chemical industry and written with an audience of industrial chemists in mind, highlights only briefly and without references or bibliography the early history of chemistry and then presents an international panorama of facts about the people, companies, processes, and products concerned in the modern chemical industry.

Another approach to teaching the history of early European chemistry has been taken on the Open University's undergraduate course, *The Rise of Scientific Europe 1500-1800*. This deals with the rise of European science in general and poses the over-arching question why science developed in Europe rather elsewhere (*e.g.* China, India or the Arab world). Within this broad remit chemistry is treated *inter alia* with other sciences and is placed in changing political, social and economic contexts from Paracelsus to the chemical revolution.[11]

In eighteenth-century Europe, public interest in science and its potential for material progress placed chemistry, along with other sciences, at the centre of informed debate and even of entertainment. This was an age of innocent enthusiasm for science, stimulated by the great programme of investigative work that was Newton's legacy. Chemistry, still in its infancy as a distinct subject, was considered capable of enlightening the public about the composition of matter and its transformations, whether natural or contrived by man.[12] As a consequence of the discoveries made by chemists from Boyle to Priestley and Lavoisier in the late eighteenth century, the fundamental revision of chemical theory, the chemical revolution, became inevitable.[13] Chemistry and the chemical industry began to claim public attention, naturally arousing popular curiosity. The reputation of chemistry as an agent of progress was less tarnished then than it has since become, and the endeavour to bring chemistry to the attention of the general non-scientific reader has been pursued assiduously ever since. Two recent books continue that tradition as they survey developments in the history of chemistry from early-modern times and strategies for improving its public relations.[14,15]

In France too, new work on the history of chemistry has appeared, partly stimulated by commemorations of the bicentenary of Lavoisier's chemical revolution, when chemistry became a French science following its domination by the German phlogiston theory during the eighteenth century. The French generally honour their important scientists as much as their political, military and literary giants, while French historians of science often pay at least as much attention to theories, language and terminology as to the experimental discoveries of individuals.[16] One effect of the chemical revolution was to concentrate attention on the ideas, nomenclature and language of chemistry, all of which required fundamental change to accommodate the ideas of the anti-phlogistic theory.[17,18] In a similar vein, Marco Beretta has published new studies of the history of chemistry from Agricola to

Lavoisier. In these, he shows how discussion of the discoveries by chemists in the eighteenth century forced the formulation of a revised theoretical structure for the discipline, leading to a new definition of chemistry and public recognition of the role of chemistry in industry and society.[19,20] Beretta argues that the reform of nomenclature had far more than theoretical implications and demonstrates the value of new perspectives on familiar subjects. The development of chemical concepts and their place in the history of chemistry also forms the basis of a Polish contribution to these new approaches.[21] Written by a practising chemist in Poland, this book discusses the rise of chemistry in broad terms that simplify the history while presenting an attractive view of the subject. It also mentions some less well known, though none the less important, Polish chemists. Another fresh approach has recently appeared in a collection of papers written between 1966 and 1992 by Trevor H. Levere, covering a group of topics important in the historiography of chemistry. The subjects covered in these papers extend from the late eighteenth-century chemical revolution to thoughts on the role of chemists and chemistry in nineteenth-century society.[22]

Among all this new writing on the history of chemistry, it is encouraging to observe the reappearance of the work of Hélène Metzger, who was a major figure in this field in the 1920s. Before publishing the work for which she was to become best known, *Newton, Stahl, Boerhaave et la doctrine chimique*, (1930), Metzger was commissioned to write a short popular history of chemistry, *La chimie*, as a contribution to a larger work, *L'histoire du monde*, edited by M. E. Cavaignac. Although Metzger completed her popular history in 1926, it was not published until 1930. After a revival of interest in 1987,[23–25] it has now been translated into English and reissued in a bid to place Metzger once more firmly amongst women scientists.[26]

Useful bibliographical reviews of chemical literature relating both to theoretical chemistry and to chemical technology, and including coverage of developments in the eighteenth century, have appeared in several publications since the late 1980s.[27–29] Chemistry is of course a practical science and its literature alone can give only a second-hand account of the experimental results. The evolution of new experimental techniques requires continual up-dating and testing of apparatus. The history of the chemical laboratory and of laboratory practices is far more elusive than that of the printed sources, but in recent years there has been growing interest in the history of the chemical laboratory, its design and equipment.[30] Lastly, it must be remarked that despite the value of these new studies for the evolution of the subject, the serious historian of chemistry will still find Partington's work indispensable for its detail and coverage. Other "classic" papers in the field are also still of interest and, recently, under the editorship of Allen Debus, the Society for the History of Alchemy and Chemistry has published a collection of reprints of notable papers originally published in its journal, *Ambix*.[31]

2.3   Biographical Studies

Biography as a literary genre has regained much of its former popularity in recent years as the private lives of prominent people, both contemporary and historical, have been exposed to public scrutiny. The reassessment of historical figures has gained in popularity and the trend has also extended to the history of science. There is a renewed interest in the special techniques of biographical writing as the broader question of the value of biography in the history of science has been reconsidered.[32] New biographies of scientists have figured largely in recent years. Robert Schofield has made a new study of Joseph Priestley and his work in pneumatic chemistry,[33] while in a new biography of Humphry Davy, David Knight reassesses the early development of post-phlogistic chemistry.[34] June Fullmer's micro-study of Davy's early years has appeared.[35] Elsewhere, Knight insists that it is necessary to consider the *whole* life of a subject, rather than merely those aspects important for the advancement of science alone. In this respect, he has reviewed Davy's life, including his later years when science played a much reduced role in his activities.[36] Knight shows that biography can provide a ready way of 'humanising' as well as popularizing scientific activity; it can be used to introduce the scientific aspects of a subject's life-work and their relationship to other equally important and perhaps better known activities.[37,38] Melhado has reviewed four recent biographical accounts of Lavoisier and his work in this light.[39]

Biographies are notoriously variable in quality and a bibliography listing the most reliable is therefore a useful guide.[40] This work includes many chemists, though it lists only substantial biographies and autobiographies published in book form. The many shorter biographical accounts in articles, chapters, obituaries and biographical dictionaries are not listed in Howsam's bibliography. These often provide accurate biographical details in concise form, although they usually include other scientists, inventers and medics as well as chemists. One recent dictionary of scientists claims to go further and to "survey the sciences through the lives of the men and women whose efforts have shaped modern science", a bold claim indeed, especially for a small volume of fewer than four hundred pages.[41] Yet there can be no doubt of the value of such a work as it breathes life into the history of science through brief accounts of the lives and works of its main exponents.

2.4   Ancient Alchemy and Chemistry

From early times, chemistry has been applied in many crafts and trades. The preparation of drugs and medicines, dyes and perfumes, bleaching and dyeing cloth, tanning hides for leather, cookery and the preserving of food, and the extraction of pure metals from their ores all use chemistry in contributing to technological and economic progress. The applications of these and other chemical processes in crafts and society can be traced back to the earliest recorded history of ancient civilizations in Egypt, Sumeria, China, India and South America. Ancient Chinese science and technology has been most fully documented in the monumental work of the late Joseph Needham.[42,43] Impressive as it was, Chinese alchemy and chemistry did not survive its fusion with early Greek and Arabic alchemy in the rise of European science. These early sources gained from Chinese experience indirectly, perhaps through travel across Asian trade routes and contact with the West by Jesuit missionaries.

In recent years interest in the history of alchemy and early chemistry has increased, perhaps encouraged in part by a burgeoning popular interest in occult phenomena. Alchemy and astrology taken together were thought to provide access to the hidden structures and processes of nature.[44] A conference on the history of alchemy held at Groningen, Holland, in 1989 gave expression to the growing scholarly interest in the history of alchemy. The proceedings of this conference include a substantial bibliography of works on the history of alchemy.[45,46] In this renewed interest in alchemy, there is a distinct European focus.[47,48] Attempts to trace formerly unknown origins[49] and fresh studies of the mystical aspects of gnosticism[50,51] and other beliefs are made to throw light on some of the spiritual ideas of the Hermetic tradition.[52] The role of alchemy in the German Rosicrucian movement of the seventeenth and eighteenth centuries may have been more important than is often recognized and further research is called for in this field.[53,54]

It has been argued that the language of alchemy was used as a tool for diplomatic mediation.[55] Its mercenary aspects supported the prospect of making gold at the courts of some European rulers, especially those of the Hapsburgs and the Holy Roman Empire,[56] as well as the court of Count Moritz of Hessen at Kassel in the first quarter of the seventeenth century.[57] In an important study of the latter, Bruce Moran has used archives at Kassel to reveal not only first-hand information on the role of alchemy but also a close examination of the controversies and the personalities involved.[58] Alchemists also travelled from one centre to another, and in an extended and not altogether valid view of the subject it has been suggested that the alchemical notion of transmutation has reappeared in modern nuclear physics.[59]

Attempts to study the philosopher's stone and, especially, the great search for ways of transmuting base metals into gold[60] ultimately extended knowledge of metallurgy.[61] Various alloys were produced, some of which were used in coins and medals.[62] With its dependence on neo-platonism, alchemy involved a large element of metaphysics and efforts to explore such philosophical aspects have always played a major role in alchemical studies. The links between alchemy and cosmology and the contributions of alchemy to an understanding of Medieval thought are also important.[63] The alchemy of Roger Bacon has attracted attention,[64,65] as has that of Ramond Lull.[66]

There is also an interesting interaction between alchemical philosophy and religious belief. Although many alchemists held strong religious beliefs, they were by no means orthodox Christians, while it is intriguing to note the extent to which Jesuits were willing to accept alchemical practices. This has been explored with respect to the reception of alchemy by seventeenth-century Jesuits.[67] In a different, but related study, the influence of ancient esotericism–the studies of alchemy, astrology and magic–on Western thought right up to the present century is traced in an effort to explain modern spiritual attitudes in the West.[68] Another wide-ranging study of the influence of Hermeticism from Alexandrian times to the present reveals the continuing importance of esoteric ideas throughout history.[69]

Symbols have always played an important role in alchemical thought and there are many ways of interpreting the alchemists' esoteric writings. What, for instance, was the significance of the iguana in David Teniers' alchemical paintings?[70] One view of alchemical symbolism uses the psycho-philosophical system of C. G. Jung, who obsessively emphasized the sexual and generative connotations of alchemical procedures intended to lead to the formation of gold from lead and other base metals.[71] The union of opposites observed in so many alchemical processes led to the symbolism of sex and gestation in Chinese as well as European alchemy.[72]Alchemical imagery and symbolism from antiquity to the seventeenth century is also portrayed in a lavishly illustrated book based on manuscript and printed sources in the British Library.[73] Lyndy Abraham has compiled a dictionary of alchemical symbolism, covering the long period from the early Christian era to the late twentieth century.[74] As an introduction and guide to symbolism in alchemy, with explanations, commentaries and detailed references to sources, Roob's *Hermetic Museum* is comprehensive and authoritative, an excellent, accessible text.[75] Since alchemical symbolism and language were intended to obscure rather than reveal procedures and so make it difficult or even impossible for the uninitiated to understand their meaning, the problems of dealing successfully with the occult ideas expressed in esoteric alchemy are considerable. Deep scholarship is required by those who would unravel its secrets and the temptation to read later scientific ideas into obscure symbolic writings is considerable, especially if it is also motivated by national pride.

One example of this type of approach is found in an attempt to explain the alchemical work of the Polish alchemist Michael Sendivogius in seventeenth century chemical terms.[76,77] In his desire to find evidence of originality and to claim priority for his Polish forbear, Szydlo makes direct links between Sendivogius' writings and seventeenth-century chemical ideas.[78] Unfortunately, the basis for such connections may not be reliable as it requires the assumption of objective attitudes and beliefs about matter, which belong to a later age, and which alchemists like Sendivogius rarely held. Nevertheless, that there is an intellectual continuity between late Medieval alchemy and seventeenth-century chemistry cannot be denied, although the precise points of demarcation are not always easy to identify.[79]

Renaissance alchemy poses philosophical problems that have often been treated in a general manner that tended to blur fine detail. There has recently been a return to study of the alchemy of this period, resulting in the revelation of many more of its details and a firmer understanding of its diversity.[80,81] As the modes of thought and experimental investigation of natural phenomena changed, the esoteric problems discussed so fervently by the alchemists gradually lost their significance, but there were seminal ideas that help to link alchemical thought to later chemical theories. For example, while most alchemical theories of transmutation were based on Platonic and Aristotelian concepts of the nature of matter, some early alchemists also held corpuscular views.[82,83] This was often based on quasi-corpuscularian leanings found in Aristotle's *Meteorologica*. A recent study examines the influences exerted by this work on Renaissance commentators on alchemy.[84]

The practical methods, chemical substances and apparatus used by alchemists proved especially important for the rise of chemistry from the sixteenth century. In their futile search for transmutation, alchemists devised chemical techniques and designs for apparatus, together with a long list of substances with known properties and uses, which passed into the corpus of chemical and medical knowledge from the time of Paracelsus and Agricola. On the practical side, the role of distillation has already been mentioned, but many other alchemical techniques also passed virtually unchanged into the work of early European chemists. Thus, not only the language and methods of the alchemists, but also their discoveries of metals, acids and other chemical substances had an important and lasting influence on the early history of chemistry. Some accounts of early alchemical and other laboratories have been studied recently, as have some archaeological remains (Chapter 9). Among these the most exciting is the discovery of a large quantity of apparatus belonging to a sixteenth-century metallurgical laboratory in Lower Austria. Investigations at this site are continuing.[85,86] How far the laboratory notebooks of a working alchemist can be used as a historical research tool has also been examined.[87]

Georgius Agricola had little time for esoteric alchemy, but the metallurgical processes he describes in *De Re Metallica* (1556), which owed much to alchemical practices, had a powerful influence on later chemistry.[88] The ordering of the metals into systems that would link alchemy to metallurgy, astronomy and other studies, or give support to the idea of transmutation was fundamental to alchemical theories as Vladímir Karpenko has shown.[89] Alchemical ideas and symbolism were by no means restricted to early periods, as Beretta shows in his study of the uses of symbolism in the transition from alchemy to chemistry.[90] Much of the philosophy and many of the practices of the alchemists passed down to early chemistry, along with the symbolism through which they were expressed. Alchemy offered ways of concealing chemical methods and discoveries.[91] This also included the various methods of distillation, for example, and the apparatus devised by alchemists to carry out such procedures.

There was also a strong element of chemical technology with its emphasis on practical methods and the necessary furnaces, apparatus and tools to carry them out.[92] In recent years, some scholars have returned to Partington's 1960 study of gunpowder manufacture and its uses in warfare, *A History of Greek Fire and Gunpowder*,[93] which has become a classic in the history of early chemistry. While in no way superseding Partington's detailed work on this subject, new studies have revived interest in it and revealed again the utility of alchemy.[94,95]

The practice of alchemy and early chemistry is commonly linked with medicine, a connection brought out in a collection of essays resulting from a colloquium held at the Warburg Institute in London in 1989.[96] This collection includes a study of chemistry teaching in the English universities and the extent to which it was based on alchemical ideas at the beginning of the eighteenth century.[97] Another example of the practical utility of alchemy can be found in the applications of "alchemical metal" in striking coins and medals.[98] Newton made good use of his alchemical studies in his work on the properties of metals for the coinage when he was Master of the Royal Mint. Newton's interest in alchemy, his large collection of books on the subject and his circle of alchemical friends is well-known, but has been revisited.[99] The subject was also studied in depth by B. J. T. Dobbs, who threw new light on the place of alchemy in Newton's world-view, its relation to his theory of matter, and to his Christian beliefs and theological works. The importance of alchemical thought for the rise of early modern chemistry is clarified in these studies. If Newton is to be regarded as the most seminal thinker in eighteenth-century science, his interest in alchemy and the part it played in forming his scientific outlook and system cannot be ignored.[100–103]

2.5   The Paracelsians

The late Walter Pagel published extensively on Paracelsian philosophy and, more recently, Allen Debus' searching work has redefined the importance of Paracelsianism for Renaissance and early seventeenth-century chemistry and medicine.[104,105] In a book published by the Folger Shakespeare Library dealing with intellectual history and the occult, Debus discusses the role of alchemy in early eighteenth-century France.[106] In a related field, concerned with Paracelsianism and iatrochemistry, a considerable amount of new work has also appeared in the past decade. An indexed catalogue of original works by and about Paracelsus, published between the early sixteenth and late eighteenth centuries, has recently been issued by Glasgow University Library. It lists many previously inaccessible works and is an invaluable resource for Paracelsian scholars.[107] Another source of selected original Paracelsian works was published at Princeton in 1995.[108] Recent Paracelsian studies have also come from several German sources where interest in the history of chemistry, always strong, has undergone expansion in recent times.[109–113]

Van Helmont's chemical work has also commanded fresh interest, as have the relationships between Paracelsus and van Helmont, both of whom exerted a powerful influence on the development of chemistry and medicine in the sixteenth and seventeenth centuries.[114,115] This gave rise to the iatrochemical movement whose members, following Paracelsus, based their medical treatments on the use of chemical substances. Among these, antimony compounds were especially popular with Paracelsus and his followers.[116] The changing interpretations of van Helmont's work from Hélène Metzger to Walter Pagel have been examined.[117] The Medieval sources for van Helmont's corpuscular theory and its role in his chemical philosophy have also been investigated.[118] Van Helmont's thought was steeped in mysticism and his contemporaries, including Boyle, found difficulty in reconciling his important discoveries with his mystical inclinations.[119] Yet van Helmont also had a more than usually open view of alchemy and attempted to reveal its secrets to make it better understood and more generally available.[120]

In these ways the shadowy details of early chemistry and the extent of its dependence on alchemy are gradually being explicated. It becomes increasingly apparent that the history of early chemistry is indeed a continuous endeavour reaching back through alchemy in an unbroken chain of attempts to solve the same problems concerning the nature of matter and its changes.[121] It is also evident that efforts were made to improve the utility of chemistry well before Roger Bacon and Robert Boyle. Paracelsian chemistry was founded upon alchemical ideas, but even in the seventeenth century there were the stirrings of a more rational, less esoteric science.[122,123] The demise of alchemy and its replacement by chemistry was greatly assisted by the work of Nicholas Lemery in Paris, where Paracelsian "analogies, similitudes and sympathies" began to be replaced by the mechanical philosophy of Descartes. John C. Powers presents a perceptive study of the early emergence of chemistry from ancient alchemical symbolism in late-seventeenth and early-eighteenth-century France.[124]

With these new approaches, chemistry began to be seen as useful in its own right, with aims going beyond gold-making and the occult fantasies of alchemy. Even if this applied mainly to its uses in medicine,[125] the new conception of the subject was valuable and prophetic. This field has been well cultivated in respect of the major workers and new studies have turned to more peripheral characters, who throw interesting light on the science of the time.[126] The relationship of alchemy and chymistry to medicine was hotly debated in the seventeenth and early eighteenth centuries.[127]

One aspect of this concerns the seventeenth-century revival of interest in the medicinal properties of spa waters.[128] The tests required to ensure that these waters were safe to drink led to important advances in chemical analysis in which Boyle figured prominently. In a recent collection of essays on Renaissance science in which some little-known workers are brought to light, it is significant that medicine figures more prominently than either alchemy or chemistry.[129] The examination of residues and deposits near the points at which mineral waters issued from the earth also initiated other forms of analysis as a means of solving wider problems in geology and related sciences.[130]

2.6   Chemistry in the Scientific Revolution

The seventeenth-century scientific revolution continues to provide new research topics for historians of science. So much had been written on this subject that the publication of a bibliography of works, in which chemistry is well represented, has been very welcome.[131] The revival of ancient particle theories, an important aspect of seventeenth-century thought about the nature of matter and its transformations, has also been revisited[132] and a new survey of the critics' of atomic theories attempts to trace the changing conception of matter from ancient times to Newton has appeared.[133] A collection of papers reprinted from *Isis,* 1970–1996, includes papers by Christoph Meinel, Thomas Kuhn, and Betty Jo Dobbs and helps to contextualize the rise of seventeenth-century chemistry within a broader scientific milieu.[134] More attention has also been devoted to chemistry and its connections with medicine than has been common in earlier studies of this period. An especially strong line of argument has been developed by Allen Debus, who finds in the Paracelsian corpus a rival to the cosmological arguments that gathered around the Copernican debate. Debus regards the Paracelsian chemical philosophy as an equal claimant with Copernicanism in defining the nature of the seventeenth-century scientific revolution.[135]

Others have attempted new interpretations of seventeenth-century chemistry, showing its relative dependency on alchemy and on the concepts and descriptive language of an earlier era. The emergence of chemistry from the shadows of esoteric alchemy under the non-scientific influences of politics, religion, and economics, together with contemporary fashions for institutionalization and open intellectual debate has been studied.[136–138] Science, especially natural philosophy, also appeared to offer the hope of future social and economic improvements, while the chemical philosophy of the seventeenth century, together with the continuing strength of alchemy, supported some of the utopian ideals of the period. Hermeticism and the persistence of pseudo-science seemed to confirm the spiritual aspects of nature and all of these attitudes had religious and political implications.[139] Allen Debus has recalled how these approaches towards the historiography of the scientific revolution have developed over the past half-century.[140]

It was in the seventeenth century that the need for corporate action in pursuit of the experimental philosophy was recognized with the beginnings of institutionalization in science. The practice of experimental chemistry in the Paris Académie des Sciences at the turn of the seventeenth century has been analysed by Ursula Klein.[141] With a more theoretical emphasis, seventeenth-century chemical ideas and the ways in which they were expressed have been examined,[142] but it is renewed interest in the foundation of the Royal Society and its early members that has marked recent historical work on the science of this period. In particular, Michael Hunter has published important studies of the early years of the Royal Society, its founders, and the efforts of early Fellows to establish the new science.[143] New studies have thrown fresh light on the early history of the Royal Society and its role in promoting science, including chemistry.[144,145]

Hunter has been most assiduous amongst those who have investigated the social and political implications of science in the seventeenth century.[146] He has produced a detailed guide to the archive of Robert Boyle's papers, housed at the Royal Society, and has also investigated some of the lesser-known aspects of Boyle's character and life which help to contextualize his scientific work.[147,148] There has been a considerable flurry of interest in this dominating figure of the scientific revolution, both in Britain and abroad. The evolution of Boyle's interpretation of Helmontian chemistry and his influence on the reception of van Helmont's ideas in England has been reconsidered.[149–154] His interpretation and use of the corpuscular theory has also been subjected to careful scrutiny, though without producing any particularly new interpretations.[155–157] There is a new study of Boyle's expressed desire to bring mechanistic science and the Biblical doctrine of creation into a mutually supportive relationship. Boyle saw in the mechanical conception of nature a denial of any innate purposive wisdom in nature herself, an attitude that suggested the proper understanding of God's supreme power over creation.[158,159] It was this belief that persuaded him of the correctness of the corpuscular view and its close connection with Christian belief.[160] The aims and nature of his scientific work, as a means of using experimental observations to support and promote theological attitudes, has been discussed.[161,162]

The claim sometimes made that Boyle was the founder of modern chemistry has also been reconsidered,[163] as has the extent to which his propensity for secrecy arose from his interest in ancient alchemy and the influence of alchemical ideas on him.[164] Much of Boyle's chemistry was strongly influenced by his efforts to integrate alchemical beliefs with the corpuscular theory and his other physico-chemical aims. In an effort to reveal the influence of such alchemical beliefs, Principe has taken a provocative new view of Boyle as an "adept".[165] Other important seventeenth-century chemists whose work has been re-examined include Lemery[166] and J. J. Becher, house physician to the Elector of Bavaria and originator of the ideas on which G. E. Stahl based his theory of phlogiston. Becher aimed to make chemistry profitable and saw it as part of the transformation of social values towards the increased importance of commerce, even at court.[167,168] In an age when the utility of science was a prime concern and chemistry under patronage was beginning to show commercial and economic promise, it seems most likely that such connections may be found, at least at an intellectual level.[169] While several studies devoted to the patronage of science in countries like Germany and Italy have appeared, the distinct nature of patronage in seventeenth-century England has also been explored.[170] New work on Boerhaave reassesses his importance in the history of chemistry.[171]

In the Royal Society, the prominence of Boyle, and later of Newton has tended to overshadow the scientific importance of other early Fellows, but among those whose contributions to the new science were far from inconsiderable Robert Hooke, the Society's first curator of experiments, cannot be ignored. A new critical account of

Hooke's work appeared in 1989.[172] Finally, in 1986 the publication of the correspondence of Henry Oldenburg, a task that had been ongoing for some years, was completed.[173] All this seventeenth-century activity culminating in Newton's work gave place to the lively experimental science of the eighteenth century in which experimental and theoretical chemistry played an increasingly important role.[174] While chemistry at the Royal Society was highly important, there were other English centres in the eighteenth century where chemistry was taught and practised. Among these, the University of Cambridge occupies a significant place, having supported a Chair of chemistry since 1702. In 2002, the tercentenary of this Chair was celebrated with a historical survey.[175] Early chemical work at the University of Oxford has also been investigated,[176] as has what constituted a laboratory in the early modern period.[177]

On a practical and commercial level, chemists made useful contributions to the burgeoning popularity of medicinal spas in the eighteenth century. The analysis of mineral waters using eighteenth-century methods and materials was not easy, but the increasing numbers of books promoting the medicinal virtues of the spas demanded improved knowledge of the chemical contents of the waters.[178,179] Studies of these mineral waters also led to attempts to classify the salts they contained.[180] This seemed to be a topic in which the chemist could make a useful contribution to medicine. There was a thriving international trade in mineral waters in the eighteenth century, including the preparation of factitious mineral waters, culminating in the establishment of mineral-water firms, such as that founded at Geneva by Jacob Schweppe in the 1790s.[181]

2.7 New Studies on Lavoisier and the Chemical Revolution

The pneumatic chemistry of Black, Cavendish,[182] Priestley, and Scheele, and especially the discovery of oxygen and the work of Antoine Laurent Lavoisier, has long fascinated historians of chemistry. From the late 1980s, however, there has been a considerable increase in work in this area, with numerous attempts to reassess the origins and nature of the changes in chemical theory and practice that were inaugurated in the late eighteenth century. A new study of Priestley reassesses his life and the influences that preceded and prepared him for his momentous discovery in 1774.[183] In another study of Priestley's wider aims, it is suggested that his scientific work was an integral part of his desire to provide a new world view under the strong influence of the eighteenth-century Dissenting movement of which he was a leading figure.[184] In provincial France too, there was a proliferation of local chemical societies, supported by physicians, apothecaries and others. The importance of the Dijon Academy has long been recognized, but there were many others. A recent study of two of these lesser Academies has revealed a wider and more dynamic public interest in chemistry throughout provincial France than has hitherto been suspected.[185,186]

Marco Beretta has published a profusely illustrated and broadly based account of the chemical revolution in which, while Lavoisier is the 'centre piece', most of those in the eighteenth century who in any way contributed to the total picture are discussed, including Priestley.[187] This is one in a series on the 'greats of science' published by the

Italian edition of *Scientific American*. It makes a fascinating story and, as there is no Italian biography of Lavoisier, this is a useful artefact. Lavishly illustrated, it serves to popularize the chemical revolution, but does it truly represent its nature?[188]

John McEvoy explores the historiography of the chemical revolution through the changing interpretations of the history of science, from the whiggish positivism which, rooted in the seventeenth century, characterized the Enlightenment and later thought until it was finally challenged in the 1960s.[189] While it is open to debate, McEvoy's thesis is worthy of consideration as a rational approach to the historiography of chemistry not only of the chemical revolution, but also more generally.

Two new biographies of Lavoisier were published in 1993.[190,191] Lavoisier's new biographers have been reviewed by Beretta,[192] who has also made a study of his library and an early expression of his new ideas.[193,194] All this activity was stimulated by celebrations of the bicentenary in 1989 of the publication of Lavoisier's *Traité Élémentaire de Chimie* and commemoration of the bicentenary of his untimely execution in 1794. It need cause no surprise to find that, in France, there was heightened interest in Lavoisier's work from the late 1980s onwards. A symposium was held in Paris in 1989,[195,196] while new studies of Lavoisier's role in the chemical revolution both in France and elsewhere proliferated.[197–200] A second symposium in 1994, consisting of a broad survey of Lavoisier's interests and achievements, marked the bicentenary of his death.[201] A brief new resumé of Lavoisier's chemical work was published by Bernadette Bensaude-Vincent,[202] followed by her study comparing the historical Lavoisier with his popular image as the sole founder of anti-phlogistic chemistry.[203] In the run-up to this period of re-evaluation, F. L. Holmes published a detailed exploration of Lavoisier's thought processes as he worked towards his new interpretation of combustion, calcination, and respiration. In a very thorough study of Lavoisier's papers, including his laboratory notebooks, early drafts of his published papers and other memorabilia, Holmes painstakingly reconstructed the stages by which Lavoisier arrived at his conclusions.[204] This work has been followed by a flurry of activity in efforts to reassess Lavoisier's work and find in this well cultivated field new evidence on which to base a re-interpretation of the events leading to confirmation of the oxygen theory.

Holmes continued his search for the sources of Lavoisier's chemical techniques,[205] and, among others in this field, Arthur Donovan has been a prolific writer. In 1988, Donovan outlined a research program for a new appraisal of Lavoisier's work in chemistry and of the chemical revolution.[206] This was followed in the same year by a collection he edited of important reinterpretative studies on Lavoisier and the chemical revolution.[207] The American Chemical Society also published a special edition of their *Bulletin for the History of Chemistry* devoted to this topic.[208] At the International Summer School in History of Science at Bolgna in 1988, F. L. Holmes gave five lectures on the theme of experimental investigation in eighteenth-century chemistry.[209] Beretta has also made a detailed study of the iconography of Lavoisier by which his scientific reputation was firmly established and enhanced in the years following his death.[210] However Lavoisier's work was interpreted, there could be no doubt that its final outcome was to change chemistry quite radically, releasing it from the phlogistic theory, which was becoming increasingly sterile, and stimulating others to think afresh. Some of the results of this, expressed in the work of his contemporaries, were explored in relation to Buffon, in 1988.[211] Donovan's work resulted in a controversy with the late Carleton E. Perrin, who also strongly advocated the search for reinterpretation. Perrin, like Holmes, thought the solution must lie in a re-evaluation of Lavoisier's original experimental results and procedures. He argued that Lavoisier had clearly articulated his idea of progress through revolution, rather than reform,[212–213] but he was constrained to proceed cautiously by the mixed reception of the new chemistry among his contemporaries.

In spite of his own preference for the notion of revolution, Lavoisier's new discoveries could be interpreted as the radical reform of eighteenth-century chemistry.[214,215] His use of physical principles in his study of gases and the fact that much of his work was carried out on the borders of physics and chemistry with emphasis on precise measurements has also led some historians of science to emphasize the role of physics in Lavoisier's chemical transformation.[216,217] These assertions relating to the suggestion that Lavoisier's use of experimental physics underlay his development of theoretical chemistry produced an exchange of views with Perrin and Evan Melhado that has wide implications for methodology.[218–220] Donovan also returned to the relationship between Lavoisier's chemistry and Newtonian physics, suggesting that Lavoisier's chemical revolution changed the very nature of chemistry by shifting its fundamental concepts away from natural history in the direction of physics. The change was away from a descriptive, classificatory science and towards the deductive, quantifiable positivism of Newtonian physics.[221] Viewed in this light, it seems that the chemical revolution may contain a more fundamental challenge to philosophical principles than has previously been thought. The new theory also assimilated the eighteenth-century study of affinity and stoichiometric principles.[222,223] Lavoisier's quantitative methods also received fresh attention from Holmes, who extended his efforts to identify the sources of Lavoisier's approach to chemical reactions in this respect.[224]

The revision of chemical nomenclature was another important product of the chemical revolution. Proposed by Guyton de Morveau, Bethollet, Fourcroy, and Lavoisier himself, the reorganization of chemical nomenclature, an essential concomitant of the chemical revolution,[225] played a significant role in organizing chemical thought and in clarifying its new concepts. A reprint of the *Méthode de Nomenclature Chimique*, with a historical introduction, has recently been published in France,[226] and the role of Guyton as its originator has been re-examined.[227] The various aspects of Lavoisier's chemistry, especially the new chemical nomenclature, are discussed in papers presented at a historiographic workshop in 1994.[228] The question of names and the use of language form the centre of various disputes of the period, which extend far wider than science itself to include social, cultural and political issues.[229]

In another study, Fourcroy is cast as the historian of the chemical revolution, a role he fulfils admirably as he was both spectator and participant in the most important chemical events of the period.[230] The importance of the language of chemistry has been stressed by Trevor Levere[231] and by Pierre Laszlo.[232] It has also been contended that the new chemistry contained the seeds of later structural concepts.[233] This may be pushing the potential of Lavoisier's anti-phlogistic chemistry and its revised nomenclature too far, yet without all these main components, the fundamental reorganization of chemistry could not have been achieved. Fresh interest was also stimulated in Lavoisier's collaborators and contemporaries. For example, the role of J.-B. Meusnier, Lavoisier's collaborator in experiments on the chemical composition of water,[234] has been re-assessed. Hassenfratz's work on combustion and respiration, often critical of Lavoisier, has also been re-examined in the light of the chemical revolution.[235] Gaspard Monge's experiments on the composition of air and water, with their relation to Lavoisier's work, have been re-examined by René Taton in the same work.[236] The part played by Mme Lavoisier as her husband's supporter, propagandist, and assistant in experimental work has also received more detailed attention.[237,238] Perrin has suggested that Mme Lavoisier studied chemistry with J. B. M. Bucquet, whose lectures at the Arsenal were probably intended for her instruction rather than of Lavoisier himself.[239]

Lavoisier's official position brought him into contact with others at many levels throughout French society. To find acceptance for his new chemical ideas, it was necessary to convince not only the scientific community, but also others outside science, particularly those whose influence extended through the world of politics and economics.[240,241] Although the links between chemistry and economics have always been prominent, its connections with politics are not at first obvious. Nevertheless, it has been argued that the chemical balance may be seen not only as a laboratory instrument, but as an abstract concept that brings together other aspects of Lavoisier's work.[242] It seems that such ideas may yet produce useful insights into the potential influence of the chemical revolution on social, political and economic change as well as the general nature of scientific change. Work along these lines was also begun in 1994 by Maurice Crosland, who has indentified some unexplored political influences upon Lavoisier and on the establishment of the new chemistry.[243,244] Lavoisier has also been regarded as the chemist who, above all others, introduced quantitative methods into chemistry in philosophical deduction as well as measurement.[245,246]

While Lavoisier undoubtedly laid the foundations for the rapid rise of chemistry in the nineteenth century, his work was supported by that of other French chemists both before and after him.[247] Without the Lavoisierian notion of the chemical element, it is unlikely that the chemical atomic theory and the laws of chemical combination would have been so widely accepted. However, a new perspective on the history of the concept of the chemical compound as the basis of modern chemistry suggests that it began long before Lavoisier and belongs to a different line of development involving the notion of 'chemical affinity'.[248–250]

Yet, in the years following 1789 many chemists continued to support the concept of phlogiston.[251] There were various reasons for this. On the one hand, theoretical chemistry had for long been based on the German phlogistic theory and initially there was strong resistance to anti-phlogistic chemistry among German chemists.[252] There were also those who had grown old in the study of phlogistic chemistry and had no desire to change. Working chemists were pragmatists for whom the new chemistry seemed to offer little advantage. The same weights and volumes of chemical substances were required to carry out experimental operations successfully and the same quantities of products were obtained whatever theory was adopted. It seemed much more important to improve chemical methods of purification, analysis and identification, all of which were problems of practice rather than theory.

It is perhaps not surprising that German chemists continued to think of their country as the homeland of chemistry, an outlook fostered by Lorenz Crell. By 1789, the chemical revolution had not even begun and there were no German advocates of Lavoisier's oxygen theory.[253] The chemical revolution enabled developments and discoveries that depended upon acceptance of the new ideas and German chemists began to fear that they were losing their predominance. Efforts were made to bolster the phlogiston theory in Germany, although after 1792 protagonists of the new chemistry had become active.[254]

A fine study of previously unpublished correspondence between Richard Kirwan and L. B. Guyton de Morveau throws fresh light on these arguments.[255] More widely in Europe too, for various reasons, anti-phlogistic notions were received with varying degrees of enthusiasm.[256–259] Edmund White has written on the reception of anti-phlogistic chemistry in America.[260] Joseph Priestley, who never wavered from his use of the phlogiston theory, continued to defend it even after his emigration to America in 1794.[261,262] The Priestley memorial volume, containing material relating to Priestley collected during the 1850s and 1860s and now in the library of the Royal Society, has been reviewed recently.[263] Comparisons between the ideas and beliefs of Priestley and Lavoisier have been drawn in new studies by Prajit K. Basu,[264,265] while Marco Berettta has examined the different attitudes adopted by Priestley and Lavoisier towards the French Revolution itself.[266]

The rejection of the traditional role of chemistry as pharmacy is discussed as an important aspect of Lavoisier's new notion of chemistry as an independent discipline in a paper by Jonathan Simon.[267] In a new critique of the development of chemical thought, F. L. Holmes considered the period following Lavoisier's enunciation of the main tenets of the new chemistry as a period of competition between two paradigms. Priestley's interpretation of the phlogiston theory, based on eighteenth-century pneumatic chemistry, is fundamentally different from the original theory propounded by Stahl; Holmes gave it the status of a rival research programme. He argued that, far from Priestley's persistent adherence to his own interpretation of the phlogiston theory being due to perversity, it was founded on his belief in a new understanding of the nature of phlogiston itself in relation to his work on 'airs'. In his studies of respiration, putrefaction, and the photosynthesis of plants as a means of restoring the 'goodness' of the air, Priestley began a conceptual revolution that for a decade dominated Europe, but was finally submerged by the stronger wave launched in Paris by Lavoisier.[268] The importance of elementary analysis to the chemistry of natural substances in Lavoisier's scheme and its broader relationship to other aspects of Enlightenment thought is also discussed in a recent paper by Jonathan Simon. He shows how Lavoisier's new system of elementary analysis contrasted with the traditional methods of solution analysis used in pharmacy. As Lavoisier's new aims and techniques began to supersede the chemical practices of doctors and pharmacists, the latter were marginalized.[269] This separation would have far reaching consequences with respect to the new chemistry.

In the early years of the nineteenth century, the new chemistry began to bear fruit on both sides of the Channel,[270] as well as in other countries, notably in Sweden. The chemical atomic theory proposed by Dalton and developed by Berzelius led to the formulation of the stoichiometric laws of chemical combination and the diligent search for accurate atomic weights. The important link between atoms and electrical charges in the early years of the nineteenth century enabled a new interpretation of chemical combination and the theory of valency. Significant improvements in analytical techniques produced other important discoveries both in organic and inorganic chemistry. One example of the latter is the persistent analytical work leading to the discovery of the rare earths. New studies in this field of chemical detection have been published, including the work of Johan Gadolin (1760–1852) and Carl Gustav Mosander (1797–1858), the discovery of cerium, and some medical uses of the rare earths.[271] In these and other ways the face of chemistry was rapidly changed from the beginning of the nineteenth century.

References

1. N. G. Coley, 'Chemistry to 1800', in *Recent Developments in the History of Chemistry*, ed. C. A. Russell, Royal Society of Chemistry, London, 1985, pp. 49–76.
2. A. G. Debus, *Science and History: A Chemist's Appraisal*, University of Coimbra Press, Coimbra, 1984.
3. A. G. Debus, 'The significance of chemical history', *Ambix*, 1985, **32**, 1–14.
4. R. Mierzecki, *The Historical Development of Chemical Concepts*, Chemists and Chemistry, Kluwer Academic, Dordrecht, 1991, vol.12.
5. W. A. Cole, *Chemical Literature 1700–1860: A Bibliography with Annotations, Detailed Descriptions, Comparisons and Locations*, Mansell, London, 1988.
6. W. B. Jensen, 'History of chemistry and the chemical community: bridging the gap?', in *Chemical Sciences in the Modern World*, ed. S. H. Mauskopf, University of Pennsylvania Press, Philadelphia, PA, 1993, pp. 262–276.
7. R. P. Multhauf, *The Origins of Chemistry*, Gordon & Breach, Philadelphia, PA, 1993.
8. J. R. Partington, *A History of Chemistry*, 4 vols, Macmillan, London, 1962–1970.
9. W. H. Brock, *The Fontana History of Chemistry*, Fontana Press, London, 1992; American edition, *The Norton History of Chemistry*, W. W. Norton, New York, 1993.
10. F. Aftalion, *Histoire de la Chimie*, Masson, Paris, 1988; trans. by O. T. Benfey; preface by P. P. McCurdy, The Chemical Sciences in Society Series, University of Pennsylvania Press, Philadelphia, PA, 1991.
11. D. C. Goodman and C. A. Russell, eds, *The Rise of Scientific Europe 1500–1800*, Hodder & Stoughton, Sevenoaks, 1991.
12. J. Golinski, *Science as Public Culture, Chemistry and Enlightenment in Britain, 1760–1820*, Cambridge University Press, Cambridge, 1992.
13. J. G. McEvoy, 'The Enlightenment and the chemical revolution', in *Metaphysics and Philosophy of Science in the 17th & 18th Centuries: Essays in Honour of Gerd Buchdahl*, ed. R.S.Woolhouse, Kluwer, Dordrecht, 1988, pp. 307–325.
14. D. M. Knight, *Ideas in Chemistry: A History of the Science*, Athlone Press, London, 1992.
15. W. B. Jensen, 'History of chemistry and the chemical community: bridging the gap?', in *Chemical Sciences in the Modern World*, ed. S. H. Mauskopf, University of Pennsylvania Press, Philadelphia, PA, 1993, pp. 262–276.

16. C. Viel, 'History and evolution of the notion of salt', *Actual. Chim.*, 1999 (1), 34–41.
17. B. Bensaude-Vincent and I. Stengers, *Histoire de la chimie*, Éditions La Découverte, Paris, 1993; trans. by D. van Dam, Harvard University Press, Cambridge, MA, 1997.
18. P. Laszlo, *La Parole des Choses ou Le Langage de la Chimie*, Hermann, Collection Savoir, Paris, 1993.
19. M. Beretta, 'The historiography of chemistry in the eighteenth century: a preliminary survey and bibliography', *Ambix*, 1992, **39**, 1–10.
20. M. Beretta, *The Enlightenment of Matter: The Definition of Chemistry from Agricola to Lavoisier*, Uppsala Studies in History of Science, 15, Science History, Canton, MA, [Uppsala University Office for History of Science, Uppsala], 1993.
21. R. Mierzecki, *The Historical Development of Chemical Concepts*, Chemists and Chemistry, Kluwer Academic, Dordrecht, 1991, vol. 12.
22. T. H. Levere, *Chemists and Chemistry in Nature and Society, 1770–1878*, Variorum, Aldershot, 1994.
23. B. Bensaude-Vincent, 'Hélène Metzger's *La chimie*: A popular treatise', *Hist. Sci.*, 1987, **25**, 71–84.
24. J. Golinski, 'Hélène Metzger and the interpretation of seventeenth century chemistry', *Hist. Sci.*, 1987, **25**, 85–97.
25. J. R. R. Christie, 'Narrative and rhetoric in Hélène Metzger's historiography of eighteenth century chemistry', *Hist. Sci.*, 1987, **25**, 99–109.
26. H. Metzger, *Chemistry*, Women in the Sciences 1, trans. and annotated by C. V. Michael, Foreword by A. J. Ihde, Locust Hill Press, West Cornwall, CT, 1991.
27. W. A. Cole, *Chemical Literature 1700–1860: A Bibliography with Annotations, Detailed Descriptions, Comparisons and Locations*, Mansell, London, 1988.
28. J. P. Swann, 'Manuscript resources in the history of chemistry at the National Library of Medicine', *Ann. Sci.*, 1989, **46**, 249–262.
29. V. Wehefritz, *Bibliography of the History of Chemistry and Chemical Technology from the 17th to the 19th Century*, Bd. 1–3/hrsd., Saur, München, 1994.
30. F. A. J. L. James, ed., *The Development of the Laboratory*, Macmillan, London, 1989.
31. A. G. Debus, ed., *Alchemy and Early Modern Chemistry: Papers from Ambix*, Society for the History of Alchemy and Chemistry, London, 2005.
32. M. Shortland and R. Yeo, eds, *Telling Lives in Science: Essays on Scientific Biography*, Cambridge University Press, Cambridge, 1996.
33. R. E. Schofield, 'The professional work of an amateur chemist: Joseph Priestley' in *Motion Towards Perfection: The Achievement of Joseph Priestley*, ed. A. T. Schwartz and J. G. McEvoy, Skinner House, Boston, 1990, pp. 1–19.
34. D. M. Knight, *Humphry Davy: Science and Power*, Blackwell Science Biographies, Blackwell, Oxford, 1992.
35. J. Z. Fullmer, *Young Humphry Davy: The Making of an Experimental Chemist*, American Philosophical Society, Philadelphia, PA, 2000.
36. D. M. Knight, 'From science to wisdom: Humphry Davy's life', in *Telling Lives in Science: Essays on Scientific Biography*, ed. M. Shortland and R. Yeo, Cambridge University Press, Cambridge, 1996, pp. 104–114.

37. D. M. Knight, 'Davy's visions', in *Fields of Influence: Conjunctions of Artists and Scientists, 1815–1860*, ed. J. Hamilton, University of Birmingham Press, Birmingham, 2001, 31–49.

38. C. Lawrence, 'The power and the glory: Humphry Davy and Romanticism', in *Romanticism and the Sciences*, ed. A. Cunningham and N. Jardine, Cambridge University Press, Cambridge, 1990, pp. 213–227.

39. E. M. Melhado, 'Scientific biography and scientific revolution: Lavoisier and eighteenth-century chemistry', *Isis*, 1996, **87**, 688–694.

40. L. Howsam, *Scientists since 1660. A Bibliography of Biographies*, Ashgate, Aldershot, 1997.

41. D. Millar, I. Millar, J. Millar and M. Millar, *The Cambridge Dictionary of Scientists*, Cambridge University Press, Cambridge, 1996, 2nd edn., rev. & enl., 2002.

42. J. Needham, *Science and Civilisation in China*, 7 vols. in 11 parts, Cambridge University Press, Cambridge, 1954–2004.

43. J. Needham, *The Grand Titration: Science and Society in East and West*, Allen and Unwin, London, 1969.

44. W. R. Newman and G. Anthony, eds, *Secrets of Nature: Astrology and Alchemy in Early Modern Europe*, MIT Press, Cambridge, MA, 2001.

45. Z. R. W. M. von Martels, ed., *Alchemy Revisited. Proceedings of the International Conference on the History of Alchemy at the University of Groningen, 1989*, Collection de travaux de l'Academie Internationale Histoire des Sciences, 33, Brill, Leiden, 1990.

46. P. M. Rattansi, 'Alchemy revisited: Proceedings of the International Conference in the History of Alchemy at the University of Groningen, 17–19 April 1989', *Ambix*, 1992, **39**, 91–95.

47. C. Kren, *Alchemy in Europe: A Guide to Research*, Garland, New York, 1990.

48. J. Font Garcia, *Historia de la Alquimia en España*, MRA, Barcelona, 1995.

49. V. Karpenko, 'The oldest alchemical manuscript in the Czech language', *Ambix*, 1990, **37**, 61–73.

50. D. Merkur, *Gnosis. An Esoteric Tradition of Mystical Visions and Unions*, State University of New York Press, Albany, NY, 1993.

51. C. Gilly and M. Afanasyeva, *In de Pelikaan. 500 Years of Gnosis in Europe*, Exhibition Catalogue of Printed Books and Manuscripts from the Gnostic Tradition, Bibliotheca Hermetica, Amsterdam, 1993.

52. D. Merkur, 'The study of spiritual alchemy: mysticism, gold-making and esoteric hermeneutics', *Ambix*, 1990, **37**, 35–45.

53. J. V. Andreä, *The Chemical Wedding of Christian Rosenkreutz*, ed. J. Godwin with Introduction by A. McLean, Magnum Opus Hermetic Source Works, 18, Phanes Press, Grand Rapids, MI, 1991.

54. C. McIntosh, *The Rose Cross and the Age of Reason. Eighteenth-Century Rosicrusianism in Central Europe and its Relationship to the Enlightenment*, Brill, Leiden, 1992.

55. P. H. Smith, 'Alchemy as a language of mediation at the Habsburg Court', *Isis*, 1994, **85**, 1–25.

56. P. H. Smith, *The Business of Alchemy: Science and Culture in the Holy Roman Empire*, Princeton University Press, Princeton, NJ, 1994.

57. B. T. Moran, 'Privilege, communication and chemiatry: the hermetic-alchemical circle of Moritz of Hessen-Kassel', *Ambix*, 1985, **32**, 110–126.

58. B. T. Moran, *The Alchemical World of the German Court: Occult Philosophy and Chemical Medicine in the Circle of Moritz of Hessen (1572–1632)*, Sudhoffs Archiv, Beihefte 29, Steiner, Stuttgart, 1991.

59. O. Bostrup, 'The travelling alchemist in Germany. 3. Darmstadt: modern and antique alchemy', *Dan. Kemi.*, 1994, **75**, 10–11 [in Danish].

60. A. De Pascalis, *Alchemy the Golden Art: The Secrets of the Oldest Enigma*, S.I., trans from Italian, Gremese International, 1995.

61. V. Karpenko, 'The chemistry and metallurgy of transmutation', *Ambix*, 1992, **39**, 47–62.

62. V. Karpenko, 'Coins and medals made of alchemical metal', *Ambix*, 1988, **35**, 65–76.

63. B. Obrist, 'Cosmology and alchemy in an illustrated 13th-century alchemical tract: Constantine of Pisa: *The Book of the Secrets of Alchemy*', *Micrologus*, 1993, **1**, 115–160.

64. W. R. Newman, 'The alchemy of Roger Bacon and the *Tres epistolae* attributed to him', in *Comprendre et Mâtriser la Nature au Moyen Age: Mélanges...Offerts à Guy Beaujouan*, Droz, Genève, 1994, pp. 461–479.

65. S. J. Linden, ed., *The Mirror of Alchimy Composed by the Thrice-Famous and Learned Fryer, Roger Bachon,* English Renaissance Hermeticism Series, Garland, New York, 1992, vol. 4.

66. M. Periera, *The Alchemical Corpus Attributed to Ramond Lull*, Warburg Institute, University of London, London, 1989.

67. M. Baldwin, 'Alchemy and the Society of Jesus in the 17th century: strange bedfellows?', *Ambix*, 1993, **40**, 41–64.

68. A. Faivre and J. Needleman, *Modern Esoteric Spirituality*, S. C. M. Press, London, 1993.

69. M. Baigent and R. Leigh, *The Elixir and the Stone: A History of Magic and Alchemy*, Viking, Penguin Books Ltd, London, 1997.

70. J. P. Davidson, ''I am the poison-dripping dragon': iguanas and their symbolism in the alchemical and occult paintings of David Teniers the younger', *Ambix*, 1987, **34**, 62–80.

71. J. Fabricius, *Alchemy: The Medieval Alchemists and their Royal Art*, Rosenkilde & Bagger, Copenhagen, 1976; rev. edn., 1989, repr. Diamond Books, London, 1994.

72. R. K. Payne, 'Sex and gestation: the union of opposites in European and Chinese alchemy', *Ambix*, 1989, **36**, 66–81.

73. G. Roberts, *The Mirror of Alchemy*, The British Library, London, 1994.

74. L. Abraham, *A Dictionary of Alchemical Symbolism*, Cambridge University Press, Cambridge, 2001.

75. A. Roob, *The Hermetic Museum: Alchemy and Mysticism*, trans. S. Whiteside, Taschen, Köln & London, 1997.

76. Z. Szydlo, 'The alchemy of Michael Sendivogius: his central nitre theory', *Ambix*, 1993, **40**, 129–146.

77. Z. Szydlo, *Water Which Does Not Wet Hands: The Alchemy of Michael Sendivogius*, Polish Academy of Science, Warsaw, 1994.

78. Z. Szydlo, 'The influence of the central nitre theory of Michael Sendivogius on the chemical philosophy of the seventeenth century', *Ambix*, 1996, **43**, 80–96.

79. B. Joly, 'Alchimie et rationalité: La question des critères de démarcation entre chemie et alchimie au XVIIe siècle', *Sci. Tech. Persp.*, 1995, **31**, 93–107.

80. J. C. Margolin and S. Matton, eds, *Alchemie et Philosophie à la Renaissance*, J. Vrin, Paris, 1993.

81. A. Perifano, 'L'alchimie entre Moyen-Age et Renaissance: Rupture ou changement de contexte?', in *Terres médiévales*, ed. B. Ribémont, Klincksieck, Paris, 1993, pp. 237–249.

82. W. R. Newman, 'Corpuscular alchemy: the transmutational theory of Eirenaeus Philalethes', *Bull. Hist. Chem.*, 1992–93, **13–14**, 19–27.

83. W. R. Newman, 'The corpuscular transmutational theory of Eirenaeus Philalethes', in *Alchemy and Chemistry in the 16th and 17th Centuries*, ed. P. Rattansi and A. Clericuzio, Kluwer Academic, Dordrecht, 1994, pp. 161–182.

84. C. Martin, 'Alchemy and the Renaissance commentary tradition on *Meteorologica IV*', *Ambix*, 2004, **51**, 245–262.

85. R. W. Soukup, S. von Osten, H. Mayer, 'Alembics, curcurbits, phials, crucibles: a 16th-century docimastic laboratory excavated in Austria', *Ambix*, 1993, **40**, 25.

86. B. Joly, 'Qu'est-ce qu'un laboratoire alchimique?' in *Les Procédures de Preuve sous le Regard de l'Historien des Science et Destechniques*, Soc. Francaise d'Histoire des Sciences et Techniques, Paris, 1992, pp. 87–102.

87. W. R. Newman and L. M. Principe, 'Chymical laboratory notebooks of George Starkey', in *Reworking the Bench: Research Notebooks in the History of Science*, ed. F. L. Holmes, J. Renn and H.-J. Rheinberger, Kluwer Academic, Dordecht, 2003, pp. 25–41.

88. M. Beretta, 'Humanism and chemistry: the spread of Georgius Agricola's metallurgical writings', *Nuncius*, 1997, **12**, 17–47.

89. V. Karpenko, 'Systems of metals in alchemy', *Ambix*, 2003, **50,** 208–230.

90. M. Beretta, 'The role of symbolism from alchemy to chemistry', in *Non-verbal Communication in Science Prior to 1900*, ed. R. G. Mazzolini, Olschki, Firenze, 1993, pp. 279–319.

91. L. M. Principe, "Robert Boyle's alchemical secrecy: codes, ciphers and concealments', *Ambix*, 1992, **39**, 63–74.

92. W. R. Newman, 'Technology and alchemical debate in the late Middle Ages', *Isis*, 1989, **80**, 423–445.

93. J. R. Partington, *A History of Greek Fire and Gunpowder*, Heffer, Cambridge, 1960.

94. V. Karpenko, 'Greek fire in a Czech alchemical manuscript', *Centaurus*, 1987, **30**, 240–244.

95. P. Pentz, 'A medieval workshop for producing 'Greek fire' grenades', *Antiquity*, 1988, **62**, 89–93.

96. P. Rattansi and A. Clericuzio, eds, *Alchemy and Chemistry in the 16th and 17th Centuries*, Kluwer, Dordrecht, 1994.

97. A. Guerrini, 'Chemistry teaching at Oxford and Cambridge, ca 1700', in *Alchemy and Chemistry in the 16th and 17th Centuries*, ed. P. Rattansi and A. Clericuzio, Kluwer, Dordrecht, 1994, pp. 183–199.

98. V. Karpenko, 'Coins and medals made of alchemical metal', *Ambix*, 1988, **35**, 65–76.

99. K. Figala and U. Petzold, 'Alchemy in the Newtonian circle: personal acquaintances and the problem of the late phase of Isaac Newton's alchemy', in *Renaissance and Revolution*, ed. J. V. Field and F. A. J. L. James, Cambridge University Press, Cambridge, 1993, pp. 173–192.

100. B. J. T. Dobbs, *The Janus Faces of Genius: The Role of Alchemy in Newton's Thought*, Cambridge University Press, Cambridge, 1991.

101. B. J. T. Dobbs, 'Newton as alchemist and theologian', in *Standing on the Shoulders of Giants: A Longer View of Newton and Halley*, ed. N. J. W. Thrower, University of California Press, Berkeley, CA, 1990, pp. 128–140.

102. B. J. T. Dobbs, 'Newton's alchemy and his 'active principle' of gravitation', in *Newton's Scientific and Philosophical Legacy*, ed. P. B. Scheurer and G. Debrock, Kluwer, Dordrecht, 1988, pp. 55–80.

103. B. J. T. Dobbs, *Alchemical Death and Resurrection: The Significance of Alchemy in the Age of Newton*, Smithsonian Institution Libraries, Washington DC, 1990.

104. A. G. Debus, *The Chemical Philosophy: Paracelsian Science and Medicine in the Sixteenth and Seventeenth Centuries*, 2 vols, Science History Publications, Canton, MA, 1977.

105. A. G. Debus, *The French Paracelsians: The Chemical Challenge to Medical and Scientific Tradition in Early Modern Europe*, Cambridge University Press, Cambridge, 1992.

106. A. G. Debus, 'Alchemy in an age of reason: the chemical philosophers in early 18th-century France', in *Hermeticism and the Renaissance: Intellectual History and the Occult in Early Modern Europe*, ed. I. Merkel and A. G. Debus, Folger Shakespeare Library, Washington DC, Associated University Presses, 1988, pp. 231–250.

107. D. Weston, *Paracelsus. A Catalogue of Works Published 1529–1793 Preserved in Glasgow University Library*, Glasgow University Library, Glasgow, 1993.

108. Paracelsus, *Paracelsus: Selected Writings*, Princeton University Press, Princeton, NJ, 1995.

109. A. M. Haas *et al.*, *Nova acta Paracelsica: Beiträge zur Paracelsus-Forschung*, B. Lang, Neue Folge 7. Hrsg. von der Schweizerischen Paracelsus-Gesellschaft, Zurich, 1993.

110. F. Rueb, *Quintessenz*, München, 1995.

111. K. Pisa, *Paracelsus in Österreich: Eine Spurensuche*, Niederösterreichisches Presshaus, St. Polten, 1991.

112. H. Dopsch, K. Goldammer and F. Kramml *et al.*, *Paracelsus (1493–1541): 'Keines andern Knecht...'* im Auftrag des Landes Salzburg, Pustet, Salzburg, 1993.

113. U. Benzenhöfer, *Paracelsus Wissenschaftlicher*, Buchgesellschaft, Darmstadt, 1993.

114. W. Pagel, *J. B. van Helmont: Reformer of Science and Medicine*, Cambridge University Press, Cambridge, 1982.

115. W. Pagel, *From Paracelsus to van Helmont: Studies in Renaissance Medicine and Science*, ed. M.Winder, Variorum, London, 1986.

116. R. T. McCallum, *Antimony in Medical History*, The Pentland Press, Edinburgh, 1999.

117. R. Halleux, 'Visages de Van Helmont, depuis Hélène Metzger jusqu'à Walter Pagel', in *Études sur Hélène Metzher. En appendice: Hélène Metzger, "Extraits de lettres, 1921–1944"*, ed. G. Freudenthal, *Corpus*, 1988, **8/9**, 35–43.

118. W. R. Newman, 'The corpuscular theory of J.B. Van Helmont and its medieval sources', *Vivarium*, 1994, **31**, 161–191.

119. B. Heinecke, 'The mysticism and science of Johann Baptista van Helmont', *Ambix*, 1995, **42**, 65–78.

120. G. Hanswille and D. Brumlich trans. & eds, *Helmont, Jan Baptista van. Alchemy Unveiled, for the First Time, the Secret of the Philosopher's Stone is being Openly Explained*, Merkur, Scarborough, Ontario, 1991.

121. W. R. Newman and L. M. Principe, *Alchemy Tried in the Fire: Starkey, Boyle and the Fate of Helmontian Chymistry*, University of Chicago Press, Chicago, 2002.

122. P. M. Rattansi, 'Recovering the Paracelsian milieu', in *Revolutions in Science: Their Meaning and Relevance*, ed. W. R. Shea, Science History Publications, Canton, MA, 1988, pp. 1–26.

123. A. G. Debus, *Chemistry, Alchemy, and the New Philosophy, 1550–1700: Studies in the History of Science and Medicine*, Variorum, London, 1987.

124. J. C. Powers, "Ars sine Arte': Nicholas Lemery and the end of alchemy in eighteenth-century France', *Ambix*, 1998, **45**, 163–189.

125. A. Clericuzio, 'From van Helmont to Boyle: a study of the transmission of Helmontian chemical and medical theories in 17th-century England', *Br. J. Hist. Sci.*, 1993, **26**, 303–334.

126. S. Clucas, 'The correspondence of a 17th-century 'chymical gentleman': Sir Cheney Culpeper and the chemical interests of the Hartlib circle', *Ambix*, 1993, **40**, 146–170.

127. A. G. Debus, *Chemistry and Medical Debate: van Helmont to Boerhaave*, Science History Publictions, Canton, MA, 2001.

128. N. G. Coley, '"Cures without care",: chymical physicians and mineral waters in seventeenth-century England', *Med. Hist.*, 1979, **23**, 191–214

129. J. V. Field and F. A. J. L. James, eds, *Renaissance and Revolution*, Cambridge University Press, Cambridge, 1993.

130. A. M. Roos, 'Martin Lister (1639–1712) and fools' gold', *Ambix*, 2004, **51**, 23–41.

131. S. A. Jayawardene, *The Scientific Revolution: An Annotated Bibliography*, Locust Hill Press, West Cornwall, CT, 1996.

132. S. Clucas, "The infinite variety of forms and magnitudes": 16th and 17th-century English corpuscular philosophy and Aristotelian theories of matter and form', *Early Sci. Med.*, 1997, **2**, 251–271.

133. A. Pyle, *Atomism and its Critics: From Democritus to Newton*, Thoemmes Press, Bristol, 1997.

134. P. Dear, ed., *The Scientific Enterprise in Early Modern Europe: Readings from Isis*, University of Chicago Press, Chicago, 1997.

135. A. G. Debus, 'A chemical key to the scientific revolution', in *Farmacia, Historia Natural y Química Intercontinentales*, ed. P. Aceves Pastrana, Univ. Autonoma Metropolitana, Unidad Xochimilco, 1996, pp. 17–33.

136. B. J. T. Dobbs, 'From the secrecy of alchemy to the openness of chemistry', in *Solomon's House Revisited: The Organisation and Institutionalisation of Science*, ed. T. Frängsmyr, Science History Publications, Canton, MA, 1990, pp. 75–94.

137. J. V. Golinski, 'Chemistry in the scientific revolution: problems of language and communication', in *Reappraisals of the Scientific Revolution*, ed. D. C. Lindberg and R. S. Westman, Cambridge University Press, Cambridge 1990, pp. 367–396.

138. J. V. Golinski, *Corpus*, 1988, **8/9**, 85–98.

139. S. A. McKnight, ed., *Science, Pseudo-Science, and Utopianism in Early Modern Thought*, University of Missouri Press, Columbia, MO, 1992. (Contributors to this anthology include A. G. Debus, B. J. T. Dobbs, Klaus Vondung, D. Walsh and W. Applebaum.)

140. A. G. Debus, 'Chemists, physicians, and changing perspectives on the scientific revolution', *Isis*, 1998, **89**, 66–81.

141. U. Klein, 'The chemical workshop tradition and the experimental practice: discontinuities within continuities', *Sci. Context*, 1996, **24**, 251–287.

142. M. Crosland, 'Changes in chemical concepts and language in the 17th century', *Sci. Context*, 1996, **9**, 225–240.

143. M. Hunter, *Establishing the New Science: The Experience of the Early Royal Society*, Boydell Press, Woodbridge, Suffolk, 1989.

144. J. V. Golinski, 'A noble spectacle: phosphorus and the public cultures of science in the early Royal Society', *Isis*, 1989, **80**, 11–39.

145. M. B. Hall, *Promoting Experimental Learning: Experiment and the Royal Society 1660–1727*, Cambridge University Press, Cambridge, 1991.

146. M. Hunter, 'The early Royal Society and the shape of knowledge', in *The Shapes of Knowledge from the Renaissance to the Enlightenment*, ed. D. R. Kelley and R. H. Popkin, Kluwer Academic, Dordrecht, 1991, pp. 189–202.

147. P. Kesaris, ed., *Letters and Papers of Robert Boyle: A Guide to the Manuscripts and Microfilm*, Intro. and guide by M. Hunter, University Pub. of America, Bethesda, MD, 1992.

148. M. Hunter, ed., *Robert Boyle Reconsidered*, Cambridge University Press, Cambridge, 1994.

149. S. Shapin and S. Schaffer, *Leviathan and the Air-Pump*, Princeton University Press, Princeton, NJ, 1985.

150. S. Shapin, 'The house of experiment in seventeenth-century England', *Isis*, 1988, **79**, 373–404.

151. S. Shapin, 'Pump and circumstance: Robert Boyle's literary technology', *Social Stud. Sci*, 1984, **14**, 481–520.

152. A. Clericuzio, 'Robert Boyle and the English Helmontians', in *Alchemy Revisited. Proceedings of the International Conference on the History of*

Alchemy at the University of Groningen, 1989, ed. Z. R. W. M. von Martels, Collection de travaux de l'Academie Internationale Histoire des Sciences, 33, Brill, Leiden, 1990, pp. 192–199.

153. A. Clericuzio, 'From Van Helmont to Boyle: a study of the transmission of Helmontian chemical and medical theories in seventeenth-century England', *Br. J. Hist. Sci.*, 1993, **26**, 303–334.

154. T. Harwood, ed., *The Early Essays and Ethics of Robert Boyle*, Southern Illinois University Press, Carbondale, IL, 1991.

155. A. Clericuzio, 'Filosofia corpuscolare e teorie chimiche' in 'Robert Boyle: Uno studio preliminare', in *Storia e fondamenti della chimica; Atti del II Convegno Nationale*, Accad. Nat. delle Scienze detta dei XL Roma, 1988, pp. 305–313.

156. A. Clericuzio, 'A redefinition of Boyle's chemistry and corpuscular philosophy', *Ann. Sci.*, 1990, **47**, 561–589.

157. Y. S. Kim, 'Another look at Robert Boyle's acceptance of the mechanical philosophy: its limits and its chemical and social contexts', *Ambix*, 1991, **38**, 1–10.

158. D. L. Woodall, 'The relationship between science and scripture in the thought of Robert Boyle', *Perspect. Sci. Christ. Faith*, 1999, **48**, 32–39.

159. J. H. Brooke and M. Hunter, 'Foreward', in R. Hooykaas, *Robert Boyle: A Study in Science and Christian Belief*, English trans. by H. Van Dyke, University Press of America Inc., Lanham, MD and Oxford, 1997; translated reprint of 1943 Dutch edition, vii–xix.

160. Robert Boyle, *A Free Enquiry into the Vulgarly Received Notion of Nature*, ed. E. B. Davis and M. Hunter, Cambridge University Press, Cambridge, 1996.

161. H. Krips, 'Ideology, rhetoric, and Boyle's new experiments', *Sci. Context*, 1994, **7**, 53–64.

162. M. Ben-Chaim, 'The value of facts in Boyle's experimental philosophy', *Hist. Sci.*, 2000, **38**, 57–77.

163. U. Klein, 'Robert Boyle-Der Begründer der neuzeitlichen Chemie?', *Phil. Natur*, 1994, **31**, 63–106.

164. L. M. Principe, 'Robert Boyle's alchemical secrecy: codes, ciphers and concealments', *Ambix*, 1992, **39**, 63–74.

165. L. M. Principe, *The Aspiring Adept: Robert Boyle and his Alchemical Quest*, Princeton University Press, Princeton, NJ, 1996.

166. M. Bougard, 'La chimie de Nicolas Lemery (1645–1715): Entre l'officine et l'amphithéâtre', in *Nouvelles Tendances en Histoire et Philosophie des Sciences*, ed. R. Halleux and A.-C. Bernès, Palais des Académies, Bruxelles, 1993, pp. 167–185.

167. M. Teich, 'Interdisciplinarity in J. J. Becher's thought', *Hist. Eur. Ideas*, 1988, **9**, 145–160.

168. P. H. Smith, *The Business of Alchemy: Science and Culture in the Holy Roman Empire*, Princeton University Press, Princeton, NJ, 1994.

169. P. H. Smith, 'Curing the body politic: chemistry and commerce at court 1664–70', in *Patronage & Institutions: Science, Technology and Medicine at the European Court, 1500–1750*, ed. B. T. Moran, Boydell Press, Rochester, NY, 1991, pp. 195–209.

170. S. Pumphrey and F. Dawbarn, 'Science and patronage in England, 1570–1625: a preliminary study', *Hist. Sci.*, 2004, **42**, 137–188.

171. R. Knoeff, 'The making of a Calvinist chemist: Herman Boerhaave, God, fire and truth', *Ambix*, 2001, **48**, 102–111.

172. M. Hunter and S. Schaffer, eds, *Robert Hooke: New Studies*, Boydell, Woodbridge, Suffolk, 1989.

173. A. R. Hall and Marie Boas Hall, eds, *The Correspondence of Henry Oldenburg*, 11 vols, University of Wisconsin Press, Madison, WI, 1965–1977; vols. 12 and 13, Taylor and Francis, London, 1986.

174. F. L. Holmes, *Eighteenth-century Chemistry as an Investigative Enterprise*, Berkeley Papers in History of Science, 12, Office for History of Science and Technology, University of California, Berkeley, CA, 1989.

175. M. Archer and C. Haley, eds, *The 1702 Chair of Chemistry at Cambridge: Transformation and Change*, Cambridge University Press, Cambridge, 2005.

176. M. Crosland, 'Difficult beginnings in experimental science at Oxford: the Gothic chemistry laboratory', *Ann. Sci.*, 2003, **60**, 399–421.

177. M. Crosland, 'Early laboratories c .1600- c. 1800 and the location of experimental science', *Ann. Sci.*, 2005, **62**, 233–253.

178. N. G. Coley, 'Physicians and the chemical analysis of mineral waters in eighteenth-century England', *Med. Hist.*, 1982, **26**, 123–144.

179. N. G. Coley, 'Physicians, chemists and the analysis of mineral waters: "the most difficult part of chemistry"', in *Med. Hist., Supplement No.10*, ed. R. Porter, 1990, 56–66.

180. M. D. Eddy, 'The "Doctrine of salts" and Rev. John Walker's analysis of a Scottish spa (1749–1761)', *Ambix*, 2001, **48**, 137–160.

181. N. G. Coley, 'The preparation and uses of artificial miner waters (ca. 1680–1825)', *Ambix*, 1984, **31**, 32–48.

182. C. Jungnickel and R. McCormmach, *Cavendish: The Experimental Life*, Bucknell University Press (distributed by Associated University Presses), Cranbury, NJ, 1999.

183. R. Schofield, *The Enlightenment of Joseph Priestley: A Study of his Life and Work from 1733 to 1773*, Pennsylvania State University Press, Philadelphia, PA, 1998.

184. D. Eshet, 'Rereading Priestley: science at the intersection of theology and politics', *Hist. Sci.*, 2001, **39**, 127–159.

185. J. Perkins, 'Creating chemistry in provincial France before the Revolution: the examples of Nancy and Metz; Part 1. Nancy', *Ambix*, 2003, **50**, 145–81.

186. J. Perkins, 'Creating chemistry in provincial France before the Revolution: the examples of Nancy and Metz; Part 2. Metz', *Ambix*, 2004, **51**, 43–75.

187. M. Beretta, 'I grandi della scienza: Lavoisier, la rivoluzione chimica', *Le Scienze, ed. Italiana di Scientific American*, anno 1, (3), March 1998.

188. F. L. Holmes, 'What was the chemical revolution about?', *Bull. Hist. Chem*, 1997, **20**, 1–9.

189. J. G. McEvoy, 'Positivism, whiggism, and the chemical revolution: a study in the historiography of chemistry', *Hist. Sci.*, 1997, **35**, 1–33.

190. J.-P. Poirier, *Antoine Laurent de Lavoisier, 1743–1794*, Pygmalion, Gérard Watelet, Paris, 1993.

191. A. Donovan, *Antoine Lavoisier: Science, Administration, and Revolution*, Blackwell, Oxford, 1993.
192. M. Beretta, 'Lavoisier and his new biographers', *Lychnos*, 1994, 153–160.
193. M. Beretta, 'The library of Antoine Laurent Lavoisier (1743–1794)', in *Bibliothecae Selectae da Usano a Leopardi*, ed. Eugenio Canone, Olschki, Firenze, 1993, pp. 457–473.
194. M. Beretta, '"A new course in chemistry": Lavoisier's first chemical paper', *Biblioteca di Nuncius: Studi e testi*, 13, Uppsala Studies in History of Science, 16, Olschki, Firenze, 1994.
195. R. Halleux, J. Jacques, M. Goupil and M. Crosland, eds, 'Actes du symposium organisé à l'occasion du bicentenaire de la publication du *Traite Élémentaire de Chimie*, par Lavoisier', *Rev. Questions Sci.*, **160**, 1989, 145–219.
196. J.-P. Poirier, 'Antoine Laurent Lavoisier (1743–1794)', *Vie. Sci.*, 1994, **11**, 197–221.
197. B. Bensaude-Vincent, 'Lavoisier: une révolution scientifique' in *Éléments d'histoire des sciences*, ed. M. Serres, Bordas, Paris, 1989, pp. 363–385.
198. B. Bensaude-Vincent, 'Lavoisier: Mémoires d'une revolution', Préface de M. Serres, *Figures de la Science*, Flammarion, Paris, 1993.
199. J.-P. Poirier, *Antoine Laurent de Lavoisier, 1743–1794*, Pygmalion, Gérard Watelet, Paris, 1993.
200. M. Beretta, in 'Échanges d'influences scientifiques et techniques entre pays Européens de 1780 à 1830', C. T. H. S., Paris, 1990, pp. 125–145.
201. C. Demeulenaere-Douyère, ed., *Actes du Colloque Organisé à l'occasion du Bicentenaire de la Mort d'Antoine Laurent Lavoisier, le 8 Mai 1794*, Technique & Documentation, Paris, 1995.
202. B. Bensaude-Vincent, 'Lavoisier et la révolution chimique', *Recherche*, 1994, **25**, 539–544.
203. B. Bensaude-Vincent, 'Between history and memory: centennial and bicentennial images of Lavoisier', *Isis*, 1996, **87**, 481–499.
204. F. L. Holmes, *Lavoisier and the Chemistry of Life; An Exploration of Scientific Creativity*, University of Wisconsin Press, Madison, WI, 1985.
205. F. L. Holmes, *Antoine Lavoisier, The Next Crucial Year; Or, the Sources of his Quantitative Method in Chemistry*, Princeton University Press, Princeton, NJ, 1998.
206. A. Donovan, 'The chemical revolution and the Enlightenment - and a proposal for the study of scientific change', in *Philosophy and Science in the Scottish Enlightenment*, ed. P. Jones, Donald, Edinburgh, 1988, pp. 87–101.
207. A. Donovan, ed., 'The chemical revolution; essays in reinterpretation', *Osiris*, *Ser.* 2, 1988, **4**.
208. W. A. Smeaton *et al.*, 'Lavoisier and the chemical revolution', *Bull. Hist. Chem.*, 1989, **5**, 4–50. (Contributors: W. A. Smeaton, A. L. Donovan, J. E. White, R. Siegfried, F. L. Holmes, A. T. Schwartz, B. B. Chastain, D. A. Davenport, K. M. Ireland and W. B. Jensen.)
209. F. L. Holmes, *Eighteenth-century Chemistry as an Investigative Enterprise*, Berkeley Papers in History of Science, 12, Office for History of Science and Technology, University of California, Berkeley, CA, 1989.

210. M. Beretta, *Imaging a Career in Science: The Iconography of Antoine Laurent Lavoisier*, Science History Publications, Canton, MA, 2001.

211. A. Donovan, 'Buffon, Lavoisier and the transformation of French chemistry', in *Actes du Colloque International pour le Bicentenaire de la Mort de Buffon*, Paris, Montbard, Dijon, 14–22 Juin, 1988. Réunis par J.-C. Beaune, S. Benoit, J. Roger, D. Woronoff, sous la direction de J. Gayon, Vrin, Paris; Institut Interdisciplinaire d'Etudes Epistémologiques, Lyon, 1992.

212. C. E. Perrin, 'Research traditions, Lavoisier and the chemical revolution', in 'The chemical revolution; essays in reinterpretation', ed. A. Donovan, *Osiris*, Ser. 2, 1988, **4**, 53–81.

213. C. E. Perrin, 'Document, text and myth: Lavoisier's crucial year revisited', *Br. J. Hist. Sci.*, 1989, **22**, 3–25.

214. C. E. Perrin, 'Revolution or reform: the chemical revolution and eighteenth-century concepts of scientific change', *Hist. Sci.*, 1987, **25**, 395–423.

215. C. E. Perrin, 'The chemical revolution: shifts in guiding assumptions', in *Scrutinising Science: Empirical Studies of Scientific Change*, ed. A. Donovan *et al.*, Kluwer Academic, Dordrecht, 1988, pp. 105–124.

216. A. Donovan, ed., 'The chemical revolution; essays in reinterpretation', *Osiris*, Ser. 2, 1988, **4**, 214–231.

217. E. M. Melhado, 'Chemistry, physics and the chemical revolution', *Isis*, 1985, **76**, 195–211.

218. C. Perrin, 'Chemistry as peer of physics: a response to Donovan and Melhado on Lavoisier', *Isis*, 1990, **81**, 259–270.

219. A. Donovan, 'Lavoisier as chemist *and* experimental physicist: A reply to Perrin', *Isis*, 1990, **81**, 270–272.

220. E. M. Melhado, 'On the historiography of science; a reply to Perrin', *Isis*, 1990, **81**, 273–276.

221. A. Donovan, 'Newton and Lavoisier: from chemistry as a branch of natural philosophy to chemistry as a positive science', in *Action and Reaction: Proceedings of a Symposium to Commemorate the Tercentenary of Newton's Principia*, ed. P. Theerman and A. F. Steff, University of Delaware Press, Newark, DE, 1993, pp. 255–276.

222. A. M. Duncan, *Laws and Order in Eighteenth-century Chemistry*, Oxford University Press, Oxford, 1995.

223. M.G. Kim, *Affinity, that Elusive Dream: A Genealogy of the Chemical Revolution*, MIT Press, Cambridge, MA, 2003.

224. F. L. Holmes, *Antoine Lavoisier - the Next Crucial Year: Or the Sources of his Quantitative Method in Chemistry*, Princeton University Press, Princeton, NJ, 1998.

225. M. Beretta, 'The grammar of matter: chemical nomenclature during the 18th century' in *Sciences et Langues en Europe*, ed. R. Chartier and P. Corsi, École des Hautes Études en Sciences Sociales, Centre A. Koyré, Paris, 1996, pp. 109–125.

226. B. Bensaude-Vincent, *Méthode de Nomenclature Chimique [Par] L. B. Guyton de Morveau A. L. Lavoisier, C. L. Berthollet et A. F. de Fourcroy*, Éditions du Seuil, Paris, 1994.

227. C. Viel, 'Guyton de Morveau, père de la nomenclature chimique (1737–1816)', in *Lavoisier et la Révolution Chimique. Actes du colloque tenu à l'occasion du Bicenntenaire de la Publication du* Traité Élémentaire de Chimie *1789*, ed. M. Goupil, P. Bret and F. Masson, SABIX, École Polytechnique, Paris, 1992, pp. 129–170.

228. B. Bensaude-Vincent and F. Abbri, eds, *Lavoisier in European Context: Negotiating a New Language for Chemistry*, Science History Publications, Canton, MA, 1995.

229. J. Riskin, 'Rival idioms for a revolutionized science and a republican citizenry', *Isis*, 1998, **89**, 203–232.

230. J. Langins, 'Fourcroy, historien de la Révolution chimique', in ref. 227, pp. 13–33.

231. T. H. Levere, 'Lavoisier, Language, instruments and the chemical revolution', in *Nature, Experiment and the Sciences*, ed. T. H. Levere and W. Shea, Kluwer, Dordrecht, 1990, pp. 207–223.

232. P. Laszlo, *La Parole des Choses ou Le Langage de la Chimie*, Hermann, Paris, 1993.

233. V. A. Kritsman, 'A. Lavoisier's contribution to the appearance of structural concepts in chemistry', *Janus*, 1986–90, **73**, 29–37.

234. P. Belin, 'Un Collaborateur d'Antoine Laurent Lavoisier à l'Hotel de l'Arsenal: Jean-Baptiste Meusnier (1754–1793)', in ref. 227, pp. 263–293.

235. E. Grison, 'Hassenfratz et Lavoisier', in ref. 227, pp. 333–353.

236. R. Taton, 'L'Oeuvre de Monge en chimie: Sa collaboration et ses relations avec Lavoisier', in ref. 227, pp. 55–89.

237. W. A. Smeaton, 'Monsieur and Madame Lavoisier in 1798; the chemical revolution and the French Revolution', *Ambix*, 1989, **36**, 1–4.

238. W. A. Smeaton, 'Madame Lavoisier, P. S. and E. I. Du Pont de Nemours, and the publication of Lavoisier's *Mémoires de chimie*', *Ambix*, 1989, **36**, 22–30.

239. C. E. Perrin, 'The Lavoisier-Bucquet collaboration: a conjecture', *Ambix*, 1989, **36**, 5–13.

240. A. Donovan, 'Lavoisier's two publics', *Trans. Int. Congr. Enlightenment*, 1992, **8**, 1181–1183.

241. A. Donovan, *Antoine Lavoisier: Science, Administration, and Revolution*, Blackwell, Oxford, 1993.

242. B. Bensaude-Vincent, 'The balance: between chemistry and politics', *Eighteenth Cent. Theory Interpr.*, 1992, **33**, 217–237.

243. M. Crosland, *In the Shadow of Lavoisier: The* Annales de chimie *and the Establishment of a New Science*, BSHS Monographs, 9, British Society for the History of Science, Faringdon, 1994.

244. M. Crosland, 'Lavoisier, the two French revolutions and the "imperial despotism of oxygen"', *Ambix*, 1995, **42**, 101–118.

245. J. Golinski, 'Precision instruments and the demonstrative order of proof in Lavoisier's chemistry', *Osiris*, 1994, **9**, 30–47.

246. J. Golinski, "The nicety of experiment'. Precision of measurement and precision of reasoning in late 18th-century chemistry', in *The Values of Precision*, ed. M. Norton Wise, Princeton University Press, Princeton, NJ, 1995, pp. 72–91.

247. M. Goupil, 'Claude-Louis Berthollet, collaborateur et continuat eur de Lavoisier' in ref. 227, pp. 35–53.

248. U. Klein, 'Origin of the concept of chemical compound', *Sci. Context*, 1994, **7**, 163–204.

249. U. Klein, 'E. F. Geoffroy's table of different 'rapports' observed between different chemical substances: a reinterpretation', *Ambix*, 1995, **42**, 79–100.

250. U. Klein, *Verbindungen und Affinität: Die Grundllegung der neuzeitlichen Chemie an der Wende von 17, zum 18, Jahrhundert*, Birkhäuser, Basel, 1994.

251. D. Allchin, 'Phlogiston after oxygen', *Ambix*, 1992, **39**, 110–116.

252. M. Engel, ed., *Von der Phlogistik zur modernen Chemie. Vorträge des Symposiums aus Anlass des 250 Geburtstages von Martin Heinrich Klaproth, Technische Universität Berlin, 29 November, 1993*, Studien and Quellen zur Geschichte der Chemie 5, Engel, Berlin, 1994.

253. K. Huffbauer, *The Formation of the German Chemical Community (1720–1795)*, University of California Press, Berkeley, CA, 1982.

254. H. G. Schneider, 'The "fatherland of chemistry"; early nationalistic currents in late eighteenth century German chemistry', *Ambix*, 1989, **36**, 14–21.

255. E. Grison, M. Goupil and P. Bret, eds, *A Scientific Correspondence during the Chemical Revoution: Louis Bernard Guyton de Morveau, and Richard Kirwan, 1782–1802*, Office for History of Science and Technology, University of California, Berkeley, CA, 1994.

256. H. A. M. Snelders, 'The New Chemistry in the Netherlands', in ref. 216, pp. 121–145.

257. A. Lundgren, 'The New Chemistry in Sweden', in ref. 216, pp. 146–168.

258. R. Gago, 'The New Chemistry in Spain', in ref. 216, pp. 169–192.

259. J. G. McEvoy, 'Continuity and discontinuity in the chemical revolution', in ref. 216, pp. 195–213.

260. J. E. White, 'The new chemistry in America', *Trans. Int. Congr. Enlightenment*, 1992, **8**, 1178–1181.

261. 'Joseph Priestley (1733–1804)', *Enlightenment & Dissent*, Commemorative issue, 1993, **2**, 1–121.

262. M. F. Conlin, 'Joseph Priestley's American defence of phlogiston reconsidered', *Ambix*, 1996, **43**, 129–145.

263. J. E. White, 'The Priestley memorial volume', *Notes Rec. R. Soc. Lond.*, 1994, **48**, 85–96.

264. P. K. Basu, 'Scientific explanation in the history of chemistry: the Priestley-Lavoisier debate', *Diss. Abstr. Int.*, 1993, 53:3937–A7.

265. P. K. Basu, 'Similarities and dissimilarities between Joseph Priestley's and Antoine Lavoisier's chemical beliefs', *Stud. Hist. Phil. Sci.*, 1992, **23**, 445–469.

266. M. Beretta, 'Chemists in the storm: Lavoisier, Priestley and the French Revolution', *Nuncius*, 1993, **8**, 75–104.

267. J. Simon, 'The chemical revolution and pharmacy: a disciplinary perspective', *Ambix*, 1998, **45**, 1–13.

268. F. L. Holmes, 'The "Revolution in Chemistry and Physics"; overthrow of a reigning paradigm or competition between contemporary research programmes?', *Isis*, 2000, **91**, 725–753.

269. J. Simon, 'Analysis and the hierarchy of nature in eighteenth-century chemistry', *Br. J. Hist. Sci.*, 2002, **35**, 1–16.
270. M. Crosland, *Science under Control: The French Academy of Sciences, 1795–1914*, Cambridge University Press, Cambridge, 1992.
271. C. H. Evans, ed., *Episodes from the History of the Rare Earth Elements*, Kluwer, Dordrecht, 1996.

CHAPTER 3

Inorganic Chemistry

W. A. CAMPBELL[†]

University of Newcastle upon Tyne

3.1 Introduction

No comprehensive history of inorganic chemistry has been written during the period since the publication in 1985 of *Recent Developments in the History of Chemistry*,[1] though two general histories of chemistry have presented useful overviews.[2,3] The subject has not been treated systematically in serial literature; instead, some half dozen topics have been worked exhaustively, usually under the impetus of a significant anniversary. For the rest, the coverage has been patchy and of uneven significance to the historian of chemistry, although one paper traces the history of inorganic chemistry from the 1870s to the 1970s.[4]

3.2 Atoms and the Periodic Table

Although the title 'inorganic chemistry' did not appear in print until the 1830s, the roots of the discipline are more than four hundred years old and have been discussed in Chapter 2. In the early years of the 19th century, two influences combined to focus attention on atomic weights. One was the atomic theory of John Dalton,[5] which stated that each element was characterized by the relative weight of its atom.[6] The other was the need felt by industrial and analytical chemists for accurate combining weights. The story of recommended atomic weights from 1882 to 1997 has been chronicled.[7] Many attempts were made to correlate chemical properties with atomic weights of elements, associated with the names of J. W. Döbereiner, J. B. A. Dumas, W. Odling and J. A. R. Newlands.[8] The periodicity of chemical properties against atomic weights was demonstrated almost simultaneously by D. I. Mendeleev and Lothar Meyer.[9–13] Inevitably, controversial questions about priority have arisen,[14–17] as have discussions about the roots of Mendeleev's discovery.[18–20] In Mendeleev's first table, the periods were vertical and the groups horizontal, an order which Lothar Meyer reversed. In 1870, Lothar Meyer published his graph of atomic volumes, plotting various quantitative properties against atomic weight and revealing periodicity in a striking manner. After that, the periodic table was quickly accepted in the USA and Britain;[21] it has since undergone many changes in form.[22,23]

Dalton's atoms were not perceived as possessing structure, but the discovery of electrons, and the distinctive phenomenon of radioactivity, inevitably generated interest in the way atoms were put together. Clearly, this had a bearing on the periodic table as any proposed structure must explain atomic weights.[24–28] The Aufbau Principle, that each element possessed one more proton in the nucleus and one more electron in the outer shell than the preceding element, effectively systematized the periodic table[29,30]

Two remaining problems associated with the periodic classification were solved in different ways. To get a good fit, gaps had to be left for elements not yet discovered, and for these Mendeleev was bold enough to predict the properties in some detail. His confidence was rewarded by the discovery of gallium in 1875, scandium in 1879 and germanium in 1886.[31,32] In the wave of nationalism that was sweeping over Europe, these elements were named after the countries in which they were discovered; fashions in naming elements have continued to change.[33]

The second problem concerned the maximum number of elements the existing table could accommodate (and therefore the number of elements still to be discovered). The work of H. G. J. Moseley in 1913 on the relation between atomic number and the frequency of characteristic X-rays finally settled this question.[34] The group of elements formerly known as the rare earths and now called lanthanides presented difficulties on two levels. The similarity of their chemical properties made separation and subsequent identification extremely troublesome; moreover, there was no obvious place for them in the periodic table.[35–38] The discovery of elements 70, 71 and 72 generated great controversy.[39] Equally difficult to place were the unsuspected gases discovered in the 1890s by William Ramsay, Lord Rayleigh and Morris Travers. These were classified first as rare gases, then as inert gases and finally as noble gases when it became clear that they were neither rare nor - following the discovery of xenon compounds by Neil Bartlett in 1961 - inert.[40–43] Recent concern about the diffusion of radon into dwellings has focused attention on the history of this member of the group,[44,45] with special emphasis on the work of Rutherford.[46,47]

Ideas on the structure of inorganic compounds were changed by the discovery of new classes of substances that did not fit easily into the older schemes. One such class was the metal carbonyls, beginning with the recognition in 1886 of $Ni(CO)_4$, the unexpectedly volatile nickel carbonyl.[48,49] A larger class of compounds defied the ordinary rules of valency. It had long been known that aqueous solutions of copper salts turned dark blue when treated with ammonia; similar coloured compounds occurred with nickel and cobalt. These were studied by C. W. Blomstrand, who proposed a chain formula to account for the structure.[50,51] Many other apparently saturated molecules were found to combine to form complexes, and these were investigated by A. Werner, who named the products coordination compounds.[52–54] Werner's work laid the foundations for modern inorganic structure theory; it is also the basis of inorganic chemical nomenclature.[55–57] In 1830, W. C. Zeise made platinum complexes with organic compounds, and from this obscure beginning came the important class or organometallic compounds.[58] Organometallic chemistry is discussed in Chapter 4. More recently this has blossomed into the field of bio-inorganic chemistry.[59]

3.3 Uranium and the Trans-uranics

Uranium was discovered in 1789 in pitchblende from Joachimisthal by M. H. Klaproth, though he isolated the oxide and not the metal.[60-63] As part of the celebration of Klaproth's discovery, his lectures have been published, together with notes taken by his student, the philosopher Schopenhauer.[64-66] At first the only use that could be found for uranium was to impart a yellow colour to glass and pottery glazes.[67]

A century after its discovery, uranium and its pitchblende source came into prominence through A. H. Becquerel's recognition of radioactivity.[68] Another fifty years were to pass before Otto Hahn and Fritz Strassman split the uranium atom into barium and krypton.[69-71] Also in 1939, the missing element 43, technetium, was discovered by C. Perrier and E. G. Segne, though a controversy soon arose with the team of W. Noddack, I. Tacke and O. Berg who, in 1927, had made short-lived isotopes.[72,73] At about the same time, Marguerite Perey had discovered the missing alkali metal francium.[74,75] Between 1934 and 1938, several workers had bombarded uranium with neutrons, but had failed to discover element 93, neptunium, largely because they expected its chemistry to resemble that of rhenium and osmium, when actually it resembled uranium.[76-78] The actinide series had started in 1917 with the independent discovery of protactinium by Liese Meitner and Otto Hahn, F. Soddy and J. Cranston and K. Fajans,[79] but the extension of the series to the transuranium elements only began with the discovery of neptunium in 1940.[80] Other elements followed in quick succession, plutonium by Glen Seaborg in 1940,[81] and americium and curium by Seaborg's team in 1944.[82-86]

3.4 Main Group Elements

Coverage of the history of the main group elements has been uneven, though individual papers of high quality have emerged. Again, anniversaries seem to have governed the choice of topics. N. N. Greenwood has reviewed the recent history of main group chemistry, seeing in it a rebirth of interest.[87] The alkali metals sodium and potassium were isolated electrolytically by Humphry Davy in 1807.[88] Of the remaining members, lithium was discovered by J. A. Arfvedson in 1817 and caesium and rubidium by R. Bunsen and G. R. Kirchhoff in 1860 and 1861, respectively using their new spectroscope.[89] Lockyer's spectroscopic work led to several "proto-elements".[90]

Chief among Group II metals, calcium in the form of limestone and lime was known to the ancient world, but the metal was isolated only in 1808 by Davy by electrolysis of its mercury amalgam.[91] Among calcium compounds, calcium montmorillonite, or fullers' earth, has been treated in great detail. Originally used for cleansing wool (fulling), its main use now is for cat litter.[92] Strontium spar was recognized as a new earth by A. Crawford in 1790, the metal being isolated by Davy in 1808.[93]

In 1702, William Homberg heated borax with iron(II) sulphate and obtained by sublimation a compound he called "sedative salt", though it was not a salt but boric acid.[94] The boron hydrides, first studied by A. Stock in 1909, have provided a rich source of new chemistry.[95] Pliny described a silvery cup, lighter than any known metal, once owned by the emperor Tiberius. Some have speculated that the Romans might have stumbled on aluminium.[96] Though alum was known in antiquity, the impure metal was isolated only in 1825 by H. C. Oersted, and prepared on a large scale by Sainte-Claire Deville in 1854.[97,98] The green spectral line of thallium was detected by William Crookes in 1861, the metal being isolated by C. A. Lamy in the following year.[99,100]

The only Group IV inorganic element to attract much attention has been silicon, isolated by J. J. Berzelius in 1824.[101] Carbon, on the other hand, has returned to centre stage with the discovery of a new allotrope C_{60}, or fullerene. A retrospective account[102] indicates a missed opportunity for its discovery in 1970. Following the first recognition in 1985, and an early report,[103] there have been many historical accounts,[104,105] one of them stressing an "epidemic spread of fullerene research"[106] and another emphasizing applications as nanotubes[107] (see Chapter 4).

Among Group V elements, the history of nitrous oxide has continued to excite interest, not least for its anaesthetic properties.[108] The discovery in 1865 of nitrogen hydroxide, or hydroxylamine, has generated a lengthy paper.[109] Phosphorus was discovered in 1669 by Hennig Brand while he was trying to make gold. It brought him little profit, though his process was exploited by J. Kunckel, G. W. Leibnitz and J. D. Krafft.[110] The somewhat sinister history of arsenic never fails to attract attention.[111]

In Group VI the history of ozone has been related by Rubin[112,113] and others in a symposium about its early history.[114] The varied fortunes of ion-hydration theories have been described for Russia, Germany and the USA[115] Other historical writing relates to the role of sulphur hexafluoride in atomic bomb production[116] and even to the humble Kipp's apparatus for generating hydrogen sulphide.[117,118]

A good deal of historical work has appeared on the halogens. The work of Scheele, Lavoisier, Berthollet, Gay-Lussac and Davy on chlorine has been briefly reviewed,[119] and a more lengthy paper has appeared on the isolation of iodine by B. Courtois in 1811.[120] Fluorine became important to the Manhattan Project as a vehicle for separating uranium isotopes.[121,122] The centenary of fluorine's isolation by H. Moissan in 1886 created much interest.[123–128]

3.5   Transition Metals

The transition metals have received considerable attention: copper,[129] mercury, including its significance in alchemy,[130] vanadium[131,132]chromium,[133] and tungsten.[134] The history of the platinum metals has received exceptional attention from two sources. One is a book[135] and the other the *Platinum Metals Review*, since July 2004 a purely electronic journal.[136] The work on platinum metals of both Werner[137] and Wilkinson[138] has been described. Attention is also given to the problems raised by a Spanish monopoly of platina ores.[139] Platinum itself was brought to Europe by Charles Wood from South America, and rendered malleable by W. H. Wollaston.[140–143] Its "secret history" is now being exposed,[144] together with the discoveries of its isotopes,[145] and even its use for glass-making at Jena.[146]

The four metals rhodium, palladium, osmium and iridium, share the same centennial and have been dealt with together.[147–149] Wollaston separated palladium from platinum ore in 1803 but concealed the identity of the metal until 1804.[150] Osmium was isolated from crude platinum by Smithson Tennant in 1804.[151] There are accounts of the discoveries of niobium (by Hatchett)[152] and ruthenium.[153,154]

References

1. C. A. Russell, 'General and inorganic chemistry,' in *Recent Developments in the History of Chemistry*, ed. C. A. Russell, Royal Society of Chemistry, 1985, pp. 77–96.
2. W. H. Brock, *Fontana History of Chemistry*, Fontana Press, London, 1992, ch. 9, p.311; ch. 15, p.570.
3. J. Hudson, *A History of Chemistry*, Macmillan, London 1992, ch. 12, 187.
4. J. A Zubieta and J. J. Zuckermann, *Chem. Eng. News*, 1978, **54**, 64–79.
5. F. Greenaway, *Chem. Br.*, 1994, **31**, 920–921.
6. W. Jansen, *Prax. Naturwiss. Chem.*, 1986, **35** (2), 34–40.
7. T. B. Coplen and H. S. Peiser, *Pure Appl. Chem.*, 1998, **70** (1), 237–257.
8. A. Arrieta, G. Ludwing and I. Beyea, *Bol. Soc. Quim. Peru*, 1991, **57** (1), 7–22.
9. E. Fluck and K. Rumff, *Chem. Z.*, 1986, **20** (4), 111–116.
10. S. Zamecki, *Kwart. Hist. Nauki Tech.*, 1988, **33** (1), 125–143.
11. O. Romanp, *Elhuyar*, 1984, **10** (4), 401–413.
12. N. E. Holden, *Chem. Int.*, 1984, (6), 19–31.
13. F. Szabadvary, *Magy. Kem. Lapja*, 1984, **39** (9), 407–411.
14. K.Borecka and A. Gorski, *Wind. Chem.*, 1984, **38**(9), 724–749.
15. B. Bensaude-Vincent, *Recherche*, 1984, **15** (159), 1206–1215.
16. F. Kober, *Chem. Ztg.*, 1985, **109** (12), 419–427.
17. D. H. Rouvray, *Chem. Br.*, 1994, **30** (5), 373–378.
18. I. S. Dmitriev, *Hist. Stud. Phys. Biol. Sci.*, 2004, **34** (2), 233–275.
19. M. D. Gordin, *Hist. Stud. Phys. Biol. Sci.*, 2002, **32** (2), 263–290.
20. M. D. Kaji, *Bull. Hist. Chem.*, 2002, **27** (1), 4–16.
21. S. G. Brush, *Isis,* 1996, **87**, 595–628.
22. L. Cerruti, *Chim. Ind.*, 1985, **67** (9), 500–507.
23. A. Arrieta, G. K. Ludwig and L. Beyerl, *Bol. Ser. Quim. Rev.*, 1991, **57** (2), 65–80.
24. H. K. Jruger, *Spectrum*, 1986, **24** (l), 16–18.
25. R. Carazza and G. P. Guidetti, *Rend. Semin. Fac. Sci.*, 1984, **54** (1), 73–86.
26. M. Davies, *Chem. Br.*, 1987, **23** (2), 118.
27. W. Kuhn, *Prax. Naturwiss. Phys.*, 1992, **41** (4), 39–41.
28. W. Kuhn, *Prax. Naturwiss. Phys.*, 1992, **41** (5), 32–33.
29. K. A. Jensen, *Dan. Kemi*, 1985, **66** (10), 276–288.
30. See also I.S. Dmitriev, *Kwart. Hist. Nauki Tech.*, 1984 , **29** (3–4), 559–567.
31. H. H. Walter, *Wiss. Fortschr.*, 1987, **37** (4), 96–99.
32. G. Ackerman, *Freiberg Forschungsh., A*, 1987, **76** (7), 9–16.
33. V. Ringnes, *J. Chem. Educ.*, 1989, **66** (9), 731–735.
34. C. M. Meyer, *Spectrum* (Pretoria), 1993, **31** (3), 14–17.
35. F. Habashi, *CIM Bull.*, 1994, **87** (976), 80–88.
36. H. D. Hardt, *Chem. Labor. Beitre.*, 1988, **39** (10), 488, 490–491.
37. F. R. Monal, *CIM Bull.*, 1990, **83** (943), 97–101.
38. F. Szabadvary, *Öesterr. Chem. Ztg.*, 1989, **90** (11), 337–338.
39. H. Kragh, in *Episodes from the History of the Rare Earth Elements*, ed. C. H. Evans, Kluwer Academic, Dordrecht, 1996, pp. 67–89.
40. C. K. Jorgensen, *Struct. Bonding*, 1990, **73**, 1–15.

41. T. Lister, *Chem. Rev.*, 1992, **1** (4), 18–20.
42. J. H. Holloway, *Chem. Br.*, 1988, **23** (7), 658, 660, 662, 664.
43. P. Laszlo, *Angew. Chem.*, 1988, **100** (4), 495–506.
44. G. B. Kaufmann, *J. Coll. Sci. Teach.*, 1988, **17** (4), 264–268.
45. W. Schüttmann, *Isotopenpraxis*, 1988, **24** (4), 158–163.
46. J. L. Marshall and V. R. Marshall, *Bull. Hist. Chem.*, 2003, **28** (2), 76–83.
47. M. F. Rayner-Canham and G. W. Rayner-Canham, *Bull. Hist. Chem.*, 2004, **29** (2), 89–90.
48. E. Abel, *Educ. Chem.*, 1992, **29** (2), 46–49.
49. W. A. Hermann, *Chem. Z.*, 1988, **22** (4), 113–122.
50. G. B. Kauffman, *Coordination Chemistry*, American Chemical Society Symposium 565, 1994, pp. 3–33.
51. L. Tansloe, *Coordination Chemistry*, American Chemical Society Symposium 565, 1994, pp. 34–40.
52. F. Kober, *Math. Naturwiss. Unterr.*, 1993, **46** (3), 157–160.
53. R. Soloniewicz, *Kwart. Hist. Nauki Tech.*, 1993, **38** (2), 79–108.
54. L. M. Venanzi, *Chimia*, 1994, **48** (1–2), 16–22.
55. G. B. Kauffmann, *Coordination Chemistry - A Century of Progress*, American Chemical Society, Washington, DC, 1994.
56. M. V. Orna, A. W. Kozlowski, A. Baskinger and T. Adams, *Coordination Chemistry*, American Chemical Society Symposium 565, 1994, pp. 65–75.
57. G. Maeueler, *Prax. Naturwiss. Chem.*, 1984, **33** (4), 104–111.
58. L. B. Hunt, *Plat. Met. Rev.*, 1984, **28** (2), 76–83.
59. R. J. P. Williams, *Coord. Chem. Rev.*, 1990, **100**, 573–610.
60. W. Schüttmann, *Kemenergie*, 1989, **32** (10), 416–420.
61. H. Huebner, *Isotopenpraxis*, 1989, **25** (9), 361–367.
62. A Hermann, *Nucl. Europ.*, 1989, **9** (9–10), 9–12.
63. L. Ninisto, *Kem.-Kemi.*, 1990, **17** (2), 132–135.
64. B. Engel, *Mitt. Ges. Deutsch. Chem. Fach. Geschichte Chem.*, 1989, **3**, 27–37.
65. M. H. Klaproth, *Vorlesungen über die Experimental-Chemie* (1789), Verlag für Wissenschafts und regional Geschichte, Berlin, 1993.
66. M. H. Klaproth, *Chemi, nach der Abschrift von Arthur Schopenhauer* (1811–12), Verlag fur Wissenschafts - und regional Geschichte, Berlin, 1993.
67. R. J. Schwankner and G. Lieckfeld, *Math. Naturwiss. Unterr.*, 1991, **44** (1), 25–32.
68. W. Schüttmann, *Naturwiss. Rundsch.*, 1988, **41** (11) 435–41.
69. A. Hermann, *lsotopenpraxis*, 1985, **21** (7), 237–240.
70. E. Amaldi, *Radioactivite Artificielle à 50 (cinquante) ans*, C. N. R. S., Paris, 1984, pp. 1–24.
71. O. Hazel, *KFK Nachr.*, 1988, **20** (4), 197–199.
72. G. Hermann, *Nucl. Phys. A*, 1989, **505** (2), 352–360.
73. P.K. Kuroda, *Nucl. Phys., A*, 1989, **503** (1), 178–192.
74. J.P. Adloff, *Actual Chim.*, 1989, (4), 127–129.
75. G. B. Kauffmann and J.P. Adloff, *Educ. Chem.*, 1989, **26**, 135–137.
76. P. H. Abelson, *Proc. Robt. A Welch Found. Chem. Res.*, 1990, **34**, 10–21.
77. K. Starke, *Isotopenpraxis*, 1990, **26** (8), 349–351.
78. U. Niese, *Isotopenpraxis*, 1990, **26** (8), 352–355.

79. R. L. Sime, *J. Chem. Educ.*, 1986, **63** (8), 653–657.
80. A. G. Maddock, *Inorg. Chim. Acta*, 1987, **139** (1–2), 7–12.
81. G. T. Seaborg, *J. Chem. Educ.*, 1989, **66** (5), 379–384.
82. G. T. Seaborg, *Americium-Curium Chem. Technol. Pap. Symp.*, 1985, 3–17.
83. R. A. Penneman, *Americium-Curium Chem. Technol. Pap. Symp.*, 1985, 25–33.
84. G. T. Seaborg, *Chem. Br.*, 1991, **27**, 200–201.
85. A. G. Maddock, in *Handbook of Hot Atom Chemistry*, ed. J. P. Adloff, VCH, New York, 1992, pp. 3–16.
86. J. Willard and D.C. Hoffman, in *Handbook of Hot Atom Chemistry*, ed. J. P. Adloff, VCH, New York, 1992, pp. 21–26.
87. N. N. Greenwood, *J. Chem. Soc. Dalton Trans.*, 1991, 565–573 (150th annual issue).
88. C. Berger and H. Frickenfrerichs, *Prax. Naturwiss. Chem.*, 1991, **40** (6), 2–9.
89. S. Boertitz and H. G. Daessler, *4th Spurenelem. Symp.*, Jena, 1983, "Lithium", pp. 10–17.
90. M. Leone and N. Robotti, *Ann. Sci.*, 2000, **57**, 241–266.
91. F. L. Wiercinski, *Biol. Bull.*, 1989, **176** (3), 195–217.
92. R. H. S. Robertson, *Fullers' Earth: a History of Calcium Montmorillonite*, Mineralogical Soc. Occasional Publication, Voltuma Press, Hythe, Kent, 1986.
93. G. C. Britton and C. H. Johnson, *School Sci. Rev.*, 1986, **68** (243), 236–244.
94. O. Bostrup, *Dan. Kemi*, 1984, **65** (6–7), 182.
95. G. Suess-Fink, *Chem. uns. Z.*, 1986, **20** (3), 90–100.
96. Anon., *Metalurgiya*, 1990, **45** (9), 24–29.
97. S. Wilkening, *Praxis Naturwiss. Chem.*, 1986, **35** (3), 11–17.
98. G. Winkhaus, *Erzmetall*, 1987, **40** (3), 115–119.
99. C. Briese, *Wiss. Z. Humboldt-Univ. Berlin Math-Natur-Wiss. Reihe*, 1985, **34** (8), 750–751.
100. A. N. Krivomazov and A. N. Kharitonova, *Voprosy Istorii estestvoenaniia i techni*, 1985, no. 1 [in Russian].
101. P. Mikulcik and R. J. Schwankner, *Prax. Naturwiss. Chem.*, 1988, **37** (5), 2–7.
102. E. Osawa, *Phil. Trans. Roy. Soc., Ser. A*, 1993, **343**, 1–8.
103. C. O'Driscoll, *Chem. Br.*, 1987, **23**, 87–88.
104. D. Spanicek, *Polimeri* (Zagreb), 1995, **16**, 117–120.
105. K. Fostiropoulos, *Int. J. Mod. Phys., B*, 1992, **8**, 3791–3800.
106. T. Braun, *Magy. Kem. Foly*, 1993, **99**, 212–215.
107. T. Ebbesen, *NATO ASI Ser. E*, 1996, **316**, 405–418.
108. S. Colás, *Sci. Culture*, 1998, **7** (3), 335–353.
109. G. Zinner, *Chem-Ztg.*, 1990, **114** (6), 197–204.
110. H. W. Prinzler, *Wiss. Z. Tech. Hochschule 'Carl Schorlemmer', Leuna-Merseburg*, 1985, **27** (l), 33–49.
111. J. Feldmann, *Chem. Br.*, 2001, **37** (1), 31–32.
112. M. B. Rubin, *Bull. Hist. Chem.*, 2001, **26** (1), 40–56, 2002, **26** (2), 81–106 *etc.*
113. M. B. Rubin, *Bull. Hist. Educ. Soc.*, 2004, **29** (2), 99–106.
114. A. R. Bandy, ed., *The Chemistry of the Atmosphere – Oxidants and Oxidation in the Earth's Atmosphere*, Special Publication no. 170, Royal Society of Chemistry, Cambridge, 1995, pp. 140–223.

115. R. E. Rice, *Bull. Hist. Chem.*, 2002, **1** (27), 17–25.
116. F. S. Preston, *Chem. Heritage*, 2003, **21** (2), 12–13, 32–36.
117. F. Habashi, *Chem. Heritage*, 2002, **20** (4), 9.
118. E. Homburg, *RSC Hist. Gp. Newsletter*, 2001 (Feb.), 14–17.
119. H. H. Grelland, *Kjemi*, 1989, **48** (10), 89–143.
120. H. Bruckner and H. Schladitz, *5th Spurenelem. Symp.*, Jena, 1986, "Iod", pp. 6–12.
121. T. Ebbesen, *NATO ASI Ser. E*, 1996, **316**, 405–418.
122. H. Goldwhite, *J. Fluorine Chem.*, 1986, **33** (1–4), 109–31.
123. O. Glemser, *J. Fluorine Chem.*, 1986, **33** (1–4), 45–63.
124. J. Flahaut, *Bull. Soc.Chim. Fr.*, 1986, (6), 856–857.
125. H. Molines, M. F. Tordeux and M. Tordeux, *Bull. Soc. Union Physic.*, 1986, **80** (688), 1403–1435.
126. C. Viel, *Nouv. J. Chim.*, 1986, **10** (11), 575–577.
127. J. Sleigh and R. Plevey, *School Sci. Rev.*, 1986, **67** (240), 488–494.
128. L. Niinisto, *Kem-Kemi.*, 1886, **13** (12), 1138–1140.
129. E. Arpaci, *Praxis Naturwiss. Chem.*, 1990, **39** (4), 2–9.
130. J. Arribas, *Quim. Ind. (Madrid)*, 1993, **39** (3), 165–169.
131. H. Kelker, *Mitt.. Ges. Dtsch. Chem. Fachgruppe Anal. Chem.*, 1991, **3**, M75–M78.
132. L. R. Caswell, *Bull. Hist. Chem.*, 2003, **28** (1), 35–41.
133. J. O. Nriagu, *Adv. Environ. Sci. Technol.*, 1988, **20**, 1–19.
134. A. Magneli, *Chem. Ser.*, 1986, **26** (4), 535–546.
135. D. McDonald and L. B. Hunt, *A History of Platinum and its Allied Metals*, Johnson Matthey, London, 1982.
136. Johnson Matthey, *Rev.*, http://www.platinummetalsreview.com/ [accessed 16 May 2005].
137. G. B. Kauffman, *Plat. Met. Rev.*, 1997, **41** (1), 34–40.
138. M. L. H. Green and W. P. Griffith, *Plat. Met. Rev.*, 1998, **42** (4), 168–173.
139. L. F. C. Vallvey, *Plat. Met. Rev.*, 1994, **38**, 22–31, 126–133.
140. H. G. Bachmann and H. Renner, *Plat. Met. Rev.*, 1984, **28** (3), 126–131.
141. L. B. Hunt, *Plat. Met. Rev.*, 1985, **29** (4), 180–184.
142. H. Renner, *Prax. Naturwiss. Chem.*, 1991, **40** (5), 2–9.
143. R. Moreno, *Plat. Met. Rev.*, 1992, **36** (1), 40–47.
144. M. C. Usselman, *Chem. Br.*, 2001, **37** (12), 38–40.
145. J. W. Arblaster, *Plat. Met. Rev.*, 2000, **44** (4), 173–178.
146. B. Fischer and K. Gerth, *Plat. Met. Rev.*, 1994, **38** (2), 74–82.
147. W. P. Griffith, *Plat. Met. Rev.*, 2003, **47** (4), 175–183.
148. W. P. Griffith, *RSC Hist. Gp. Newsletter*, 2004 (Jan.), 20–27.
149. W. P. Griffith, *Chem. World*, 2005, **2** (3), 50–53.
150. I. R. Cottington, *Plat. Met. Rev.*, 1991, **35** (3), 141–151.
151. H. Renner, *Metallurgy*, 1990, **44** (7), 687–690.
152. W. P. Griffith and P. J. T. Morris, *Notes Rec. Roy. Soc.*, 2004, **57** (3), 299–316.
153. K. R. Seddon, *Plat. Met. Rev.*, 1996, **40** (3), 128–134.
154. V. N. Pitchkov, *Plat. Met. Rev.*, 1996, **40** (4), 181–188.

CHAPTER 4

Organic Chemistry

COLIN A. RUSSELL

History of Chemistry Research Group, The Open University

4.1 General

No one who has studied the history of organic chemistry for a long time can fail to be impressed by considerable changes in the historiography of the subject over the last twelve years or so. In *Recent Developments in the History of Chemistry* the chapter on organic chemistry[1] had almost three-hundred references to literature from the previous quarter century, of which most applied to what might be termed 'classical organic chemistry', *i.e.* the subject in pre-electronic days. However, a trend noted in Chapter 1 of this volume appears to be even more pronounced for organic chemistry than for other branches. This is a relocation of interest from the classical to the modern period, and especially to the very recent past. To be sure there have been major studies of Liebig, Kekulé, Kolbe, Frankland, Hofmann and a few others, but they have demonstrated how much there is still to be learned about a whole range of different approaches to organic chemistry and about inter-personal relationships between organic chemists. Where, for instance, is any biographical treatment of B. F. Duppa, a pioneer in acetoacetic ester chemistry and much else in related fields? Above all, the nuances of theoretical schemes are only beginning to emerge from such studies and from the all too rare attempts to study in depth salient concepts and ideas.

A second feature, also noted earlier, is the interest of many living chemists in the history of their science, often manifested in some autobiographical form. A third characteristic of many of the works cited below, also of general occurrence, is the strong emphasis on obviously useful organic chemistry. There is, therefore, much activity in the history of medical chemistry (Chapter 8) and also in the studies of natural products and of synthesis that must precede any usefulness. Polymers come well to the fore.

While a modern history of organic chemistry remains to be written, there is a useful book of essays devoted solely to the subject.[2] Several general histories of chemistry have recently appeared (Chapter 1), though one has almost nothing specific on organic chemistry, one devotes less than 30 pages to it, and another has an excellent text marred by a profusion of misprinted formulae. A rare venture into very recent history has been prompted by the centenary of the RSC's *Annual Reports*, an account of 'Advances in organic chemistry over the last 100 years'.[3] Several papers have addressed a number of questions about organic chemistry in general. A brief essay is concerned with the rise of organic chemistry from animal chemistry.[4] There is, for example, the problem of conventionalism and how far a realist view can be applied to organic structures that we have accepted. A paper on this theme includes a general survey of the growth of organic theory up to the emergence of structure theory.[5]

Another paper speaks of contributions to the formation of structural organic chemistry by the Russian chemist Butlerov and by Kekulé.[6] To illustrate the related problem of language in science, a recent paper shows how the different usages of Dalton's 'atom' and Avogadro's 'molecule' can be perceived in the organic chemistry of Berzelius, Dumas, Laurent, Gerhardt, and others in the first half of the 19th century.[7] Then the question arises as to how far the methodology of organic chemistry proclaimed by its practitioners has been anything more than rhetoric uttered for political purposes; the difficulties in resolving such a question have been well expounded in a paper that concludes that we do have to take at least some of the pronouncements at face value.[8] One may also ask whether scientific world-views can be influenced by the wider philosophies of their holders, as in the case of the organic chemist H. Kolbe; in this case there is little doubt that they can.[9]

The circumstances attending discoveries in organic chemistry, and especially those that may prevent them, are the subject of a perceptive paper that focuses on the diene synthesis of Diels and Alder and the orbital symmetry conservation rules.[10] Still on the matter of scientific discovery, the role of serendipity (or accidental discovery) has been outlined for a series of polymers, drugs, sweeteners and dyes, including Perkin's mauve and Duisberg's benzopurpurin 4B.[11] The other kind of accident familiar to every organic chemist who has lived is that of fire and explosion. A paper shows how a glass-blower's nightmare developed from a series of explosions during attempted syntheses of 1,6-dimethylcyclodecapentene, when sodium iodide, used in detosylations, was accidentally replaced by sodium iodate – a confusion probably caused by the use of Latin names on bottles.[12]

Interesting work has been done on organic chemistry in various national contexts. A complete book is devoted to the development of organic chemistry from 1875 to 1955 in the USA,[13] while a chapter in another volume relates the formation of a specialist community of physical organic chemists in the same country.[14] At the other geographical extreme is the small German state of Hesse; a short paper describes the development of organic chemistry by chemists from Hesse,[15] with special reference to three men who made widely different contributions to the science: Liebig, Hofmann and Staudinger. Several accounts have appeared of the 'provincial cradle of Russian organic chemistry', the University of Kazan. They include a paper on Aleksandr Zaitsev and his students, including E. E. Wagner and S. N. Reformatsky,[16] and another that discusses the life at Kazan university of the Russian chemist N. Zinin, whose discoveries in aromatic chemistry included the reduction of nitrobenzene to aniline.[17] Two further papers[18, 19] and a book chapter[20] look at this cradle of Russian organic chemistry. Amongst twentieth-century studies are accounts of the development of organic chemistry in Mexico (paying particular attention to the role of companies like

Syntex and its founders R. R. Marker, E. Somlo and F. A. Lehmann)[21, 22] and a review of forty years of organic chemistry at the Polish Academy of Sciences.[23] Other organic chemical communities to be encountered later in this chapter include many in America and others in Germany, Sweden, Poland and Croatia.

One topic of general and perennial interest to organic chemists is that of nomenclature. Its history is much more than a story of changing names, for every move in this area reflects complex relationships between individuals, institutions, and even nations, as well as new understanding of organic structure. These complexities are well illustrated in a brief paper,[24] a longer essay,[25] and in two whole books devoted to the subject. One was written by P. E. Verkade (1891–1979), for thirty-seven years chairman of the International Commission for the Nomenclature of Organic Chemistry, and records the Geneva nomenclature, the Liège nomenclature, the work of IUPAC, contributions of Combes, Istrati, Siboni, and many others.[26] The other embodies the proceedings of a meeting celebrating the centennial of the Geneva Conference.[27] Another account of the Geneva Conference and the following fifty years summarizes the circumstances leading to the conference, its main achievements, and the ways in which some rules of nomenclature were retained while others were revised or abandoned.[28] The interactions between chemical journals and nomenclature reforms are related for the period from 1892 to the Liège conference of 1930.[29] In the more recent period some personal observations have been supplied by K. L. Loening[30] and V. Prelog.[31] The important work of the Chemical Abstracts Service, with its Registry, Indexes, and other services relating to organic nomenclature, has been described,[32] and also the special problems in this matter faced by the Laboratory of the Government Chemist in London.[33]

4.2   People

The role of biography in the history of chemistry has been discussed in Chapter 1. Organic chemists have been particularly well represented by this genre. To write an effective biography today necessitates great familiarity not only with primary as well as secondary sources, but also with written archives, and not merely printed papers and books. A sample of this kind of treatment has been offered in relation to the chemist T. Curtius (notable for his research on organic nitrogen compounds) and colleagues at the University of Heidelberg where a small collection of their manuscripts has been salvaged from the university archives.[34]

Several major representatives of past organic chemistry have been given extended biographical treatment. A modern major biography of Liebig, written by W. H. Brock, discusses his work in not only organic chemistry but also commerce, agriculture, and the chemistry of food.[35] One of the founders of organic chemistry was J. J. Berzelius. New books with a strong biographical flavour include a collection of nine essays that endeavour to set Berzelius' work within the contemporary context of Europe, especially Sweden.[36] Four of these are papers previously published (by Lindroth, Eriksson, Lundgren and Brooke), all revised in varying degrees and, in the case of the first three, available in English for the first time. A chapter by Melhado touches briefly on the Berzelian concept of species in organic chemistry,[37] but the most significant new material on organic chemistry comes in a chapter on Berzelius' animal chemistry from 1805 to 1814. It stresses the 'fundamental ambivalence' of Berzelius to issues of vitalism, scorning on one hand the speculative excesses of *Naturphilosophie,* then rampant in German lands, and, on the other hand, a materialistic philosophy of complete reductionism both in practice and in principle.[38]

A man who followed closely in the Berzelian tradition of electrochemistry, and had little patience with those who held opposed views, was Hermann Kolbe. He also is entitled to be considered one of the great founders of organic chemistry and has now at last received his first substantial biography. This is one of those rare examples that may be honestly called 'definitive'.[39] The full-length study by Alan Rocke has the admirable characteristic of relying extensively on manuscript sources (many of them new) and of endeavouring to view the contemporary scene through his subject's eyes. In addition to a portrait of a leader of chemical thought who became, eventually, an irascible and eccentric conservative, we are provided with much information on German institutions where organic chemistry was nourished, especially Marburg and Leipzig. The subject of organic chemistry itself is seen in its rapidly changing forms, of which one may specially note the rise and transformation of structure theory and the immense progress in organic synthesis, not forgetting the electrolytic coupling reaction and the carboxylation of sodium phenoxide, both reactions later to bear Kolbe's name.

One of Kolbe's early colleagues was the English chemist Edward Frankland.[40] A new scientific biography of Frankland also depends heavily on new manuscript sources, and has some relevance to the history of organic chemistry. It emphasizes Frankland's first researches with dialkylzincs and the subsequent emergence of what he was the first to call 'organometallic chemistry'. These have also been described in a recent paper.[41] Frankland's labours with aliphatic acids and esters led him to synthetic methods of great power, to a first recognition of what might be legitimately called 'structure and mechanism', to pioneering studies on acetoacetic ester, and to a deserved reputation as one of the great synthetic organic chemists of the 19th century. He was a master of chemical techniques, and these have featured in a general volume relating to the history of chemical instrumentation.[42] Frankland also promoted the modern system of organic notation with atoms linked to each other by 'bonds', another term he invented.

The year 1992 saw the centenary of the death of A. W. von Hofmann, and following it two major publications. The first is remarkable for its detailed descriptions of the laboratories at the Chemical Institute in the Friedrich-Wilhelms Universität in Berlin, and at the Technischen Hochschule in Charlottenburg, and also for biographical details (often with portraits) of nearly 600 German chemists in academia and industry in the late 19th century.[43] The second is a collection of papers given at the centenary symposium in Berlin.[44]

Other figures from the classical period include H. E. Armstrong, chiefly remembered for his heuristic approach to chemical education, his dyspeptic criticism of other chemists, notably those of the Ingold/Robinson schools, and for his own research on naphthalene derivatives. He is commemorated in a recent short article.[45] A chemist who has attracted more interest for his other activities than for his science is the musician Aleksandr Borodin. Several articles in 1987 marked the centenary of his death,[46–48] and a longer one examined his chemistry in greater detail, notably his work on aldehydes, and the opposition encountered from Kekulé.[49] An English translation of a biography, from the Russian original of 1850, is chiefly memorable for its evocation of Soviet values at the height of the Cold War.[50] There have been many biographical accounts of the lives of distinguished organic chemists in the 20th century. Some, specially identified with one particular part of organic chemistry, will be encountered in the appropriate section. Others, with a wide variety of interest and achievement, are mentioned below. The order is alphabetical to avoid any suggestions as to relative importance or merit.

The English chemist and Nobel Prize winner Sir Derek Barton contributed to the chemistry of free radicals, was one of the founders of conformational analysis, the inventor of important new synthetic methods and one of the few Britons so far to be honoured with an autobiographical place in the American Chemical Society's series 'Profiles, Pathways and Dreams'.[51] Another autobiography in the same series is that of D. J. Cram, whose work has ranged from cyclophanes to the stereochemistry of carbanions, mechanism studies and host/guest inclusion compounds.[52] M. J. S. Dewar, apostle of the MO methodology in organic chemistry, aptly entitled his autobiography *A Semi-Empirical Life*.[53]

Perhaps as influential, though in a very different way, was Henry Gilman (1893-1986), memorialized in a brief obituary notice.[54] One of the USA's leading organic chemists, best known through his multi-volume *Organic Chemistry*, Gilman researched widely, but most notably within organometallic chemistry. The next three chemists have reminisced in the 'Profiles, Pathways and Dreams' series. Egbert Havinga[55] is a Dutch chemist who has worked in a wide variety of fields including stereochemistry, vitamin photochemistry, peptides and proteins. W. S. Johnson[56] developed syntheses of oestrone and of steroids generally by cyclization of polyunsaturated aliphatics, while John D. Roberts[57] is specially noted for his studies of carbocations such as $C_4H_7^+$, NMR, benzyne etc.

Our last example is Sir Robert Robinson, well served with a biography by Trevor Williams,[58] though it contains few details of Robinson's organic chemistry. Robinson's own essay in autobiography was not written until he was over eighty and totally blind.[59] His contribution to organic chemistry was prodigious, with important synthetic work in natural product chemistry (steroids, penicillin *etc.*), and the foundation of his own influential version of theoretical organic chemistry. His work in organic chemistry and much else is reported in a special issue of *Natural Product Reports*.[60] The papers cover Robinson's work on alkaloid chemistry,[61] on anthocyanins, brazilin, and related compounds,[62] steroids,[63] and penicillin.[64] Other papers deal with his work in physical organic chemistry (Chapter 5).

Finally, it may be appropriate to mention a number of short articles on individuals appearing in *Chemistry in Britain* and now its successor *Chemistry World*, including articles on: Berzelius,[65] Fischer,[66] Perkin,[67] Frankland,[68] van't Hoff,[69] Beckmann,[70] and Williamson.[71] Though making no claims to new knowledge, they may be useful for first introductions.

4.3 Organic Compounds

One difference between a historian of organic chemistry and a laboratory worker is that the latter frequently focuses on individual compounds, the former rarely so. The reasons are complex, but the history of – let us say – malonic ester is so liable to cut across major issues of general interest that these can only be encountered sporadically and cannot easily be subjected to any kind of sustained analysis. Instead one moves in an irregular trajectory from one set of problems to another, linked only by fortuitous association with this one compound. Moreover, the technical difficulties of much of the science are likely to obscure the issues the historian likes to address, even assuming he or she can understand the detailed chemistry. Yet it need not be like that and some noble exceptions exist in which chemical and historical questions are faced with equal determination. A study of even simple substances like malonic ester is replete with opportunities for understanding new aspects of organic theory, questions of priority, development of experimental techniques, management of laboratories, scientific communication, and even personal ambition. However, in the period of this review, not many opportunities of this kind have been taken. We shall notice a few that have.

4.3.1   Simple Organic Compounds

A history of ether and etherification is a welcome, and now rare, focus on an individual compound.[72] It covers work by Berzelius, Gerhardt, Hennell, Kolbe, Liebig, and of course Williamson. Acetoacetic ester has received detailed historical notice in a biography,[73] as have salicylic acid and the salicylates.[74] Apart from natural products, few heterocyclic substances have been recently the subject of historical enquiry. An impressive exception is that of pyrrole, a simple molecule explored by Dippel, Reichenbach, Runge and others, and manufactured by Du Pont.[75] There is also an account of the structural problems posed by piperidine.[76] Accounts have been given of the discovery of aniline from 'crystallin' (a product of the thermal decomposition of indigo),[77] of the history of phenol over the last two centuries,[78] and of organic nitrates and their uses in medicine.[79]

Most other simple compounds whose history has been examined have been those of obvious industrial interest. Thus, fluorinated compounds have been discussed, especially the now controversial chlorofluorocarbons (CFCs).[80] Another aliphatic chemical that has received exhaustive historical treatment, chiefly in the context of its industrial use, is lactic acid.[81] The development of urea as a fertilizer has been studied,[82] and a history provided of the synthesis of methanol.[83]

4.3.2   Organometallic Chemistry

The origins of organometallic chemistry have been traced in the early work of Frankland[84] and a brief account has appeared of the life of Victor Grignard.[85] An autobiography of one of its later researchers has been published.[86] On organometallic and related topics, attention is given to Otto Roelen (1897–1993) and his work on the Fischer–Tropsch reaction, hydroformylation, and polyethylene.[87] The histories have been recounted of triethylaluminium,[88] of the production of organosilicon compounds in Russia,[89] and of the use of palladium compounds in organic syntheses.[90] Recently, a series of historical papers by D. Seyferth has appeared in the journal *Organometallics*, including studies of the anion of Zeises's salt,[91] arsenicals related to cacodyl,[92] dimethyldichlorosilane,[93] (cyclobutadiene)iron tricarbonyl,[94] bis(benzene)chromium,[95] tetraethyllead,[96] and uranocene.[97]

4.3.3   Early Dyestuffs

It is impossible to separate the business of dye manufacture in the 19th century from a growing understanding of the underlying organic chemistry. A monumental work on British dye-makers will be of great value for years to come,[98] as will another book with a more European coverage.[99] There is also a paper on early-19th-century efforts to synthesize ultramarine artificially.[100] An account has appeared of Perkin's 'mauve' and its descriptions in Japanese chemistry books.[101] A traditional structure attributed to mauve has been recently corrected,[102] taking into account the presence of aniline and *o*- and *p*-toluidines needed in the reaction mixture (though in a popular exposition incorrect structures are given for *o*- and *m*-toluidines).[103]

Artificial dyes in the laboratory of the Lancashire calico-printer John Lightfoot included his own invention of aniline black; he also pioneered new methods for mordanting and the use of vanadium in aniline black printing.[104] A general account has been given of the role of rosaniline in the development of the synthetic dye industry.[105] A paper on quinones focuses chiefly on the case of anthraquinone and the synthesis of alizarin from anthracene.[106]

4.3.4   Calixarenes

Much interest has centred on the branch of cyclophanes known as calixarenes. They are polyphenol systems that can act as 'hosts' in the formation of inclusion compounds, where a small 'guest' molecule resides completely in a cavity within a single host; they are cone-shaped 'cavitands'. Several accounts have appeared of their history. The discovery by Baeyer of a formaldehyde/phenol resin led to Bakelite and to the work of A. Zincke and E. Ziegler, who gave to the first oligomer a tetrameric structure of a calix[4]arene. Later syntheses by Gutsche (1978) led to calixarenes with 4, 6 or 8 phenol residues.[107–109]

4.4   Reactions and Syntheses

Many of the substances in the previous section, most notably the organometallic compounds, were of course often employed as synthetic tools. Other reagents and reactions have also been studied historically.

The discovery of Fehling's reagent, and the use of Cu^{2+} ions by H. A. Vogel in sugar chemistry, has been described.[110] The Cambridge chemist, H. Fenton (1854–1929), his reagent, and his research into oxidation of aliphatic hydroxy compounds have been appropriately commemorated in a journal concerned with the chemistry of free radicals.[111] The production of glycidic esters from base-catalysed condensation of ketones with esters of *α*-halogenated acids is the reaction associated with Georges Darzens, whose time in Kazan has been described.[112, 113] Articles have appeared on the Friedel–Crafts reaction[114] and the Diels–Alder reaction.[115] A feature of modern organic chemistry is the frequent recourse to free radicals in mechanistic explanation and also as intentionally chosen agents to achieve desired synthetic results. Two books with an autobiographical emphasis relate to this theme, each covering a half-century time span. They are by D. H. R. Barton[116] and C. Walling.[117]

On the question of synthesis in general, not much has appeared recently on classical methods of synthesis, though an exception must be the work of W. H. Perkin Sr. His life and achievements have been described.[118] Synthesis in general in the 19th century was pursued for a wide variety of reasons: theoretical, industrial, and even ideological.[119] The recent use of a 'combinatorial library' in developing new methods of synthesis has historical precedents in the three methods employed: mixed reactant method, portioning mixing method, and light-directed synthesis.[120] Amongst the pioneers of organic synthesis of modern times, A. J. Birch has produced an autobiography[121] and R. B. Woodward has been commemorated with a booklet to accompany an exhibition at the Beckman Center in 1992,[122] an analysis by his daughter of the aesthetic aspects of his research programme,[123] and a collection of his most important papers, with careful historical commentary.[124]

4.5   Aromatic Chemistry

If one organic compound has dominated the historical literature of the last few years, that compound must be benzene. Most probably, this is because its structure in some respects marks a transition from the most austere form of classical organic chemistry, in which carbon was tetravalent and tetrahedral, to a continuing series of changes from oscillating molecules, through partial valencies to MO descriptions, and Hückel's rules of aromaticity. It is the case *par excellence* of a single substance whose history intersects all major streams of chemical theory – except perhaps the periodic law – and which also has enormous industrial and economic importance.

4.5.1   Kekulé and Benzene

A concise historical survey of the structure of benzene has been provided.[125] The role of Kekulé was obviously crucial and an essay celebrating the centenary of Kekulé's death examines the institutional and intellectual contexts of his theoretical achievements.[126] An amusingly illustrated biography of Kekulé gives a good brief account of his life and work with benzene.[127] Kekulé's work has been the subject of two recent controversies, each pursued with unusual energy, and sometimes with a degree of personal commitment rarely seen in current history of science. The first relates to Kekulé's famous 'dream' of intertwining snakes. Was it, or was it not, the real origin of his hexagon theory? Kekulé's claim to the dream has been discussed [128, 129] and refuted,[130] while a whole symposium has devoted itself largely to the issue.[131] The part played by experiment and hypothesis in the formulation of Kekulé's benzene theory has been judiciously assessed[132] and a paper relates spectroscopic studies to the structure of benzene.[133]

The other contentious issue is what might be termed the Kekulé/Loschmidt controversy. Here Kekulé is said to have been anticipated by J. Loschmidt in his proposal of a ring structure for benzene.[134, 135] Indeed one of the authors has proceeded to repeat the charge and, on the strength of researches by Kekulé's biographer,

Anschütz, to claim that Kekulé had also been anticipated by A. S. Couper in the idea of catenation of 4-valent carbon atoms.[136] Although the last claim has long been discussed, the first is more controversial and has been vigorously contested.[137] It has been well-recognized for many years that, though Couper and Loschmidt may have had unusual insights into aspects of organic chemistry, and into benzene in particular, neither offered any empirical evidence for their proposal and neither played a part in subsequent developments. The argument has been elaborated in an article stressing the human character of all scientific endeavour.[138] The real achievements of Loschmidt have been celebrated by a postage stamp in Austria.[139] Another little-known author of a benzene formula was the Belgian chemist Paul Havrez.[140]

Turning from the polemics surrounding the Kekulé case, it is refreshing to read of real experiments and genuine problems in the laboratory. The complexities associated with early attempts to reduce benzene to cyclohexane by HI included the production of methylcyclopentane, whose boiling temperature was nearly that of hexane, and which originated in an unsuspected rearrangement.[141]

4.5.2 Aromatic Substitution

Apart from some routine references in chemistry textbooks, little has been written about the history of the great problems posed by the complexities of aromatic substitution. However, one discussion of the early theories of substitution took the subject from a series of empirical rules, through the Crum Brown/Gibson rule, to the theories of Flürscheim and Holleman, and the threshold of electronic understanding.[142] Such substitutions were, of course, nearly always electrophilic in character, but a recent note[143] describes the renaissance of aromatic nucleophilic substitution in the 1980s. The latter had traditionally involved replacement of an electronegative group, as Cl, SO_2OH *etc.*, and only rarely of hydrogen. Following experiments with palladium-catalysed substitutions, metallations, and new fluorination methods, new techniques have been developed for 'vicarious nucleophilic substitution', *i.e.* direct replacement of hydrogen atoms.

4.5.3 Aromaticity and Non-benzenoid Aromatics

The concept of aromaticity continues to engage modern organic chemists, but it has a long history. Amongst the compounds to which it is dubiously applicable is the highly reactive cyclo-octatetrene, whose history has been described,[144] together with a note on Walter Reppe, who achieved its early synthesis.[145] And some molecules, though not benzenoid, are manifestly aromatic in their chemical behaviour. They include tropolone and other non-benzenoid aromatics, which were studied by T. Nozoe, whose autobiography spans forty years of organic research, at Formosa from 1926 to 1948 and at Tohoku from then to 1966.[146] Indeed the 1960s and 1970s saw intense interest in non-benzenoid aromatics, often based on the assumption that the necessary delocalization of π-electrons is best realized in planar molecules. An excellent example is [18]annulene (Sondheimer 1962). The existence of aromatic transition metal complexes was first recognized with ferrocene (1952), and analogues of other metals (such as chromium) quickly followed. The aromatic complexes of chromium had been known since their first synthesis by F. Hein in 1919, but their structure was not determined until the 1960s.[147] The complexing power of non-benzenoid aromatics was explored, further synthetic methods developed and much research conducted on the effects of transition metals on ligand reactivity.[148]

4.5.4 Fullerenes

One outcome of this strong interest in aromaticity and non-benzenoid aromatics was a search for molecules that might – contrary to prevailing views – have a 3-dimensional p-electron delocalization over a surface of high symmetry. These might have such lowering of energy as to be termed 'superaromatics', a term previously applied by Gilman and others to molecules like furan, which are more susceptible to electrophilic substitution than benzene. A molecule known as corannulene (Barth and Lawton, 1966), shown by X-rays to be bowl-shaped, appeared to an example. The Japanese chemist E. Osawa saw the further possibility of a completely spherical analogue, prompted by recognition of the pattern of the corannulene molecule in his son's football. Though predicted in 1970, it became a story of a missed opportunity for discovering C_{60}, which is told with charm and frankness in a paper that reprints, in English, part of the original description.[149]

In 1985, a stable C_{60} species was detected in a laser-vaporized graphite by mass spectrometry. Harry Kroto called it 'buckminsterfullerene' after the designer of geodesic structures Buckminster Fuller, other cyclic forms of carbon being called simply 'fullerenes'. Admitting the role of serendipity in his discovery, and commenting on the great stability of C_{60}, with 12498 more resonance forms than benzene, Kroto observed that 'perhaps after Kekulé's advance from a chain to a ring, C_{60} provides the next step – to a sphere'.[150] By the late 1980s, the search for these novel forms of carbon had led to an 'epidemic spread of fullerene research'.[151] This kind of synthesis more than any other has attracted the attention of contemporary historians, with reflections on the prehistory of 1985–90,[152] including an account describing work on carbon clusters in soot in the 1980s leading to the isolation of solid C_{60} by the physicists W. Krätschmer and D. R. Huffman.[153] There is also a well-referenced review of the discovery, physical, and chemical characteristics of fullerenes and their occurrence in inter-stellar dust.[154]

There are also numerous brief reviews[155, 156] and a book[157] describing the course of discovery. Later work, in the 1990s, has included studies of addition reactions to C_{60} (it behaves like a 'super-alkene'), the production of metal inclusion derivatives as La@C_{60}, and the synthesis of lower and higher homologues such as C_{28} and C_{76} (the first chiral fullerene).[158] A further development is the discovery of 'bucky tubes', elongated pipelines of hexagonal carbon faces, more elegantly known as carbon nanotubes.[159] It is, of course, problematic how far these novel allotropic forms of carbon are proper subjects for organic chemistry, but they would appear to be so on account of their structure, reactions and above all their history.

4.5.5 Arynes

Finally, reference must be made to benzyne and other arynes. Something of their history may be gleaned from the autobiography of R. Huisgen, whose research on benzyne chemistry was followed by other work of great importance on 1,3- and 1,4-dipolar cycloadditions and other electrocyclic reactions.[160]

4.6 Stereochemistry

Pasteur's famous experiments with tartrate crystals have been described,[161] the sesquicentenary of their start being marked by a paper that examined missed opportunities in stereochemistry and, in particular, the work by Mitscherlich which nearly anticipated Pasteur.[162] The centenary in 1995 of Pasteur's death prompted publication of several papers.[163–165] His observations, and those of le Bel and van't Hoff, are described in a paper outlining the reasoning that led to the momentous recognition of the tetrahedral carbon atom.[166] The reception of a tetrahedral carbon atom was varied. Its application to the relatively new Kekulé structure for benzene was made by van't Hoff himself and more particularly by Wilhelm Körner.[167] Eventual support came from Wislicenus, a widely ranging organic chemist, noted specially for his work on isomerism.[168]

There have been several detailed analyses of the early development of conformational analysis[169, 170] and a short historical review dedicated to one of the key participants, Sir Derek Barton.[171] Charlotte Roberts, a young woman teacher who took a doctoral degree in the subject at Yale and wrote a textbook on stereochemistry in 1896, has been profiled.[172] The now outdated 'screw theory' of Moeller features in a paper on stereochemistry and the 'ether'.[173]

There have been several papers offering overviews of the route from crystalline to universal dissymmetry,[174–176] one including a brief discussion of its importance for medicine.[177] The need for asymmetric syntheses in drug preparation has been discussed with reference to their historical development.[178] Industrial processes using steroids include fermentation techniques and also resolution of racemic mixtures of drugs where one enantiomer may have harmful effects.[179]

Lives of individuals who have distinguished themselves in stereochemistry include E. L. Eliel[180] and V. Prelog.[181] Prelog's ninetieth birthday was appropriately celebrated in *Croatica Chemica Acta*, since Zagreb was home to his School of Organic Chemistry from 1935.[182] Its work continued after his departure for Zurich in 1941 and included the syntheses of adamantane and analogues of quinine. The Cornforths have related the circumstances in which they and Prelog established the absolute configuration of (+)- and (−)- linalool (1960).[183]

4.7 Analysis and Techniques

Amongst the general techniques much used in organic chemistry is, of course, distillation. Its development has recently been described for France, England and Germany in the 19th century, during which the rectification column was invented.[184] The distillation of tar commenced in Germany in the 1840s and a commemorative paper on its founder, E. Sell, has appeared, with appropriate emphasis on the relevance for the German dyestuffs industry.[185] In micro-scale manipulations, two valuable tools have been the Willstätter 'nail' and the Schwinger microfilter.[186]

A history of the all-important techniques of elemental combustion analysis attributes their origin to Lavoisier, their early development to the little-known Cumbrian chemist Richard Rigg (1799–1864) and to William Prout, and their refinement on to a micro-scale to Fritz Pregl. However, little is said of subsequent changes involving semi-micro techniques, Unterzaucher and other methods for oxygen *etc.*[187] Liebig's combustion analysis has been investigated.[188, 189] Another article correctly identifies this procedure as 'the key to organic chemistry'.[190] The determination of nitrogen by the Kjeldahl method has been familiar to generations of organic chemists; a brief history has been given of this ubiquitous technique.[191]

Rather more specialized are the chemical methods for the determination of acetylcholine, but a short account has been given of their historical development in a brief correspondence in 1968 between Sir Henry Dale, D. J. Jenden, and B. Holmstedt, who wrote it up for publication.[192]

Chromatography could (wrongly) be said to have originated with the early years of the petroleum industry. However, work by D. T. Day, C. Engler and J. F. Gilpin on differential adsorption of petroleum led to their technique of adsorptive filtration and could thus be seen as a precursor of chromatography – though not an early example.[193] The use of paper chromatography has been described as a preliminary technique in the journey to steroidal drug discovery.[194] P. Fischer has described a quarter century of automated protein sequencing.[195] Automation is of course intrinsic in the modern but almost universally used procedure of mass spectrometry.[196]

4.8 Natural Products

The study of natural products has attracted more attention from historians than any other aspect of organic chemistry. Many of the autobiographies mentioned earlier related to chemists who included natural product research as one of their main interests. In some cases it has been an all-consuming one, as in the case of K. Nakanishi.[197] Many recent historical accounts apply to specific groups of organic compounds.

4.8.1 Aliphatic Amines, Amino Acids, Peptides and Proteins

A detailed study of choline, neurine, and betaïne constituted an early interest of the German Chemical Society.[198] An account of hydroxy- and amino-acid incorporation includes some historical data.[199] An autobiography of B. Merrifield,[200] who received a Nobel Prize in 1984, describes his work on peptides, chiefly in association with the Rockefeller Institute and University. The discovery of glycopeptides is described.[201] The long haul of M. A. Ondetti at the Squibb Institute for Medical Research led from peptides to peptidases, and ultimately to the α-methyl analogue of 3-mercapto-propanoyl-L-proline, or captopril, an inhibitor of ACE, now used in treatment of heart failures.[202] An account is also given of George Cohen's research on β-galactosidase and its impact on molecular biology.[203] The phenomenon of protein folding is of crucial importance for many areas of structural biology and its recent history has been recorded.[204]

Personal reminiscences in macromolecular and protein chemistry have been recorded.[205] Both D. E. Koshland[206] and J. T. Edsall[207] have reported their own experiences in early protein research, the former stressing his induced fit theory and the latter describing his long career at Harvard after two years with Hopkins at

Cambridge. An account has been given of the work of T. B. Osborne (1859–1928), noted for his analyses of amino acids from seed proteins.[208] The idea that enzymes might be proteins was a matter of heated debate among chemists from about 1915, including Willstätter and James Sumner who, in 1926, isolated the enzyme urease and claimed it to be a protein. The argument and its resolution are described.[209] The first enzyme whose structure was disclosed by X-ray diffraction methods was lysozyme. Louise Johnson has given a vivid account of this discovery in 1965 by D. Phillips and the events leading up to it.[210]

4.8.2   Nucleic Acids and DNA

Underlying the revolution in molecular biology in the late 1940s was a new under-standing of the concept of biochemical specificity, namely the realization that the *sequence* of subunits – proteins and nucleic acids – was crucially important. This was largely due to Frederick Sanger and Erwin Chargaff.[211] At the same time it was slowly becoming clear that genes were made not of protein but of DNA. It was, in fact, genes that determined the structure of proteins. The difficulties in accepting these ideas are vividly conveyed by one participant in the debate, Max Perutz.[212] A short history of the path to the double helix refers specially to the work of Rosalind Franklin, Maurice Wilkins and Erwin Chargaff,[213] while ten years' work on DNA before recognition of the double helix are described.[214] There is also a recent account, aimed at a general readership, which considers the role of Rosalind Franklin.[215] Special attention is paid to the work at Nottingham in the years before 1953, especially by J. M. Gulland (killed in a railway accident in 1947),[216] together with that of his colleague D. O. Jordan.[217] A brief account of the discovery of the double helix has been given,[218] together with two autobiographical notes by Francis Crick himself.[219, 220] A complete volume of the New York Academy of Science is dedicated to DNA, and commences with facsimile reprints of three original papers and includes many portraits of the early workers in this field.[221] Subsequent reflec-tions on the genetic role of DNA have not excluded a measure of historical analy-sis.[222, 223] Work since the 1960s on nucleotide modification and base conversion in RNA is described in an autobiographical account by S. Nishimura.[224]

4.8.3   Steroids

The oldest steroid known, cholesterol, has been studied from the second half of the 18th century. Its history has been briefly recounted.[225] Little was done on steroid research until its medical implications became clear in the 1940s. An excellent review by Sir Ewart Jones tells the story of early steroid researches in England.[226] The trigger to an explosion of research activity seems to have been the discoveries of vitamin D in fish liver oils and of oestrogens in pregnancy urine. Notable research in the new field was made by Ian Heilbron at Liverpool and by Otto Rosenheim at the National Institute of Medical Research in London, to say nothing of Jones' own significant work at Imperial College, Manchester and Oxford. A different kind of stimulus came to A. J. Birch in wartime Oxford, where he was encouraged in gen-eral by Robert Robinson and in particular by a rumour that Luftwaffe pilots were dosed on cortical hormones to promote resistance to shock.[227] During this work, he devised a technique for the partial hydrogenation of aromatic systems, using sodium in liquid ammonia with ethanol as a proton donor. Birch's autobiography has appeared.[228] The application by others of what became known as the 'Birch reduction' led to the development of oral contraceptives. Prominent among these investigators was C. Djerassi,[229, 230] whose partial syntheses of steroids and later studies in marine steroids, paved the way to his celebrated work in optical rotatory dispersion and circular dichroism. Another development from this early research on steroids was Barton's pioneering studies in conformational analysis. Further reminiscences include those by Fried on the discovery of the 9α-fluorosteroids at the Squibb Institute for Medical Research in New Jersey.[231]

4.8.4   Carbohydrates

It is just over a century ago that Emil Fischer suggested his famous 'lock-and-key' theory for the specificity of enzymatic action, particularly on carbohydrates. This metaphor has been celebrated as acknowledgement of the debt of organic chemistry to the biological world and of the centrality of molecular recognition, essentially different from post-synthesis Darwinian selection.[232] The motivation behind Fischer's enunciation of the theory has been probed and his slightly earlier work on elucidating the configuration of the aldohexoses, including D-glucose, discussed.[233] The derivation from Fischer's 'lock-and-key' theory of the related 'induced fit theory' is described by its originator.[234] The specific accomplishments of Fischer in determining the constitution of glucose were accompanied by a change in his view of the philosophy of chemistry.[235] A further paper recounts and analyses the reactions to Fischer's carbohydrate work of a French chemist, Louis Simon (1867–1925).[236]

An autobiography of the distinguished sugar chemist R. U. Lemieux has appeared,[237] and also an account of industrial aspects of sugar chemistry in Louisiana.[238]

4.8.5   Cyclodextrins

The cyclodextrins (cycloamyloses) are torus-shaped molecules that can form crystalline inclusion compounds, recently attracting much attention as enzyme-site models. Their history has been seen in three phases. From 1891 to 1935 they were known as natural products, but with no recognition of their exact chemical structure. This recognition emerged in the second period, to about 1970, when most of their characteristics were also elucidated. The period from 1970 to the present has seen considerable research into their industrial use and production.[239] Their inclusion compounds or complexes have found employment in such diverse fields as explosives, insecticides, pharmaceutical products, rust-prevention agents, and even baking powder.

4.8.6   Natural Pigments

A description is given of a discovery in 1833 by Bartolomeo Bizio, a Venetian chemist, of the origin and chemical properties of Tyrian Purple. This was subsequently shown in 1911 by Friedländer to be 6,6′-dibromindigo.[240] Going from animal to plant pigments, a book on carotenoids notably devotes considerable space to their history.[241] There has been recent historical work discussing the enduring importance of natural dyestuffs well into the era of synthetics.[242–244]

4.8.7 Rubber

This substance has a long history.[245] Its structure was established by Staudinger on the basis of decomposition products and molecular weight determinations through viscosity.[246] A paper on the early development of the Polish rubber industry stresses the difficulties faced in the 19th century, the existence of two factories before 1914, and the rapid growth after 1918.[247]

4.8.8 Alkaloids

Amongst specific alkaloids whose history has been discussed are strychnine[248] and cocaine. The latter covers the ground from the establishment of its constitution by Willstätter and determination of its absolute configuration to work since 1980 on addiction and the development of new antagonists.[249] Quinine has attracted much attention with a 'short history'[250] and two substantial papers on quinine and related alkaloids.[251, 252] An excellent account has appeared of the development of alkaloid chemistry in Hungary, not least with respect to the manufacturing of morphine, codeine, papaverine and other alkaloids of medical importance.[253] Liebig's alkaloid work has been discussed[254] and the work of his pupil A. W. Hofmann has been investigated.[255]

4.8.9 Vitamins

The complex molecule vitamin B_{12} has been studied for forty years. Its structure as a cobalt-containing corrin was established in 1952 by Dorothy Hodgkin, who would win the Nobel Prize for Chemistry in 1964. Since then the work of Battersby and many others has led ultimately to the understanding of all the stages of its biosynthesis: 'nature's route to vitamin B_{12} has finally been mapped out'.[256] In 1995, a paper about vitamin C celebrated the 200th anniversary of an Admiralty decision to issue every sailor with a daily ration of lime-juice as a preventative against scurvy and the 25th anniversary of Pauling's controversial advocacy of the vitamin to ward off the common cold and many other illnesses.[257]

4.8.10 The Chemistry of Life

The development of biochemistry as a subject may be traced to Lavoisier's work on the chemistry of life[258] and to Prout and Berzelius.[259] The age-old question about the nature of life has left organic chemistry with a legacy of thinking about vitalism which, despite claims to the contrary, is still associated in the popular mind with the synthesis of urea by Wöhler in 1828, by which it is alleged vital forces were banished for ever from organic chemistry.[260] Yet another article on the Wöhler synthesis has appeared, this time an examination of the ways in which it has been reported in textbooks.[261] An unusual aspect to the history is the rise and breakdown of vitalism in 19th-century South Slavic chemical literature, in another essay on the chemistry of a region.[262] The influence of stereochemistry on vitalist and other views on the nature of life is discussed in a paper that touches on the chemistry of Crum Brown, Japp, and Pope.[263]

Since then, many advances have been made and accounts have been given of the Nobel Prize winners Martin and Synge,[264] and of another Nobel Laureate, Sanger, and his colleague, Chargaff.[265] Less conspicuous than their work, but of great importance, was work on the tricarboxylic cycle[266] and the synthesis of bovine insulin.[267]

On the history of organic chemistry as applied to plants, F. Skoog gives a personal account of work related to cytokinin and plant hormones, chiefly by himself, at Wisconsin, and Miller, at Indiana, in a race to isolate a naturally occurring cytokinin.[268] Another reminiscence relates to studies of ponasterones, which are insect molting hormones from plants.[269] There is a historical survey of biochemical herbicides, including polyurethanes, triazines, bipyridines, and nitrophenols. It starts with the use of phenylurethane by O. Warburg in 1920 and concludes with the 1980s, 'the decade of the chloroplast'.[270] A Chinese paper relates an unusual aspect of the history of aspirin in discussing its recognition as a plant regulator, that is, a phytohormone.[271]

The application of organic chemistry specifically to human medicine is discussed in Chapter 8. It is, however, worth noting that the recognition by Pauling[272] in 1949 that sickle-cell anaemia was 'a molecular disease' may be said to have put down the clinical roots of molecular biology and thus of the organic chemistry on which it is founded.[273] The course of research on β-lactam antibiotics over forty years has been charted by G. N. Rolison, one of the chief investigators.[274] A recent bibliography on the history of pharmacy has much of relevance to the development of natural products chemistry.[275]

4.9 Polymers

There are some book-length treatments of the history of polymers[276] and of the lives of some of the most important pioneers,[277] together with some shorter papers.[278] An account has been given of the major meetings on macromolecules, held in Prague, mostly under the auspices of IUPAC.[279] A brief account of early polymer studies describes experiments on addition polymerization and investigations of rubber, starch, and cellulose, suggesting that the 'colloid school' of thought inhibited real progress.[280] The same opinion is voiced in a fuller review that focuses on the very concept of a 'macromolecule', stressing the fruitless attempt to understand 'large' molecules within the concepts of aggregation and complex formation derived from classical colloid chemistry. The very different contributions of H. F. Mark and The Svedberg are also emphasized,[281] the latter being featured in another paper that focuses on his discovery of protein macromolecules.[282] An Austrian, Mark became one of the major figures of American polymer chemistry, developing with Wulff a process for catalytic dehydrogenation of ethylbenzene to styrene and founding the

Polymer Research Institute at Brooklyn Polytechnic.[283] In addition to his autobiography, a memorial symposium in 1993 included a short account of his work by a colleague, Charles Price,[284] and a further tribute was later written by Morawetz.[285] Another indefatigable worker at this time was C. S. Marvel.[286] The macromolecular hypothesis was triumphantly confirmed by Staudinger in the 1920s, leading to subsequent discoveries of radical polymerization mechanisms.[287-290] A further advance came with the discovery by Natta of the stereospecific polymerization of α-olefins and of structure determinations of stereoregular polymers.[291]

Meanwhile, however, semi-synthetic plastics had been made for decades before anyone knew anything of macromolecular science, amongst them the ubiquitous cellulose nitrate, or celluloid.[292] It was followed early in the 20th century by the phenol-formaldehyde resin known as Bakelite.[293] Once polymer science had taken root, other synthetic polymers followed rapidly. From Staudinger's research came a new understanding of polystyrene and PVC plastics, both appearing in the late 1920s,[294] as well as a detailed study of poly(vinyl acetate).[295] A new synthetic rubber, a polychlorobutadiene later named neoprene, was discovered in 1931 by W. Carothers,[296] who also invented the polyamide nylons and conducted other research at Du Pont.[297] His death through suicide was brought on by acute depression and alcoholism;[298] he is the subject of a recent biography.[299] Polyethylene (polythene) was discovered at ICI, where it was manufactured on a small scale from 1938.[300] Polymerization of ethylene at low pressures was discovered by Ziegler in 1953, as a witness reports.[301] A historical account has been given of a much more recent development, the production of alternating olefin/carbon monoxide thermoplastics.[302] The rise and growth of synthetic rubber industries in Germany and the USA form a major part of a paper that does not avoid the chemical issues underlying decisions taken by politicians.[303] The work of Ambros and Reppe has also been investigated.[304]

The specific employment of polymers as fibres has a relatively long history. Cellulose acetate, manufactured for this purpose since 1865, is one of several dating to the second half of the 19th century.[305] Rayon soon followed, beginning with 'Chardonnet silk'[306] and cuprammonium rayon,[307] whose production by the viscose process has been well-documented.[308] A symposium paper examines the work of Carothers in synthetic fibre production.[309] A very full historical account has been presented of the preparation, properties and uses of acrylonitrile fibres up to 1983[310] and there is a description of acrylic fibre production generally in the second half of the 20th century.[311] In addition to synthetic polymers a whole range of miscellaneous manmade fibres has appeared, including inorganic and proteinaceous fibres, their history having been reviewed.[312] Other historical accounts deal with fibres from polyethylene[313] and polypropylene,[314] glycol-derived polyesters as terylene,[315] and with 'aramid' fibres (polyamides where at least 85% of the amide linkages are directly connected to aromatic rings).[316, 317] The history has also been written of spandex elastomeric fibres[318] and of fibre-production by the melt spin-orientation process.[319]

Polymer mixtures have attracted some historical interest,[320] as have liquid crystalline polymers.[321] Several papers have examined the history of conducting polymers[322] and of organic superconductors,[323, 324] the first having been discovered by Jerome in 1980.

References

1. J. H. Brooke, 'Organic chemistry', in *Recent Developments in the History of Chemistry*, ed. C. A. Russell, Royal Society of Chemistry, London, 1985, pp. 97–152.
2. J. G. Traynham, ed., *Essays on the History of Organic Chemistry*, Louisiana State University Press, Baton Rouge, LA, 1987.
3. C. A. Russell, *Annu. Rep. Prog. Chem., Sect. B*, 2004, **100**, 3–31.
4. F. Szabadvary, 'Die Organiker I: Von der Tierchemie zur organischen Chemie', *Österr. Chem.*, 1993, **94**, 118–119 [no references].
5. A. J. Rocke, 'Convention versus ontology in nineteenth century organic chemistry', in ref. 2, pp. 1–20.
6. V. I. Kuznetsov and A. Shamin, 'Butlerov's and Kekulé's contributions to the formation of structural organic chemistry', in *The Kekulé Riddle: A Challenge for Chemists and Psychologists*, ed. J. H. Wotiz, Glenview Press, Carbondale, IL, 1993, pp. 211–220.
7. M. G. Kim, 'The layers of chemical language, II: stabilizing atoms and molecules in the practice of organic chemistry', *Hist. Sci.*, 1992, **30**, 397–437.
8. J. H. Brooke, 'Methods and methodology in the development of organic chemistry', *Ambix*, 1987, **34**, 147–155.
9. A. J. Rocke, 'Kolbe versus the "transcendental chemists": the emergence of classical organic chemistry', *Ambix*, 1987, **34**, 156–168.
10. J. A. Berson, 'Discoveries missed, discoveries made: creativity, influence, and fame in chemistry', *Tetrahedron*, 1992, **48**, 3–17.
11. H. Ferles, 'Unusual causes of discoveries in organic and pharmaceutical chemistry', *Chem. Listy*, 1992, **86**, 773–779 [in Czech].
12. J. Dale, 'What is beautiful, is also good', *Croat. Chem. Acta*, 1996, **69**, 423–425 [in English].
13. D. S. Tarbell and A. T. Tarbell, *Essays on the History of Organic Chemistry in the United States, 1875–1955*, Folio Publishers, Nashville, TN, 1986.
14. L. Gortler, 'The development of a scientific community: physical organic chemistry in the United States, 1925–1950', in ref. 2, pp. 95–113.
15. H. Kammerer, 'The place of Hessian chemists in the development of organic chemistry: Liebig, Hofmann and Staudinger', *Chem. Z.*, 1982, **106**, 13–18.
16. D. E. Lewis, 'Aleksandr Mikhailovich Zaitsev (1841–1910)', *Bull. Hist. Chem.*, 1995, **17/18**, 21–30.
17. N. M. Brooks, 'Nicolai Zinin at Kazan University', *Ambix*, 1995, **42**, 129–142.
18. D. E. Lewis, 'The University of Kazan – provincial cradle of Russian organic chemistry; Part I: Nicolai Zinin and the Butlerov school', *J. Chem. Educ.*, 1994, **71** (1), 39–42.
19. D. E. Lewis, 'The University of Kazan – provincial cradle of Russian organic chemistry; Part II: "Aleksandr Zaitsev and his students"', *J. Chem. Educ.*, 1994, **71**, (2), 93–97.
20. N. M. Brooks, 'The evolution of chemistry in Russia during the eighteenth and nineteenth centuries', in *The Making of the Chemist in Europe: The Social*

History of Chemistry in Europe, 1789–1914, ed. D. M. Knight and H. Kragh, Cambridge University Press, Cambridge, 1998, pp. 163–176.

21. F. Y. López, 'El desarrollo de la química orgánica en México', *Ciencia* (Mexico City), 1994, **45**, 141–145 [no references].

22. P. A. Lehmann, 'Early history of steroid chemistry in Mexico: the story of three remarkable men', *Steroids*, 1992, **57**, 403–408.

23. M. Chmielewski, 'Instytut Chemii organicznej pan 1954–1994', *Wiad. Chem.*, 1995, **49**, 623–631 [no references].

24. H. A. Smith, Jr., 'The centennial of systematic organic nomenclature', *J. Chem. Educ.*, 1992, **69**, 863–865.

25. J. G. Traynham, 'The familiar and the systematic: a century of contention in organic chemical nomenclature', in ref. 2, pp. 114–126.

26. P. E. Verkade, *A History of the Nomenclature of Organic Chemistry*, D. Reidel, Dordrecht, 1985.

27. M. V. Kisakürek, ed., *Organic Chemistry: Its Language; Its State of the Art*, Verlag Helvetica Chim. Acta, Basel, 1993.

28. J. G. Traynham, 'Organic nomenclature: the Geneva Conference, 1892, and the following fifty years', in ref. 27, pp. 1–8.

29. M. V. Kisakürek, 'Chemistry journals and nomenclature 1892–1930', in ref. 27, pp. 55–75.

30. K. L. Loening, 'Organic nomenclature: the Geneva Conference and the second fifty years: some personal observations', in ref. 27, pp. 35–45.

31. V. Prelog, 'My "nomenclature years"', in ref. 27, pp. 47–54.

32. M. P. Giles, Jr. and W. V. Metanomski, 'The history of Chemical Abstracts nomenclature at Chemical Abstracts Service (CAS)', in ref. 27, pp. 173–196.

33. E. W. Godly, 'Chemical nomenclature in the Government Laboratory', in ref. 27, pp. 116–132.

34. I. Howarth, 'A modern view of the past: verifying the work of pioneering organic chemists through examination of remnants of their sample archives', *J. Chem. Educ.*, 1994, **71**, 726–729.

35. W. H. Brock, *Justus von Liebig: The Chemical Gatekeeper*, Cambridge University Press, Cambridge, 1997.

36. E. M. Melhado and T. Frängsmyr, ed., *Enlightenment Science in the Romantic Era: The Chemistry of Berzelius in its Cultural Setting*, Cambridge University Press, Cambridge, 1992.

37. E. M. Melhado, 'Novelty and tradition in the chemistry of Berzelius (1803–1819)', in ref. 36, pp. 132–170.

38. A. J. Rocke, 'Berzelius's animal chemistry: from physiology to organic chemistry (1805–1814)', in ref. 36, pp. 107–131.

39. A. J. Rocke, *The Quiet Revolution: Hermann Kolbe and the Science of Organic Chemistry*, University of California Press, Berkeley, 1993.

40. C. A. Russell, *Edward Frankland: Chemistry, Controversy and Conspiracy in Victorian England*, Cambridge University Press, Cambridge, 1996.

41. D. Seyferth, *Organometallics*, 2001, **20**, 2940–2955.

42. C. A. Russell, 'Chemical techniques in a pre-electronic age: the remarkable apparatus of Edward Frankland', in *Instruments and Experimentation in the History of Chemistry*, ed. F. L. Holmes and T. H. Levere, MIT Press, Cambridge, MA, 2000.

43. M. Engel and B. Engel, *Chemie und Chemiker in Berlin die Ära August Wilhelm von Hofmann 1869–1892*, M. Engel, Berlin, 1992.

44. C. Meinel and H. Scholz, ed., *Die Allianz von Wissenschaft und Industrie: August Wilhelm von Hofmann 1818–1892*, VCH, Weinheim, 1992.

45. W. Brock, 'A colourful character', *Chem. Br.*, 1996, **32**, 37–39.

46. C. B. Hunt, 'Aleksandr Borodin: chemist and composer', *Chem. Br.*, 1987, **23**, 547–550.

47. A. D. White, 'Alexander Borodin: full-time chemist, part-time musician', *J. Chem. Educ.*, 1987, **64**, 326–327.

48. G. B. Kauffman, I. D. Rae, I. Solov'ev, 'Borodin - chemist and composer', *Chem. & Eng. News*, 1987, **65**, 28–35.

49. I. D. Rae, 'The research in organic chemistry of Aleksandr Borodin', *Ambix*, 1989, **36**, 121–137.

50. N. A. Figurovskii and Y. I. Solov'ev, *Aleksandr Porfir'evich Borodin: A Chemist's Biography,* trans. C. J. Steinberg and G. B. Kauffman, Springer-Verlag, Berlin, 1988.

51. D. H. R. Barton, *Some Recollections of Gap Jumping,* American Chemical Society, Washington DC, 1991 (in the series 'Profiles, Pathways and Dreams: Autobiographies of Eminent Chemists').

52. D. J. Cram, *From Design to Discovery*, American Chemical Society, Washington DC, 1990 (in the series 'Profiles, Pathways and Dreams: Autobiographies of Eminent Chemists').

53. M. J. S. Dewar, *A Semi-Empirical Life*, American Chemical Society, Washington DC, 1992 (in the series 'Profiles, Pathways and Dreams: Autobiographies of Eminent Chemists').

54. J. D. Roberts, *Org. Synth.*, 1987, **66**, xv–xvii.

55. E. Havinga, *Enjoying Organic Chemistry, 1927–1987*, American Chemical Society, Washington DC, 1991 (in the series 'Profiles, Pathways and Dreams: Autobiographies of Eminent Chemists').

56. W. S. Johnson, *A Fifty-Year Love Affair with Organic Chemistry*, American Chemical Society, Washington DC, 1996 (in the series 'Profiles, Pathways and Dreams: Autobiographies of Eminent Chemists').

57. J. D. Roberts, *The Right Place at the Right Time*, American Chemical Society, Washington DC, 1990 (in the series 'Profiles, Pathways and Dreams: Autobiographies of Eminent Chemists').

58. T. I. Williams, *Robert Robinson, Chemist Extraordinary*, Clarendon Press, Oxford, 1990.

59. R. Robinson, *Memoirs of a Minor Prophet: 70 Years of Organic Chemistry*, Elsevier, London, 1976, vol. 1.

60. *Nat. Prod. Rep.*, 1987, **4** (1).

61. K. W. Bentley, 'Sir Robert Robinson - his contributions to alkaloid chemistry', ref. 60, p. 13.

62. R. Livingstone, 'Anthocyanins, brazilin and related compounds', ref. 60, pp. 25–33.
63. Sir John Cornforth, 'Steroids and synthetic oestrogens', ref. 60, pp. 35–40.
64. E. P. Abraham, 'Sir Robert Robinson and the early history of penicillin', ref. 60, pp. 41–46.
65. C. A. Russell, 'A chemical colossus', *Chem. Br.*, 1998, **34** (9), 36–38.
66. M. Engel, 'A projection on Fischer', *Chem. Br.*, 1992, **28**, 1106–1109.
67. D. H. Leaback, 'Perkin's pioneering enterprises', *Chem. Br.*, 1988, **24**, 787–790.
68. C. A. Russell, 'The Frankland enigma', *Chem. Br.*, 1999, **35** (9), 43–45.
69. C. A. Russell, 'From tetrahedra to thermodynamics', *Chem. Br.*, 2001, **37** (6), 44–45.
70. C. A. Russell, 'One thing leads to another', *Chem. Br.*, 2003, **39** (11), 36–38.
71. C. A. Russell, 'Ethereal philosopher', *Chem. World*, 2004, **1** (3), 46–49.
72. C. Priesner, 'Spiritus aethereus - formation of ether and theories of etherification from Valerius Cordus to Alexander Williamson', *Ambix*, 1986, **33**, 129–152.
73. C. A. Russell, *Edward Frankland: Chemistry, Controversy and Conspiracy in Victorian England*, Cambridge University Press, Cambridge, 1996, pp. 262–270.
74. A. J. Rocke, *The Quiet Revolution: Hermann Kolbe and the Science of Organic Chemistry*, University of California Press, Berkeley, 1993, pp. 291–297, 365–366, 446–447.
75. H. J. Anderson, 'Pyrrole: from Dippel to Du Pont', *J. Chem. Educ.*, 1995, **72**, 875–878.
76. E. W. Warnhoff, 'When piperidine was a structural problem', *Bull. Hist. Chem.*, 1998, **22**, 29–34.
77. O. Bostrup, '"Crystallin" and the discovery of aniline', *Dan. Kemi*, 1995, **76**, (6/7), 43 [in Danish].
78. H. Uenaka, 'Stories of phenol', *Soda to Enso*, 1996, **47**, 280–289 [in Japanese].
79. W. Sneader, 'Organic nitrates', *Drug News Perspect.*, 1999, **12** (1), 58–63.
80. R. Powell, 'The rise and fall of CFCs', *Chem. Rev.* (Deddington), 1996, **6**, (1), 18–23.
81. H. Benninga, *A History of Lactic Acid Making*, Kluwer Academic, Dordrecht, 1990.
82. T. Jojima, 'History of the development of urea as a fertiliser', *Kagakushi Kenkyu*, 1993, **20**, 161–200 [in Japanese].
83. M. Esaki and M. Ohtani, 'History of methanol synthesis', *Kagakushi Kenkyu*, 1994, **21**, 250–277 [in Japanese].
84. C. A. Russell, *Edward Frankland: Chemistry, Controversy and Conspiracy in Victorian England*, Cambridge University Press, Cambridge, 1996, pp. 262–270.
85. D. Hodson, 'Victor Grignard (1871–1935)', *Chem. Br.*, 1987, **23**, 141–142.
86. F. G. A. Stone, *Leaving No Stone Unturned: Pathways in Organometallic Chemistry*, American Chemical Society, Washington DC, 1993 (in the series 'Profiles, Pathways and Dreams: Autobiographies of Eminent Chemists').
87. B. H. Cornils, W. A. Kohlpainter and W. Christian, 'Otto Roelen: father of organometallic homogeneous catalysis', *Nachr. Chem., Tech. Lab.*, 1993, **41**, 544, 546–548, 550 [in German].

88. W. Kubak, 'Triethylaluminium: a milestone in organometallic chemistry', *Prax. Naturwiss., Chem.*, 1995, **44**, (4), 42–45 [in German].

89. M. A. Ezerets, ' The latest developments of organosilicon compounds production in Russia', *Khim. Prom-st.* (Moscow), 1995, 711–717 [in Russian].

90. M. J. H. Russell, 'The advantageous use of palladium compounds in organic syntheses: formation of carbon-carbon bonds', *Plat. Met. Rev.*, 1989, **33**, 186–192.

91. D. Seyferth, *Organometallics*, 2001, **20**, 2–6.

92. D. Seyferth, *Organometallics*, 2001, **20**, 1488–1498.

93. D. Seyferth, *Organometallics*, 2001, **20**, 4978–4992.

94. D. Seyferth, *Organometallics*, 2002, **21**, 1520–1530; 2800–2820.

95. D. Seyferth, *Organometallics,* 2003, **22**, 2–20.

96. D. Seyferth, *Organometallics*, 2003, **22**, 2346–2357; 5154–5178.

97. D. Seyferth, *Organometallics*, 2004, **23**, 3562–3583.

98. M. R. Fox, *Dye Makers of Great Britain, 1856–1976: A History of Chemists, Companies, Products and Changes*, Imperial Chemical Industries plc, Manchester, 1987.

99. A. S. Travis, *The Rainbow Makers: The Origins of the Synthetic Dyestuffs Industry in Western Europe*, Lehigh University Press, Bethlehem, PA, 1993.

100. J. Mertens, 'The history of artificial ultramarine (1787–1844): science, industry and secrecy', *Ambix*, 2004, **51**, 219–244.

101. Y. Hiyoshi, 'W. H. Perkin, discoverer of the first artificial purple dye, mauveine, and following studies', *Kagakushi Kenkyu*, 1994, **21**, 345–357.

102. O. Meth-Cohn and M. Smith, 'What did W. H. Perkin actually make when he oxidised aniline to make mauveine?', *J. Chem. Soc., Perkin Trans.* I, 1994, 5–7.

103. O. Meth-Cohn and A. S. Travis, 'The mauveine mystery', *Chem. Br.*, 1995, **31**, 547–549.

104. A. S. Travis, 'Artificial dyes in John Lightfoot's Broad Oak laboratory', *Ambix*, 1995, **42**, 10–27.

105. D. Fauque, 'History of synthetic dyes of the 19th century: the role of rosaniline', *Bull. Union Physic.*, 1994, **88**, 1753–1774.

106. S. Bittner, 'Quinones - reality and myth', *Kim., Handasa Kim.*, 1994, **19**, 11–13 [in Hebrew].

107. Z. Asfari and J. Vicens, 'Les calixarènes: de la Bakélite aux édifices supramoléculaires', *Actual. Chim.*, 1995, **1**, 10–16.

108. J. Szejtli, 'Historical background' in *Comprehensive Supramolecular Chemistry*, ed. J. Atwood, Pergamon, Oxford, 1996, vol. 3, pp.1–3.

109. T. Kappe, 'The early history of calixarene chemistry', *J. Inclusion Phenom. Mol. Recognit. Chem.*, 1994, **19**, 3–15 (reprinted in *Calixarenes*, ed. Z. Asfari, J. Vicens and J. McB. Harrowfield, Kluwer Academic, Dordrecht, 1995).

110. P. M. Boll, 'Revisiting Fehling and discovering Vogel', *J. Chem. Educ.*, 1994, **71**, 220–221.

111. W. H. Koppenol, 'The centennial of the Fenton reaction', *Free Rad. Biol. Med.*, 1993, **15**, 645–651.

112. G. Bram and P. Laszlo, 'Young Darzens via translation', *New J. Chem.*, 1995, **19**, 767–768.

113. P. Laszlo, 'Georges Darzens (1867–1954): inventor and iconoclast', *Bull. Hist. Chem.*, 1994, **15/16**, 59–64.

114. X. Bataille and G. Bram, 'La découverte de la réaction de Friedel et Crafts', *Compt. Rend. Ser. 2c*, 1998, **1**, 293–296.

115. B. Kruiswijk, 'Een fascinerend gereedschap voor de organisch chemicus', *Chem. Mag.*, 1998, 311–312.

116. D. H. R. Barton with S. I. Parekh, *Half a Century of Free Radical Chemistry*, Cambridge University Press, Cambridge, 1993.

117. C. Walling, *Fifty Years of Free Radicals*, American Chemical Society, Washington DC, 1995 (in the series 'Profiles, Pathways and Dreams: Autobiographies of Eminent Chemists').

118. S. M. Edelstein, 'Sir William Henry Perkin', *Am. Dyest. Rep.*, 1992, **81**, 117–118, 122–124, 127–129.

119. C. A. Russell, 'The changing role of synthesis in organic chemistry', *Ambix*, 1987, **34**, 169–180.

120. A. Furka, 'History of combinatorial chemistry', *Drug. Dev. Res.*, 1995, **36**, 1–12.

121. A. J. Birch, *To See the Obvious*, American Chemical Society, Washington DC, 1995 (in the series 'Profiles, Pathways and Dreams: Autobiographies of Eminent Chemists').

122. M. E. Bowden and T. Benfey, *Robert Burns Woodward and the Art of Organic Synthesis*, Beckman Center for the History of Chemistry, Philadelphia, PA, 1992.

123. C. Woodward, 'Role of aesthetic pleasure or art in organic chemistry in the work of R. B. Woodward', *Actual. Chim.*, 1993, **6**, 63–70.

124. *Robert Burns Woodward, Architect and Artist in the World of Molecules*, ed. O. T. Benfey and P. J. T. Morris, Chemical Heritage Foundation, Philadelphia, PA, 2001.

125. R. Norman, 'The structure of benzene', *Chem. Rev.* (Deddington), 1991, **1**, (1), 7–10.

126. W. H. Brock, 'August Kekulé (1829–96): theoretical chemist', *Endeavour*, 1996, **20**, 121–125.

127. W. Caesar, 'Friedrich August Kekulé (†1896)', *Dtsch. Apoth. Ztg.*, 1996, **136** (28), 25–28.

128. G. Somsen, 'Kekulé dreamed a splendid story', *Chem. Mag.* (Rijswijk), 1994, **9**, 359–360.

129. H. L. Kretzenbacher, 'Closed chains and writhing snakes - the metaphorical representation of the benzene formula', in *Sprache Chemie*, 2nd Erlenmeyer-Kolloq. Philosophie der Chemie, Koenigshausen & Neumann, Würzburg, 1996, pp. 187–196.

130. J. H. Wotiz and S. Rudofsky, 'The unknown Kekulé', in ref. 2, pp. 21–34.

131. J. H. Wotiz, ed., *The Kekulé Riddle: A Challenge for Chemists and Psychologists*, Glenview Press, Carbondale, IL, 1993.

132. A. J. Rocke, 'Hypothesis and experiment in the early development of Kekulé's benzene theory', *Ann. Sci.*, 1985, **42**, 355–381.

133. M. A. Sutton, 'Linking the bands to the bonds: spectroscopy and the structure of benzene', in ref. 131, pp. 139–152.

134. C. Noe and A. Bader, 'Facts are better than dreams', *Chem. Br.*, 1993, **29**, 126–128.

135. C. Noe and A. Bader, 'Josef Loschmidt', in ref. 131, pp. 221–245.

136. A. Bader, 'A chemist turns detective', *Chem. Br.*, 1996, **32**, 41–42.

137. A. J. Rocke, 'Waking up to the facts', *Chem. Br.*, 1993, **29**, 401–402.

138. J. Baggott, 'Who discovered the structure of benzene?' *Chem. Rev.* (Deddington), 1995, **5**, (1), 30–34.

139. H. G. Winkler 'Chemistry and philately, Josef Loschmidt', *Chem. Labor. Biotech.*, 1996, **47**, 66–68.

140. E. Heilbronner and J. Jacques, 'Paul Havrez (1838–1875) et sa formule du benzène', *Compt Rend., Ser. 2c*, 1998, **1**, 387–396.

141. E. W. Warnhoff, 'The curiously intertwined histories of benzene and cyclo-hexane', *J. Chem. Educ.*, 1996, **73**, 494–497.

142. C. A. Russell, 'Early concepts of aromatic substitution', in ref. 131, pp. 153–175.

143. S. M. Brown and M. C. Bowden, 'New twists on an old tale', *Chem. Ind. (London)*, 1993, 143–147.

144. P. J. T. Morris, 'The technology-science interaction: Walter Reppe and cyclo-octatetrene chemistry', *Br. J. Hist. Sci.*, 1992, **25**, 145–167.

145. P. J. T. Morris, 'An industrial pioneer', *Chem. Br.*, 1993, **29**, 38–40.

146. T. Nozoe, *Seventy Years in Organic Chemistry*, American Chemical Society, Washington DC, 1991 (in the series 'Profiles, Pathways and Dreams: Autobiographies of Eminent Chemists').

147. E. Uhlig, 'Chromium aromatic complexes: the long pathway from their first synthesis to the determination of their structures', *Mitt.-Ges. Dtsch. Chem., Fachgruppe Gesch. Chem.*, 1993, **9**, 56–67.

148. P. L. Pauson, 'Aromatic transition complexes - the first 25 years', *Pure Appl. Chem.*, 1977, **49**, 839–855.

149. E. Osawa, 'The evolution of the football structure for the fullerene (C_{60}) mol-ecule: a retrospective', *Phil. Trans. R. Soc., Ser. A*, 1993, **343**, 1–8.

150. H. Kroto, 'Giant fullerenes', *Chem. Br.*, 1990, **26**, 40–42, 45.

151. T. Braun, 'Epidemic spread of fullerene research', *Magy. Kem. Foly.*, 1993, **99**, 212–215 [in Hungarian].

152. H. Moellendal, 'C_{60} fullerenes - round surprises with many unexpected prop-erties', *Kjemi*, 1993, **53** (4), 20–23 [in Norwegian].

153. K. Fostiropoulos, 'Discovery and isolation of solid fullerene', *Int. J. Mod. Phys.*, B, 1992, **6**, (23–4), 3791–3800.

154. J. V. Yakhmi, 'Fullerenes', *Phys. News (Bombay)*, 1991, **22** (3–4), 75–87.

155. C. O'Driscoll, 'Designs on C_{60}', *Chem. Br.*, 1996, **32** (9) 32–36.

156. R. Stevenson, 'Bouncing to a Nobel Prize', *Chem. Br.*, 1996, **32** (11) 22–23.

157. J. Baggott, *Perfect Symmetry: The Accidental Discovery of Buckminsterfullerene*, Oxford University Press, London, 1994.

158. C. O'Driscoll, 'A round up of carbon chemistry', *Chem. Br.*, 1996, **32** (9) 34–35.

159. T. Ebbesen, 'Carbon nanotubes: past, present and future', *NATO ASI Ser. E*, 1996, **316**, 405–418.

160. R. Huisgen, *The Adventure Playground of Mechanisms and Novel Reactions*, American Chemical Society, Washington DC, 1994 (in the series 'Profiles, Pathways and Dreams: Autobiographies of Eminent Chemists').

161. A. Beck, H. Kurtz and J. Mauch, 'Famous chemists and their experiments. Part 2. Stereochemistry yesterday and today: Pasteur's experiments with tartrate crystals (1850)', *Prax. Naturwiss., Chem.*, 1992, **41** (5), 34–37.

162. G. B. Kauffman, I. Bernal and H.-W. Schütt, 'Overlooked opportunities in stereochemistry, Part IV. Eilhard Mitscherlich's near discovery of conglomerate crystallization: on the sesquicentennial of Pasteur's resolution of sodium ammonium racemate', *Enantiomer*, 1999, **4**, 33–45.

163. Qi Wu, 'L. Pasteur and molecular dissymmetry', *Huaxue Tongbao*, 1995, **46**, 55–57 [in Chinese].

164. A. Carneiro and A. M. N. Dos Santos, 'Louis Pasteur (1822–1895): homenagem no centenário da sua morte', *Química (Lisbon)*, 1995, **59**, 8–15.

165. A. M. Amorim da Costa, 'Ainda a propósito do Centenário da Morte de Pasteur: Simetria e Quiralidade Moleculares', *Química (Lisbon)*, 1996, **60**, 33–40.

166. D. E. Drayer, 'The early history of stereochemistry: from the discovery of molecular asymmetry and the first resolution of a racemate by Pasteur to the asymmetrical chiral carbon of van't Hoff and le Bel', in *Clin. Pharmacol. vol.18, Drug Stereochemistry* (2nd edn.), D. E. Drayer and I. W. Weiner, Dekker, New York, NY, 1993, pp. 1–24.

167. L. Paoloni, 'Stereochemical models of benzene, 1869–1875: the conflicting views of Kekulé, Körner, le Bel and van't Hoff', *Bull. Hist. Chem.*, 1992, **12**, 10–24.

168. P. J. Ramberg, 'Johannes Wislicenus, atomism and the philosophy of chemistry', *Bull. Hist. Chem.*, 1994, **15/16**, 45–54.

169. C. A. Russell, 'The origins of conformational analysis', in *van't Hoff-Le Bel Centennial*, ed. O. B. Ramsay, ACS Symposium Series No. 12, Washington DC, 1975, pp. 159–178.

170. O. B. Ramsay, 'The early history and development of conformational analysis' in *Essays on the History of Organic Chemistry*, ed. J. G. Traynham, Louisiana State University Press, Baton Rouge, LA, 1987, 54–77.

171. M. Beugelmans-Verrier, 'Aperçu historique sur l'analyse conformationnelle', *Actual. Chim.*, 1993, **5**, 87–90.

172. M. R. S. and T. M. Creese, 'Charlotte Roberts and her textbook on stereochemistry', *Bull. Hist. Chem.*, 1994, **15/16**, 31–36.

173. A. Greenberg, 'Stereochemistry and the "ether" in the evolution of molecular structure theory: the musings of a chemist on Moeller's supplanted 'screw theory', beginning with a view of the obnoxious and equally outdated cod-liver oil', *J. Chem. Educ.*, 1993, **70**, 284–286.

174. S. F. Mason, 'From molecular morphology to universal dissymmetry', in ref.2, pp. 35–53.

175. S. F. Mason, 'From Pasteur to parity nonconservation: theories of the origin of molecular chirality', in *Circular Dichroism*, ed. K. Nakanishi, N. Berova and R. W. Woody, VCH, New York, 1993, pp. 39–57.

176. E. L. Eliel, 'Chemistry in three dimensions', *Chemical Structure,* Springer, Berlin, 1993, pp. 1–8.

177. C. Gros and G. Boni, 'The world of chirality', *Actual. Chim.*, 1995, **2**, 9–15.

178. P. Eiden, 'Ariadne's thread: from Socrates to asymmetric syntheses', *Pharm. Ztg.*, 1994, **139**, 9–12, 14, 16, 18–19.

179. E. L. Eliel, 'Louis Pasteur and modern industrial steroids', *Croat. Chem. Acta*, 1996, **69**, 519–533.
180. E. L. Eliel, *From Cologne to Chapel Hill*, American Chemical Society, Washington DC, 1990 (in the series 'Profiles, Pathways and Dreams: Autobiographies of Eminent Chemists').
181. V. Prelog, *My 132 Semesters of Chemistry Studies*, American Chemical Society, Washington DC, 1991 (in the series 'Profiles, Pathways and Dreams: Autobiographies of Eminent Chemists').
182. R. Seiwerth, 'Prelog's Zagreb School of Organic Chemistry (1935–1945)', *Croat. Chem. Acta*, 1996, **69**, 379–397.
183. R. T. Cornforth and J. W. Cornforth, 'How to be right and wrong', *Croat. Chem. Acta*, 1996, **69**, 427–433.
184. L. Deibele, 'Development of distillation techniques during the 19th century', *Chem.-Ing.-Tech.*, 1994, **66**, 809–818 [in German].
185. D. Wagner, 'Ernst Sell, founder of tar distillation in Germany, on the occasion of the 150th anniversary of the first tar distillation in Germany', *Mitt.-Ges. Dtsch. Chem., Fachgruppe Gesch.*, 1993, **9**, 39–44 [in German].
186. J. T. Stock, 'The Willstätter 'nail' and the Schwinger microfilter', *J. Chem. Educ.*, 1992, **69**, 822.
187. D. T. Burns, 'Precursors to and evolution of elemental organic tube combustion analysis over the last two hundred years', *Anal. Proc.*, 1993, **30**, 2722–2725.
188. M. C. Usselman, 'Liebig's alkaloid analyses: the uncertain route from elemental to molecular formulae', *Ambix*, 2003, **50**, 71–89.
189. M. C. Usselman, A. J. Rocke, C. Reinhart and K. Foulser, 'Restaging Liebig: a study in the replication of experiments', *Ann. Sci.*, 2005, **62**, 1–55.
190. H. Kelker, 'Two hundred years of analysis. Chapter 4. Organic elemental analysis, the key to organic chemistry', *Git. Fachz. Lab.*, 1992, **36**, 454, 457–458, 461–462, 464–466, 469–470.
191. N. B. Kiriakidis, 'Determination of nitrogen by the Kjeldahl method', *Chim. Chron. Genike Ekdose*, 1996, **58**, 401–404.
192. B. Holmstedt, 'Chemical methods for the determination of acetylcholine', *Life Sci.*, 1996, **58**, 1917–1919.
193. L. S. Ettre, 'Early petroleum chemists and the beginnings of chromatography', *Chromatographia*, 1995, **40**, 207–216.
194. A. Zaffaroni, 'From paper chromatography to drug discovery', *Steroids*, 1992, **57**, 642–648.
195. P. Fischer, 'Twenty-five years of automated protein sequencing', *Nachr. Chem. Tech. Lab.*, 1992, **40**, 963–964, 966, 968, 970–971.
196. A. A. Polyaskova, 'Mass spectrometry and its role in organic analysis', *Ross. Khim. Zh.*, 1994, **38**, 47–53 [in Russian].
197. K. Nakanishi, *A Wandering Natural Products Chemist*, American Chemical Society, Washington DC, 1991 (in the series 'Profiles, Pathways and Dreams: Autobiographies of Eminent Chemists').
198. H. Teichmann, 'Cholin - neurin – betaïn: ein kapitel naturstoff-chemie aus der gründungszeit der Deutschen chemischen gesellschaft', *Mitt.-Ges. Dtsch. Chem., Fachgruppe Gesch. Chem.*, 1994, **10**, 31–45.

199. A. M. Shkrob, 'Reaction of hydroxy- and amino-acid incorporation', *Bioorg. Khim.*, 1994, **20**, 100–113 [in Russian].

200. B. Merrifield, *Life During a Golden Age of Peptide Chemistry: The Concept and Development of Solid-Phase Peptide Synthesis*, American Chemical Society, Washington DC, 1993 (in the series 'Profiles, Pathways and Dreams: Autobiographies of Eminent Chemists').

201. R. C. Yao and L. W. Crandall, 'Glycopeptides: classification, occurrence and discovery', *Drugs. Pharm. Sci.*, 1994, **63**, 11–27.

202. M. A. Ondetti, 'From peptides to peptidases: a chronicle of drug discovery', *Annu. Rev. Pharmacol. Toxicol.*, 1994, **34**, 1–16.

203. G. N. Cohen, 'Studies on β-galactosidase and their impact on molecular biology: a personal view', *Structure*, 1994, **2**, 569–570.

204. R. L. Baldwin, 'Protein folding from 1961 to 1982', *Nat. Struct. Biol.*, 1999, **6**, 614–617.

205. C. Tanford, 'Macromolecules', *Protein Sci.*, 1994, **3**, 857–861.

206. D. E. Koshland, 'The joys and vicissitudes of protein science', *Protein Sci.*, 1993, **2**, 1364–1368.

207. J. T. Edsall, 'Memories of early days in protein science, 1926–1940', *Protein Sci.*, 1992, **1**, 1526–1530.

208. J. S. Fruton, 'Thomas Burr Osborne and chemistry', *Bull. Hist. Chem.*, 1995 **17/18**, 1–8.

209. A. B. Costa, 'James Sumner and the urease controversy', *Chem. Br.*, 1989, **25**, 788–790.

210. L. N. Johnson, 'The early history of lysozyme', *Nat. Struct. Biol.*, 1998, **5**, 942–944.

211. H. F. Judson, 'Frederick Sanger and Erwin Chargaff, and the metamorphosis of specificity', *Gene*, 1993, **135** (1–2), 19–23.

212. M. F. Perutz, 'Before the double helix', *Gene*, 1993, **135** (1–2), 9–13.

213. J. Barciszewski, 'The DNA double helix', *Biotechnologia*, 1993, **3**, 138–144 [in Polish].

214. R. D. Hotchkiss, 'DNA in the decade before the double helix', *Ann. N. Y. Acad. Sci.*, 1995, **758**, 55–73.

215. B. Maddox, *Rosalind Franklin: The Dark Lady of DNA*, HarperCollins, London, 2002.

216. K. L. Manchester, 'British Rail and the discovery of the double helical structure of DNA', *S. Afr. J. Sci.*, 1993, **89**, 525–527.

217. H. Booth and M. J. Hey, 'DNA before Watson and Crick - the pioneering studies of J. M. Gulland and D. O. Jordan at Nottingham', *J. Chem. Educ.*, 1996, **73**, 928–931.

218. T. Travis, 'The twisted molecule', *Educ. Chem.*, 1993, **30**, 152–155.

219. F. Crick, 'DNA: a co-operative discovery', *Ann. N. Y. Acad. Sci.*, 1995, **758**, 198–199.

220. F. Crick, 'Looking backwards: a birthday card for the double helix', *Gene*, 1993, **135** (1–2), 15–18.

221. *Ann. N. Y. Acad. Sci.*, 1995, **758**.

222. G. S. Stent, 'The aperiodic crystal of heredity', *Ann. N. Y. Acad. Sci.*, 1995, **758**, 25–31.

223. M. McCarty, 'A fifty-year perspective on the genetic role of DNA', *Ann. N. Y. Acad. Sci.*, 1995, **758**, 48–54.

224. S. Nishimura, 'Studies of modified nucleosides in RNA: past and future reflections on my work for the last three decades', *Biochimie*, 1994, **76**, 1105–1108.

225. B. M. Ganzer, 'Cholesterol - the history of a controversial steroid', *Pharm. Ztg.*, 1993, **138** (5), 9–14, 16–17.

226. E. R. H. Jones, 'Early English steroid chemistry', *Steroids*, 1992, **57**, 357–362.

227. A. J. Birch, 'Steroid hormones and the Luftwaffe: a venture into fundamental strategic research and some of its consequences: the Birch reduction becomes a birth reduction', *Steroids*, 1992, **57**, 363–377.

228. A. J. Birch, *To See the Obvious*, American Chemical Society, Washington DC, 1995 (in the series 'Profiles, Pathways and Dreams: Autobiographies of Eminent Chemists').

229. C. Djerassi, *Steroids Made It Possible*, American Chemical Society, Washington DC, 1990 (in the series 'Profiles, Pathways and Dreams: Autobiographies of Eminent Chemists').

230. C. Djerassi, *The Pill, Pigmy Chimps and Degas' Horse,* HarperCollins, New York, 1992.

231. J. Fried, 'Hunt for an economical synthesis of cortisol: discovery of the fluorosteroids at Squibb (a personal account)', *Steroids*, 1992, **57**, 384–391.

232. A. Eschenmoser, 'One hundred years of the lock-and-key principle', *Angew. Chem. Int. Ed. Engl.*, 1994, **33**, 2363.

233. F. W. Lichtenthaler, '100 years "Schlüssel-Schloss-Prinzip": what made Emil Fischer use this analogy?', *Angew. Chem. Int. Ed. Engl.*, 1994, **33**, 2364–2374.

234. D. E. Koshland, Jr., 'The key-lock theory and the induced fit theory', *Angew. Chem. Int. Ed. Engl.*, 1994, **33**, 2375–2378.

235. M. Engel, 'On the 100th anniversary of the determination of the constitution of glucose by Emil Fischer. Comments on a change in paradigms', *Mitt.-Ges. Dtsch. Chem., Fachgruppe Gesch. Chem.*, 1991, **6**, 44–55.

236. M. Engel, 'The chemistry of carbohydrates and Emil Fischer in the view of a French chemist', *Mitt.-Ges. Dtsch. Chem., Fachgruppe Gesch. Chem.*, 1994, **10**, 46–50.

237. R. U. Lemieux, *Explorations with Sugars: How Sweet It Was*, American Chemical Society, Washington DC, 1990 (in the series 'Profiles, Pathways and Dreams: Autobiographies of Eminent Chemists').

238. J. A. Heitmann, 'A new science and a new profession: sugar chemistry in Louisiana, 1885–1895', in ref. 2, pp. 78–94.

239. J. Szejtli, 'Historical background', in *Comprehensive Supramolecular Chemistry*, ed. J. Atwood, Pergamon, Oxford, 1996, vol. 3, pp. 1–3.

240. F. Ghiretti, 'Bartolomeo Bizio and the rediscovery of Tyrian purple', *Experientia*, 1994, **50**, 802–807.

241. C. H. Eugster, 'History: 175 years of carotenoid chemistry', in *Carotenoids*, Birkhaeuser, Basel, 1995, pp. 1–12.

242. A. Nieto-Galan, 'Calico-printing and chemical knowledge in Lancashire in the early nineteenth century: the life and "colours" of John Mercer', *Ann. Sci.*, **54**, 1997, 1–28.

243. R. Fox and A. Nieto-Galan, ed., *Natural Dyestuffs and Industrial Culture in Europe, 1750–1880*, Science History Publications, Canton, MA, 1999.

244. A. Nieto-Galan, *Colouring Textiles: A History of Natural Dyestuffs in Industrial Europe*, Boston Studies in the Philosophy of Science, Kluwer Academic, Dordrecht, 2001.

245. W. Botsch, 'History of rubber', *Prax. Naturwiss., Chem.*, 1992, **41**, 17–20.

246. A. Tanaka, 'H. Staudinger's research and the birth of the polymer industry in Germany. II. Staudinger's research on the constitution of rubber and the path to synthetic rubbers', *Kagakushi Kenkyu*, 1992, **19**, 247–261 [in Japanese].

247. L. Ciechanowicz, 'The early development of the Polish rubber industry', *Polimery (Warsaw)*, 1994, **39** (3), 133–135.

248. J. W. Nicholson, 'The story of strychnine', *Educ. Chem.*, 1993, **30**, 46–47.

249. G. Fodor, 'New chemical results in the fight against cocaine', *Magyar Kém. Lapja*, 1995, **3**, 93–101.

250. J. K. Borchardt, 'A short history of quinine', *DN&P*, 1996, **9**, 116–120.

251. F. Eiden, 'Chinin und andere Chinaalkaloide', *Pharm. Z.*, 1998, **27**, 257–271.

252. F. Eiden, 'Chinin und andere Chinaalkaloide', *Pharm. Z.*, 1999, **28**, 11–20, 74–86.

253. C. N. Szantay, 'History of alkaloid chemistry in Hungary', *Magyar Kém. Foly.*, 1994, **100**, 423–442 [in Hungarian].

254. M. C. Usselman, 'Liebig's alkaloid analyses: the uncertain route from elemental to molecular formulae', *Ambix*, 2003, **50**, 71–89.

255. M. N. Keas, 'The nature of organic bases and the ammonia type', in *Die Allianz von Wissenschaft und Industrie: August Wilhelm von Hofmann 1818–1892*, ed. C. Meinel and H. Scholz, VCH, Weinheim, 1992.

256. L. R. Milgrom, 'Vitamin B_{12}: the view from the summit', *Chem. Br.*, 1994, **30**, 923–927.

257. J. Emsley, 'A life on the high Cs', *Chem. Br.*, 1995, **31**, 946–948.

258. F. L. Holmes, *Lavoisier and the Chemistry of Life: An Exploration of Scientific Creativity*, University of Wisconsin Press, Madison, WI, 1985.

259. C. Garcia *et al.*, 'Contributions of Prout and Berzelius to the birth of biochemistry', *Quim. Ind.* (Madrid), 1993, **40**, 43–46.

260. P. J. Ramberg., 'The death of vitalism and the birth of organic chemistry: Wöhler's urea synthesis and the disciplinary identity of organic chemistry', *Ambix*, 2000, **47**, 170–195.

261. P. S. Cohen and S. M. Cohen, 'Wöhler's synthesis of urea: how do the textbooks report it?', *J. Chem. Educ.*, 1996, **73**, 883–886.

262. I. Sencar-Cupovic, 'The rise and breakdown of vitalism in nineteenth-century South Slavic chemical literature', *Hist. Phil. Life Sci.*, 1984, **6**, 183–198.

263. P. Palladino, 'Stereochemistry and the nature of life: mechanist, vitalist and revolutionary perspectives', *Isis*, 1980, **81**, 44–67.

264. S. Konstantinovic, 'The British biochemists Archer John Porter Martin and Richard Laurence Millington Synge shared the 1952 Nobel Prize for chemistry', *Hem. Pregl.*, 1994, **35**, 76–78.

265. H. F. Judson, 'Frederick Sanger, Erwin Chargaff, and the metamorphosis of specificity', *Gene*, 1993, **135**, 19–23.

266. R. Bentley, 'A history of the reaction between oxaloacetate and acetate for citrate biosynthesis: an unsung contribution to the tricarboxylic acid cycle', *Perspect. Biol. Med.*, 1994, **37**, 362–383.

267. C.-L. Tsou, 'Chemical synthesis of crystalline bovine insulin: a reminiscence', *Trends Biochem. Sci.*, 1995, **20**, 289–291.

268. F. Skoog, 'A personal history of cytokinin and plant hormone research', in *Cytokinins, Chemistry, Activity, and Function*, ed. D. W. S. Mok and M. C. Mok, CRC Press, Boca Raton, FL, 1994, pp. 1–14.

269. K. Nakanishi, 'Past and present studies with ponasterones, the first insect molting hormones from plants', *Steroids*, 1992, **57**, 649–657.

270. D. E. Moreland, *Z. Naturforsch., C: Biosci.*, 1993, **48**, 121–131.

271. Q. Shi, 'Aspirin. An old medicine and a new phytohormone', *Guangxi Shifan Daxue Xuebao, Ziran Kexueban*, 1993, **11**, 69–74 [in Chinese].

272. A. Rich, 'Linus Pauling: clinical and molecular biologist', *Ann. N. Y. Acad. Sci.*, 1995, **758**, 74–82.

273. P. Heller, 'Historic reflections on the roots of molecular biology', *Ann. N. Y. Acad. Sci.*, 1995, **758**, 83–93.

274. G. N. Rolison, 'Forty years of β-lactam research', *J. Antimicrob. Chemother.*, 1998, **41**, 580–603.

275. J. H. Gregory and E. C. Stroud, *The History of Pharmacy: An Annotated Bibliography*, Garland, New York, 1995.

276. H. Morawetz, *Polymers: The Origins and Growth of a Science*, Wiley, New York, 1985.

277. P. J. T. Morris, *Polymer Pioneers*, Beckman Center for the History of Chemistry, Philadelphia, PA, 1986.

278. E. Pöcksteiner, 'Polymers from scientific curiosities to commodities', *J. Polym. Sci.*, Polymer Symposium no. 75, *Polymers to the Year 2000 and Beyond*, 1993, 137–142.

279. P. Cefelin, 'Prague meetings on macromolecules', *Chem. Int.*, 1996, **18**, 163–165.

280. H. Morawetz, 'A few comments on early polymer studies', *Macromol. Symp.* no. 98, 1995, 1163–1171.

281. H. Eisenberg, 'Birth of the macromolecule', *Biophys. Chem.*, 1996, **59**, 247–257.

282. B. Raanby, 'Svedberg – discoverer of protein macromolecules', *Macromol. Symp.* no. 1995, 1227–1245.

283. H. Mark, *From Small Organic Molecules to Large: A Century of Progress*, American Chemical Society, Washington DC, 1993 (in the series 'Profiles, Pathways and Dreams: Autobiographies of Eminent Chemists').

284. C. C. Price, 'Herman Mark and polymer history', *J. Polym. Sci.*, Polymer Symposium no. 75, *Polymers to the Year 2000 and Beyond*, 1993, 209–212.

285. H. Morawetz, 'Herman Mark: life and accomplishments', *Macromol. Symp.* no. 98, 1995, 1173–1184.

286. H. K. Hall Jr., 'Professor Carl S. "Speed" Marvel', *Macromol. Symp.* no. 98, 1995, 1185–1198.

287. P. R. Dvornic, 'The macromolecular hypothesis - birth of a science', *Hem. Pregl.*, 1992, **33**, 111–117 [in Serbo-Croat].

288. E. Martuscelli, 'The history of polymers: Staudinger's 'macromolecular' hypothesis', *Chim. Ind.* (Milan), 1996, **78**, 735–738 [in Italian].

289. P. R. Dvornic, 'Polymer chemistry 1930–1960: the founding of a new science: Part I: free radical and step growth polymerisation reactions', *Hem. Pregl.*, 1992, **33**, 117–125 [in Serbo-Croat].

290. P. R. Dvornic, 'Polymer chemistry 1930–1960: the founding of a new science: Part II: polysiloxanes and co-ordination, ionic and ring-opening polymerisations', *Hem. Pregl.*, 1993, **34**, 15–21 [in Serbo-Croat].

291. P. Corradini, 'The impact of the discovery of stereoregular polymers in macromolecular science', *Macromol. Symp.* no. 89, 1995, 1–11.

292. S. T. Mossman, 'Parkesine and celluloid', in *The Development of Plastics*, ed. S. T. I. Mossman and P. J. T. Morris, Royal Society of Chemistry, Cambridge, 1994, pp. 10–25.

293. P. Reboul, 'Britain and the bakelite revolution', in *The Development of Plastics*, ed. S. T. I. Mossman and P. J. T. Morris, Royal Society of Chemistry, Cambridge, 1994, pp. 26–37.

294. A. Tanaka, 'H. Staudinger's research and the birth of the polymer industry in Germany. I. The foundation of industrial production of polystyrene and polyvinyl chloride plastics', *Kagakushi Kenkyu*, 1996, **23**, 1–14, 147–166 [in Japanese].

295. A. Tanaka, 'H. Staudinger's research and the birth of the polymer industry in Germany. The research and development on polyvinylacetate by Staudinger and chemists in German chemical industries', *Kagakushi Kenkyu*, 1993, **20**, 243–258 [in Japanese].

296. J. K. Smith, 'The ten-year invention: neoprene and Du Pont research, 1930–1939', *Technol. Cult.*, 1985, **26**, 34–55.

297. J. K. Smith and D. A. Hounshell, 'Wallace H. Carothers and fundamental research at Du Pont', *Science*, 1985, **229**, 436–442.

298. M. E. Hermes, 'The ineluctable fate of Carothers', *Chem. Ind. (London)*, 1996, 209.

299. M. E. Hermes, *Enough for One Lifetime*, American Chemical Society and Chemical Heritage Foundation, Washington DC, 1996.

300. G. D. Wilson, 'Polythene: the early years', in *The Development of Plastics*, ed. S. T. I. Mossman and P J. T. Morris, Royal Society of Chemistry, Cambridge, 1994, pp. 70–86.

301. R. Magri, 'Forty years after a great chemical discovery, in the recollection of a witness', *Chim. Ind. (Milan)*, 1993, **75**, 582–584 [in Italian].

302. C. E. Ash, 'Alternating olefin/carbon monoxide polymers: a new family of thermoplastics', *J. Mater. Educ.*, 1994, **16**, 1–20.

303. P. J. T. Morris, 'Synthetic Rubber: Autarky and War' in *The Development of Plastics*, ed. S. T. I. Mossman and P. J. T. Morris, Royal Society of Chemistry, Cambridge, 1994, pp. 54–69.

304. P. J. T. Morris, 'Ambros, Reppe, and the emergence of heavy organic chemicals in Germany, 1925–1945', in *Determinants in the Evolution of the European Chemical Industry, 1900–1939: New Technologies, Political Frameworks, Markets and Companies*, ed. A. S. Travis, H. G. Schröter, E. Homburg, P. J. T. Morris, Kluwer Academic, Dordrecht, 1998, pp. 89–122.

305. M. Raheel, 'History of cellulose acetate fibers', pp. 142–168, in *Manmade Fibers: Their Origin and Development.* Proceedings of an international symposium, ed. R. B. Seymour and R. S. Porter, Elsevier, London, 1993.

306. G. Kauffman, 'Rayon: the first semi-synthetic fiber product', *J. Chem. Educ.*, 1993, **70**, 887–893.

307. G. Kauffman, 'A brief history of cuprammonium rayon', in ref. 305, pp. 63–71.

308. T. A. Summers, B. J. and J. R. Collier and J. L. Haynes, 'History of viscose rayon', in ref. 305, pp. 72–90.

309. M. E. Hermes, 'Synthetic fibers from "pure science": Du Pont hires Carothers', in ref. 305, pp. 227–243.

310. M. Raheel, 'History of acrylonitrile fibers', in ref. 305, pp. 183–207.

311. H. J. Koslowski, 'Fifty years of acrylic fibres', *Chem./Text.*, 1993, **43**, E69, 443.

312. R. B. Seymour, 'History of miscellaneous manmade fibers', in ref. 305, pp. 429–442.

313. R. S. Porter and T. Kanamoto, 'Development of polyethylene fibers', in ref. 305, pp. 295–310.

314. M. Wishman and J. Scruggs, 'History of polypropylene fibers' in ref. 305, pp. 311–314.

315. N. W. Hatman and F. S. Smith, 'Major advances in polyester 2GT technology 1941–1990', in ref. 305, pp. 363–394.

316. S. L. Kwolek and H. H. Yang, 'History of aramid fibers', in ref. 305, pp. 315–336.

317. K. F. Mulder, 'The other aramid fibers', in ref. 305, pp. 337–360.

318. A. J. Ultee, 'History of spandex elastomeric fibers', in ref. 305, pp. 278–294.

319. H. D. Noether, 'History of fiber production by melt spin-orientation processes', in ref. 305, pp. 267–277.

320. V. N. Kuleznev, 'Polymer mixtures: history and perspectives', *Komp. Tashkent Politekhn. In-t*, 1991, 3–14 [in Russian].

321. X. Li and M. Huang, 'Historical development in liquid crystalline polymers', *Gaofenzi Cailiao Kexue Yu Gongcheng*, 1993, **9** (4), 1–8 [in Chinese].

322. H. Krueger, 'Conducting polymers', *Spectrum (Pretoria)*, 1994, **32**, 4–5.

323. H. Mori, 'Overview of organic superconductors', *Int. J. Mod. Phys. B*, 1994, **8**, (1/2), 1–45.

324. K. Murata, 'Organic superconductors: a role of high pressure physics for the understanding of the properties of organic superconductors', *Koatsuryoku no Kagaku to Gijutsu*, 1993, **2**, 206–211 [in Japanese].

CHAPTER 5

Physical Organic Chemistry

JOHN SHORTER

Formerly Department of Chemistry, University of Hull

5.1 Introduction

5.1.1 The Nature of Physical Organic Chemistry

The term physical organic chemistry is commonly attributed to Louis Hammett, who used it in the title of a book in 1940.[1] Previously, in the 1930s books dealing with essentially the same area of chemistry had referred to physical aspects of organic chemistry[2] and modern theories of organic chemistry.[3] According to Hammett, the term implied the investigation of the phenomena of organic chemistry by quantitative and mathematical methods. He noted that one of the chief directions that the development of the subject had taken had been the study by quantitative methods of the mechanism of reactions and of the related problem of the effect of structure and environment on reactivity.

Hammett's view of the scope of the subject is summarized in the rarely mentioned sub-title of his book: 'Reaction Rates, Equilibria, and Mechanisms'. His conception of the subject still defines its core, but requires amplifying; certain other topics are now usually deemed part of physical organic chemistry. Thus the rationalization of the experimental results of studies of reaction rates, equilibria, and mechanisms involves the application of the electronic theory of the structures and reactions of organic molecules, either in its early forms as developed by Robinson, Ingold, and others on the basis of the electron-pair covalent bond, or in its later forms involving quantum mechanical treatments.

The application of physical techniques to the study of the structures of organic molecules may also be regarded as part of physical organic chemistry. Such techniques include X-ray and electron diffraction; infrared, ultraviolet, and visible spectroscopy; dipole moment measurements; nuclear magnetic resonance spectroscopy (various nuclei); electron spin resonance spectroscopy; and other techniques. Applications of some of these techniques are of course widespread as aids to organic synthesis and in identifying organic compounds. To that extent all organic chemists are now physical organic chemists. It is, however, possible to distinguish the use of these techniques as tools from their application to the fundamental study of the structures, chemical properties, and physical properties of organic compounds.

It is appropriate to quote the statement of 'Aims and scope', that appears inside the front cover of each issue of the *Journal of Physical Organic Chemistry*:

> The *Journal of Physical Organic Chemistry* provides an international forum for the rapid publication of original scientific papers dealing with physical organic chemistry in its broadest sense. As such, it is aimed at scientists working in this field, not only in chemistry itself, but also in biochemistry and pharmaceutical chemistry, materials and polymer science, and in other areas of research in which the approaches used in physical organic chemistry are now applied. The *Journal of Physical Organic Chemistry* will devote particular attention to the following fields: Organic chemistry; Bio-organic chemistry; Organometallic chemistry; Theoretical chemistry; Catalytic chemistry; Photochemistry; Supramolecular chemistry. Among the topics covered within the fields will be: Reaction mechanisms; Reactive intermediates; Novel structures; Spectroscopy; Chemistry at interfaces; Stereochemistry; Conformational analysis; Quantum chemical studies; Structure/reactivity relationships; Solvent, isotope and solid-state effects; Long-lived charged, sextet or open-shell species; Magnetic, non-linear optical and conducting molecules; Molecular recognition.
>
> [Quoted by kind permission of John Wiley and Sons Ltd., Chichester.]

The changing nature of what is regarded as physical organic chemistry during the last thirty years or so is also well shown in the editorial introduction to a special issue of the *Journal of Physical Organic Chemistry*, which contains reviews and papers collected at the European Symposium on Organic Reactivity (ESOR VI, Louvain-la-Neuve, July 1997),[4] intending to illustrate the diversity and the dynamism of recent physical organic chemistry, its new ambitions and applications in the numerous fields of the chemical sciences.

5.1.2   The Scope of this Chapter

In *Recent Developments in the History of Chemistry*, physical organic chemistry was dealt with incidentally in a long chapter on the history of organic chemistry in general.[5] For this reason, the present chapter will not be restricted solely to contributions to the history of physical organic chemistry written in the last twenty years or so. Further, a rather broad view will be taken as to what constitutes historical writing relating to physical organic chemistry, including obituary notices and other biographical articles and books, and also review articles, monographs and textbooks of some antiquity, in particular those written before World War II, or in the 1940s. It is important to record the existence of the latter categories of material, much of which is now disappearing from the open shelves of libraries, either to remote storage or, in the extreme, for disposal in skips. This material is quasi-historical in character, in so far as it records the development of topics to their states at particular times, usually with copious references to the primary literature.

5.2 Robinson and Ingold: Electronic Theories and Reaction Mechanisms

The centenaries of the births of Sir Robert Robinson (1886–1975)[6] and Sir Christopher Ingold (1893–1970)[7] have stimulated much historical activity. The Royal Society of Chemistry commemorated the Robinson Centenary in 1986 with a tribute symposium organized by the Historical Group at the Annual Chemical Congress. Versions of the various talks occupied an issue of *Natural Product Reports*.[8] The selection of this journal for the publication reminds us of Robinson's main interest in organic chemistry, even though he wrote in his autobiography that the development of an electronic theory constituted to his mind his most important contribution to knowledge.[9] In the symposium the relevant talks were: 'Theoretical Organic Chemistry before Robinson' (Russell),[8a] 'The Development of Sir Robert Robinson's Contributions to Theoretical Organic Chemistry' (Saltzman),[8b] and 'Electronic Theories of Organic Chemistry: Robinson and Ingold' (Shorter).[8c,10] Saltzman also wrote a centennial tribute to Robinson for *Chemistry in Britain*.[11] Shorter's symposium contribution dealt with the controversies of the 1920s involving Robinson, Lapworth, Ingold, and others, the outcome of which rankled with Robinson for the rest of his life. Robinson's own extended account of these matters is in Chapter 11 of his autobiography. Chapter 7 of Trevor Williams' biography of Robinson is devoted to 'The Electronic Theory of Reaction: The Ingold Controversy'.[12] As long ago as 1980, Saltzman wrote on 'The Robinson-Ingold Controversy: Precedence in the Electronic Theory of Organic Reactions'.[13] The topic also appears in W. H. Brock's one-volume history of chemistry, Chapter 14 is 'Structure and Mechanism in Organic Chemistry'.[14] M. J. Nye's *From Chemical Philosophy to Theoretical Chemistry* deals with the development of electronic theories of organic reactions, and the Robinson–Ingold controversy in Chapter 7.[15]

The early development of the electronic theory of organic reactions is also dealt with at length in three works that are only available as copies deposited in certain libraries. G. N. Burkhardt (1900–1991) was on the chemistry staff at the University of Manchester from the 1920s and was thus a junior colleague of Robinson and Lapworth during the controversies with Ingold. In the 1970s, he put together his memoirs of fifty years earlier and a great deal of related scientific material.[16] More recently, K. Schofield (Emeritus Professor of Chemistry at the University of Exeter) has written an extensive account of the growth of physical organic chemistry to the middle of the 20th century and this contains much on the development of the electronic theories in the 1920s and 1930s.[17] A Chemistry Part II thesis of the University of Oxford was written in 1972 by Jennifer M. Seddon[18] on the development of electronic theory in organic chemistry. It was based largely on a very thorough examination of the relevant primary literature, mainly of the 1920s and earlier. The author had the benefit of conversations with J. C. Smith (Section 5.3) and other colleagues of Robinson, who was still alive in 1972 and whose autobiography had not then appeared. A historical account of the electronic (electrochemical) theory of the course of organic reactions has appeared in Japanese.[19]

Perkin Division and the Historical Group of the RSC arranged a one-day symposium on the work of Sir Christopher Ingold to take place in University College,

London on the day of the centenary of his birth, 28th October 1993. The main historical paper was given by Schofield, and this formed the basis of a subsequent publication on the development of Ingold's system of organic chemistry.[20] While this naturally involves some account of the development of Ingold's version of the electronic theory of organic reactions (including the dispute with Robinson), it is mainly concerned with Ingold's application of kinetic studies to the elucidation of reaction mechanisms during the 1930s and 1940s, *i.e.* the distinction between the S_N1 and S_N2 mechanisms, *etc.* Schofield subsequently dealt with these matters at length.[21] An article by Maccoll on 'A mechanistic pioneer' also appeared in the month of the centenary.[22]

The Historical Group had already held a commemoration of Ingold during the Annual Chemical Congress of the RSC at the University of Southampton in April 1993. The venue was appropriate because Ingold had been an undergraduate at Hartley University College in Southampton from 1911–1913. In the main the papers given at the half-day symposium were not concerned directly with Ingold's work in physical organic chemistry, but were biographical or devoted to other aspects of Ingold's chemical work or career. They are available only as summaries,[23] but an expanded version of the paper by G. K. Roberts on Ingold as department head and educator was later published.[24]

A more extended commemoration of the Ingold centenary was held by the Division of the History of Chemistry of the American Chemical Society at the ACS National Meeting in Chicago during August 1993. Speakers at the symposium included Ingold's son Keith U. Ingold, and several of his former students and colleagues. A journalist's impression of the symposium appeared shortly afterwards.[25] The main publication appeared in 1996:[26] 'Ingold as educator and department head' (G .K. Roberts),[26a] 'The Faraday Society Discussions of 1923, 1937, and 1941' (Davenport),[26b] 'Ingold's influence as historian and educator' (Benfey),[26c] 'Ingold's development of the concept of mesomerism' (Saltzman),[26d] 'Physical organic terminology after Ingold' (Bunnett),[26e] 'Ingold, Robinson, Winstein, Woodward, and I' (Barton),[26f] 'The beginnings of physical organic chemistry in the United States' (J. D. Roberts),[26g] 'The Paris school of reaction mechanisms in the 1920s and 1930s' (Nye),[26h] 'Base hydrolysis of cobalt(iii) ammines' (Basolo,[26i] dealing with Ingold's venture into the reactions of octahedral complexes), 'Reactions in micelles and the scope of the Hughes–Ingold solvent theory' (Bunton),[26j] 'A personal history of the benzidine rearrangement' (Shine).[26k]

The later part of the chapter on 'Structure and Mechanism in Organic Chemistry' in Brock's one-volume history of chemistry outlines important features of Ingold's work in the 1930s on the kinetics and mechanism of substitution and elimination reactions.[14] He also mentions the first proposal in 1939 of a non-classical carbonium ion by Ingold's former student and University College, London colleague, C. L. Wilson. The circumstances surrounding this proposal were put on record by Saltzman in 1980, with the aid of Wilson himself, a few years before he died.[27] Chapter 8 of the book by Nye[15] is devoted to 'Reaction mechanisms: Christopher Ingold and the integration of physical and organic chemistry, 1920–1950'. This account (and indeed the whole book) is very much written from the standpoint of a professional historian rather than the standpoint of a professional chemist. There are quite a few chemical inaccuracies, including erroneous formulae, nomenclature, and equations. Furthermore, there is an overemphasis of Ingold's role in the development of physical organic chemistry at the expense of the important work of other pioneers.

Other articles relating to Ingold include a version of a lecture by Bunton on 'C. K. Ingold. A Chemical Revolutionary' delivered at the IUPAC conference on physical organic chemistry in Padua, Italy in 1994[28] and a version of a talk by Davenport, 'On the comparative unimportance of the invective effect'.[29] The latter presents amusing sidelights on the controversy with Robinson and on other situations in which Ingold's 'invective effect' was manifested, for example in controversies with H. C. Brown and with William Taylor.

K. T. Leffek of Dalhousie University, Halifax, Nova Scotia, has written a biography of Ingold.[30] Leffek was a student at University College, London during the last part of Ingold's career there. The book is a substantial work and contains much detail of personal, family, and scientific matters; there is also a good collection of photographs. Leffek makes his own attempt to assess the dispute between Ingold and Robinson. An account of the dispute has also been published in German to make it more accessible to chemistry students for whom that language is native.[31]

Both Robinson and Ingold married chemists –Gertrude Maude Walsh (1886–1954) and Edith Hilda Usherwood (1898–1988), respectively. At various times both wives acted as research assistants (usually unpaid) to their husbands. Their important contributions in this way have been discussed.[32]

Much of Ingold's work from 1930 onwards was carried out in association with E. D. Hughes (1906–1963), who will be discussed more particularly in Section 5.3. It would, however, be appropriate to mention here the obituary notice of Hughes written by Ingold himself,[33] and also that written by Ingold on B. J. Flürscheim (1874–1955),[34] whose pre-electronic theory of chemical affinity exerted a major influence on Ingold during the development of his approach to the electronic theory of organic chemistry.[35,36] Finally, we should mention the obituary notice of D. R. Boyd (1872–1955), who taught Ingold as an undergraduate in Southampton.[37] Ingold owed much to Boyd's influence. Boyd himself made pioneering studies of the kinetics of organic reactions in solution.

5.3 Physical Organic Chemistry in the United Kingdom

Saltzman has written a good summary of the development of physical organic chemistry in the United States and the United Kingdom, pointing out various parallels and contrasts.[38] In particular, he emphasizes that the subject developed more vigorously in the UK than in the USA between 1919 and 1939, even though in both countries there had been considerable work done in the kinetics and mechanisms of organic reactions before World War I. He suggests various reasons for this difference. As far as the UK is concerned, Saltzman mentions most of the key figures in the development of physical organic chemistry before 1939, and we may take his exposition as a guide in presenting the relevant bibliography.

One of the leading people in British chemistry during late-Victorian and Edwardian times was H. E. Armstrong (1848–1937).[39] It would be an exaggeration to describe Armstrong as an early physical organic chemist, but he had a very strong interest in the theoretical organic chemistry of the time. Two of his pupils at the Central Technical College in London, however, certainly exerted a considerable effect on the development of physical organic chemistry in the UK: Arthur Lapworth (1872–1941)[40,41] and Thomas Martin Lowry (1876–1936).[42] Their interest in the mechanisms of chemical change was probably stimulated by their experience of the complexities of camphor chemistry while working with Armstrong.

Lapworth is well known for establishing the mechanisms of the formation of cyanohydrins from ketones (1903), of the acid-catalysed bromination of acetone (1904), and of various other reactions before World War I, work which is well summarized by Saltzman.[43] Subsequent interest in Lapworth has been stimulated by his association with and influence on Robinson. Lapworth came to Manchester in 1909 as a Senior Lecturer when Robinson was a junior staff member there. Robinson left Manchester in 1912, but their close association was resumed in the 1920s when both were Professors at Manchester. Lapworth had devised a pre-electronic theory of alternating polarities for reactions in organic chemistry and discussions with Lapworth were of great help to Robinson in developing his electronic theory. They participated together in the controversy with Ingold.[44] The unpublished memoirs of G. N. Burkhardt[45] are of special value in connection with Lapworth, since Burkhardt was one of Lapworth's last students. Schofield's unpublished volume[46] deals extensively with Lapworth's work. He has also published a more generally accessible account.[47] Lapworth also features in the books by Brock and by Nye, and also in Robinson's autobiography.[48]

Lowry occupied various positions in London colleges, before becoming the first professor of physical chemistry at Cambridge in 1920. He applied the skills of both the physical chemist and the organic chemist in work on such subjects as catalysis, stereochemistry, acid-base theory, and valency theory. In the early 1920s he became particularly interested in the applications of electronic theory to organic reactions. In 1923 he was the organizer of a Faraday Society symposium on the electronic theory of valency at Cambridge.[49] G. N. Lewis gave the introductory talk, and Lapworth, Robinson, and Lowry all presented papers dealing with their applications of Lewis electron-pair theory to organic chemistry. In the mid-1920s, Lowry was a major participant in the controversies involving Robinson, Lapworth, and Ingold.[50] Saltzman has discussed Lowry's concept of the mixed multiple bond.[51] After this period, Lowry devoted much effort to the investigation of optical rotatory dispersion. Lowry encouraged W. A. Waters to write his early book on physical aspects of organic chemistry. The book had originally been planned by Lowry; Waters was later invited to be co-author, but ultimately Lowry had to withdraw from the project.[52]

Saltzman[53] rightly refers to K. J. P. Orton (1872–1930)[54,55] as a very significant figure in the development of physical organic chemistry in the UK. Orton studied at Cambridge and Heidelberg, and at the latter obtained a Ph.D. with K. von Auwers. He acted for some years as assistant to F. D. Chattaway[56,57] at St. Bartholomew's Hospital Medical College before becoming professor in the small Chemistry Department of the University College of North Wales at Bangor in 1903. Orton had been introduced by Chattaway to working on the chemistry of N-halogeno-aromatic amines and related compounds. At Bangor, Orton was drawn into studies of the kinetics and mechanisms of their transformations.[58] Orton's conclusion that these molecular rearrangements were inter- and not intra-molecular brought him into conflict with Chattaway, who had a somewhat negative attitude to physical chemistry, and led to a quarrel of long duration between the two men.[59]

From 1919 to Orton's death in 1930, the Chemistry Department at Bangor was one of the major centres for physical organic chemistry in the UK. At various times during this period, the principal members of Orton's staff were H. B. Watson (1894–1975),[60] A.E. Bradfield (1897–1953),[61] and F. G. Soper (1898–1982),[62] all of whom were involved in research in the kinetics and mechanism of organic reactions. The staff were assisted by an able group of students, some of whom became distinguished physical organic chemists. The group included Brynmor Jones (1903–1989),[63] Gwyn Williams (1904–1955),[64] and E. D. Hughes (1906–1963).[65,66] Orton's death in 1930 largely ended the Bangor school of physical organic chemistry, because he was succeeded by the terpene chemist J. L. Simonsen. Some residual physical organic activity remained in the hands of Soper. A revival occurred when Hughes held the Chair of Chemistry at Bangor from 1943 to 1948 and physical organic chemistry has flourished there again much more recently under C. J. M. Stirling.

In 1930 Watson became head of the Chemistry Department of the Cardiff Technical College. With various collaborators, Watson continued studies of the kinetics and mechanism of organic reactions, with a particular interest in structure–reactivity correlations and wrote the important monograph to which reference has already been made.[67] In 1936, Soper went to the Chair of Chemistry at Otago, New Zealand, where studies in physical organic chemistry continued for many years, apart from an interval in World War II. Soper's contributions to physical organic chemistry ceased with his appointment as Vice-Chancellor of Otago University in 1953.

Hughes moved to University College, London in 1930 to work as a post-doctoral fellow under Ingold, who had just moved there from Leeds. After a series of temporary appointments, Hughes became an established lecturer at UCL and, after a few years back in Bangor, was a professor at UCL from 1948 to 1963. Much of Hughes' research work, although by no means all, over 33 years was carried out in association with Ingold. Hughes' expertise in chemical kinetics, as acquired under Orton and Watson, was doubtless of great value to Ingold in getting established in this field. Saltzman has revisited the single short paper Hughes published in the *Journal of the American Chemical Society* in 1935 and has endeavoured to answer the question: Why did Hughes choose to publish this paper entitled 'Hydrolysis of secondary and tertiary halides' in an American journal when all the rest of his publications appeared in British Journals?[68] In 1929 Brynmor Jones moved to Cambridge and worked on optical rotatory dispersion with Lowry. In 1931 he went to Sheffield, and extended work on the kinetics and mechanism of halogenation of anilides and phenolic ethers begun with Orton and Bradfield. These studies were continued in Hull, where he occupied the Chair of Chemistry from 1947 to 1956, when he became Vice-Chancellor (1956–1972). Gwyn Williams joined the staff of Kings College, London and ultimately became Professor at Royal Holloway College. He did much distinguished work, particularly on the mechanism of nitration.

In September 1941, under wartime conditions, the Faraday Society organized a one-day discussion in London on 'Mechanism and Chemical Kinetics of Organic Reactions in Liquid Systems'.[69] It proved to be a Bangor reunion, because Watson,

Bradfield, Gwyn Williams, Hughes, and Brynmor Jones were all present. Ingold concluded his 'Introductory Remarks' as follows:

> Finally, you will not fail to observe that more than half of the reading matter we are to consider has come from the pens of five distinguished pupils of the late Professor Kennedy Orton. Those who remember him must well appreciate the enthusiasm with which he would have participated in a discussion, whose motive was his own, and whose official title might appropriately have been applied to his own life's work. A great leader and a pioneer of the movement we are here to further, it is appropriate to notice the large part which, through the first generation of his successors, he has taken in our proceedings.

By the time Robinson went to Oxford in 1930, the main controversies with Ingold were over and Robinson largely devoted his energies from then onwards to the chemistry of natural products. The university and college laboratories of Oxford, however, had their own traditions of interest in the physical aspects of organic chemistry. These were closely connected to a long-standing interest in chemical kinetics, which may be traced back to A.G. Vernon Harcourt (1834–1919),[70–72] who devoted many years to studying the basic principles of kinetics, but appears never to have made the connection between kinetics and mechanism. One of his pupils, however, realized early in his career the importance of using physical chemistry to study organic reactions. This was N. V. Sidgwick (1873–1952), who sought to broaden his experience of physical chemistry by working in the laboratory of Ostwald at Leipzig and his experience of organic chemistry by working in the laboratory of von Pechmann at Tübingen.[73,74] In the period 1900–1914, Sidgwick and his students carried out various studies of the kinetics of organic reactions in the laboratory of Magdalen College. In 1910, he wrote a book on the organic chemistry of nitrogen, which devoted rather more space to physical aspects of organic chemistry than was usual in organic textbooks at that time.[75,76] His most important work came after World War I when he had become interested in the electronic theory of valency, about which he wrote an outstanding book.[77] Gavroglu and Simoes have written on 'Preparing the ground for quantum chemistry in Great Britain: the work of the physicist, R. H. Fowler, and the chemist, N. V Sidgwick'.[78] An article on Sidgwick has also been written recently by Russell to commemorate the 50th anniversary of his death.[79]

In the 1920s, the Balliol-Trinity Laboratory became a strong centre for studies in chemical kinetics under C. N. Hinshelwood.[80–82] Initially, interest was mainly in gas kinetics, but studies of the kinetics of organic reactions in solution started around 1930, when Hinshelwood was joined by E. A. Moelwyn-Hughes (1905–1978),[83] who had worked on solution kinetics with W. C. McC. Lewis (1885–1956)[84] at Liverpool. The present contributor has written a study of the work that Hinshelwood and Moelwyn-Hughes carried out in the early 1930s and its continuation by Hinshelwood after Moelwyn-Hughes left Oxford in 1933 to work with Bonhoeffer in Frankfurt, before establishing himself in Cambridge, where he remained for the rest of his life.[85] This work was carried out by people who regarded themselves as physical chemists, rather than organic chemists, but in its nature it may be regarded as physical organic chemistry. In the early 1930s, interest in solution kinetics in the Balliol-Trinity

Laboratory was strengthened by the return of R. P. Bell (1907–1996)[86–88] from a period of working with J. N. Brønsted in Copenhagen (Section 5.4). While he always regarded himself as a physical chemist, much of Bell's work on the kinetics of acid-base catalysed reactions and isotope effects is appropriately regarded as belonging to physical organic chemistry. In 1941 the Balliol-Trinity Laboratory was replaced by the University Physical Chemistry Laboratory; the work on kinetics of organic reactions in solution continued there until Hinshelwood retired in 1964 and Bell became the founder Professor of Chemistry at the University of Stirling in 1967.

The Oxford centre of organic chemistry, the Dyson Perrins Laboratory (the DP), developed its own interests in physical organic chemistry in the 1920s, well before Robinson's arrival in 1930. The history of this laboratory has been described in considerable detail by J. C. Smith,[89] who joined it in 1928 after a period of working with Robinson in Manchester. Smith lists[90] the physical organic chemists on the staff of the DP when Robinson arrrived as: Sidgwick (who had transferred from the Magdalen Laboratory), D. Ll. Hammick (1887–1966),[91] T. W. J. Taylor (1895–1952),[92] and L. E. Sutton (1906–1992).[93–95] (Smith evidently did not regard himself as a physical organic chemist, although some of his work verged on the field.) The physical organic chemists of the DP carried out a great variety of work, including on kinetics, but were particularly interested in the applications of dipole moment measurements to problems of molecular structure, including stereochemistry. This followed the visit of Sutton to Peter Debye's laboratory in Leipzig in 1929, during which he learnt the experimental technique. Around 1934, Sutton also learnt the technique of electron diffraction from L. O. Brockway in the USA and used it in Oxford. He later began the critical compilation of bond lengths and angles.[96]

The interest of these members of the DP in physical organic studies continued through the 1930s and into the 1940s. It gradually declined with the involvement of Sidgwick in other matters, the interruptions of World War II, the departure of Taylor in 1946 to be the first Principal of the University College of the West Indies, and the transfer of Sutton to the Physical Chemistry Laboratory in 1942. Hammick continued studies in physical organic chemistry until some years after his formal retirement from the DP in 1952. By that time, a new interest in physical organic chemistry had appeared in the DP with the arrival of Waters in 1946. He had spent a long period in Durham after leaving Cambridge. Also, Robinson attempted to revive his personal role in electronic theory: in 1946 he began a collaboration with M. J. S. Dewar to write an exposition of the Lapworth–Robinson system of organic chemical theory, as he termed it. For various reasons the collaboration broke down and Dewar was the sole author of the ultimate book.[97] The book was certainly rather different from what Robinson had envisaged (for example, the Schrödinger equation appears on p. 1!), but distinct traces of the Lapworth–Robinson system remain. Since most of Dewar's career was subsequently pursued in the USA, there will be further reference to him in Section 5.5. This also applies to A. R. Katritzky, whose career in chemical research began in the DP around 1950, although not at first involving work in physical organic chemistry.

Ingold was Professor of Organic Chemistry at the University of Leeds from 1924 to 1930.[98] At the time he went to Leeds, he was antipathetic to the new electronic theories, and favoured the alternating affinity approach of Flürscheim.[99] This is what the controversy with Robinson and Lapworth between 1923 and 1927 was all about.[100] By the time it was over, Ingold had taken over the electronic theory and was increasingly interested in the physical aspects of organic chemistry. Thus his Leeds period marks Ingold's conversion to physical organic chemistry. The Chemistry Department at Leeds already had a foothold in physical organic chemistry. Ingold's predecessor at Leeds was J. B. Cohen (1859–1935),[101] who was a prolific writer of textbooks of organic chemistry. By comparison with other textbooks of those times, Cohen's books contained an unusual amount about the physical aspects of organic chemistry. For example, Part I (of three parts) of the 4th edition (1923) of his *Organic Chemistry for Advanced Students*[102] had long chapters on 'Valency of carbon', 'Nature of organic reactions', 'Dynamics of organic reactions', and 'Abnormal reactions'.

The other already existing contact of Leeds with physical organic chemistry lay in the research interests of the Professor of Physical Chemistry, H. M. Dawson (1876–1939),[103] who for many years had worked on the kinetics of the iodination of ketones. Dawson was prominent in developing the concept of general acid-base catalysis. There is no doubt that Dawson helped to turn Ingold's interests towards kinetic studies and Ingold later paid tribute to Dawson's influence. There is an account of the history of the School of Chemistry at Leeds written by F. Challenger.[104] After Ingold's departure to University College, London in 1930, research in physical organic chemistry continued at Leeds in the hands of several of Ingold's students who had been appointed to the Leeds staff. These included J. W. Baker (1898–1967),[105] who introduced the concept of hyperconjugation in 1935.[106] He also wrote a monograph on *Tautomerism*.[107] An article by Challenger shows how the traditions of Dawson and of Ingold were still effective at Leeds in the early 1950s.

When Brynmor Jones moved to Sheffield in 1931, G. M. Bennett (1892–1959)[108] had recently become Firth Professor of Chemistry there, having been Lecturer in Organic Chemistry at Sheffield since 1924. By 1931, Bennett's interests had for several years been oriented towards physical organic chemistry, particularly the influence of sulphur-containing substituents on reactivity. He continued such work through the 1930s, some of it in collaboration with Brynmor Jones and also with Samuel Glasstone (1897–1986),[109] well known to several generations of students as an author of textbooks of physical chemistry. In 1938 Bennett became Professor at Kings College, London, where he pursued the study of nitration, particularly in collaboration with Gwyn Williams. His work in physical organic chemistry largely came to an end with his appointment as Government Chemist in 1945. He was succeeded at Kings by D. H. Hey. Under Hey's leadership, the Chemistry Department at Kings became a leading centre for physical organic chemistry in which Victor Gold (1922–1985)[110] played a key role. Gold moved from Ingold's school at UCL to a junior position at Kings in 1945, ultimately succeeding Hey when the latter retired in 1971. Gold's work included studies of the kinetics and mechanism of a great variety of organic reactions, with a particular emphasis on the study of kinetic solvent isotope effects.

Major contributions to the application of stereochemistry as a tool in the investigation of reaction mechanisms were made by Joseph Kenyon (1885–1961).[111] He was associated for many years with R. H. Pickard (1874–1949)[112] at the Blackburn Municipal Technical School. Together they developed very useful methods for the separation of optical isomers and investigated the relationship between optical rotation and chemical constitution. In 1920 Pickard was appointed Principal of Battersea Polytechnic; Kenyon became Head of the Chemistry Department there, and led a flourishing research group in physical organic chemistry at Battersea until well after his retirement in 1951. The reactions to which stereochemical methods were applied included nucleophilic substitution, the hydrolysis of esters, and molecular rearrangements.

A little remembered or recognized figure in the development of physical organic chemistry is J. J. Sudborough (1869–1963).[113] He studied under Victor Meyer at Heidelberg, working on steric hindrance in reactions of di-ortho-substituted benzoic acids. He was Professor of Chemistry at University College, Aberystwyth 1901–1911, and Professor of Organic Chemistry at the Indian Institute of Science in Bangalore 1911–1925. His main work from 1901 to 1925 was on rates of esterification of organic acids and of hydrolysis of esters. He accumulated much information about structure–reactivity relationships in such systems, but it was work before its time, because much of it could not then be interpreted. About 1950, R. W. Taft in the USA used many of Sudborough's data in an analysis of carboxylic acid and ester reactivity (Section 5.5).

To mark the sesquicentenary in 1978 of Wöhler's discovery that ammonium cyanate may be converted spontaneously into urea, the present contributor wrote 'The conversion of ammonium cyanate into urea–a saga in reaction mechanisms'.[114] The saga began in Dundee in 1895 with work done by James Walker (1863–1935)[115] and his student, F. J. Hambly. Kinetic studies on the reaction in water showed it to be of the second order. The precise mechanistic significance of this finding was not clearly established until the second half of the 20th century. The saga involved contributions from numerous authors world-wide, including many in the UK. The subject has been discussed again by M. J. ten Hoor.[116]

The pioneering contributions of Walker, Lapworth, Sidgwick, Sudborough, Orton, and Dawson to the kinetics and mechanism of organic reactions have been discussed by Shorter.[117]

During the closing years of Lapworth's time at Manchester and after Robinson's departure from there in 1928, physical organic chemistry continued there in other hands. G. N. Burkhardt[118] essentially carried on the Lapworth tradition through the 1930s and for some years beyond, until other responsibilities drew him away from active participation in research. In the meantime, the Manchester staff had acquired D. H. Hey, M. G. Evans (1904–1952),[119] and Michael Polanyi (1891–1976).[120,121] In Manchester, there was strong interaction between Evans and Polanyi in work that may be regarded as physical organic chemistry, although both would have regarded themselves as physical chemists. They were much interested in transition state theory from the standpoint of thermodynamics and quantum mechanics, and in various aspects of the kinetics of organic reactions, including polymerization. After a spell as professor of physical chemistry at Leeds, largely during World War II, Evans returned to Manchester as Professor in 1948, following Polanyi's departure from chemistry to be a full-time philosopher. However, Evans' tenure was terminated by his untimely death in 1952.

We, finally, mention the names of certain other British chemists who at various times did some work of a physical organic nature. T. Patterson (1872–1949),[122] after study under Victor Meyer at Heidelberg, served the University of Glasgow for almost forty years. He carried out much research on optical activity, including the studies of the influence of solvents on the rotations of optically active compounds. A. W. Stewart (1880–1947)[123,124] held various appointments before occupying the Chair of Chemistry at Queens University, Belfast from 1919 to 1944. He did pioneering work in the spectroscopy of organic compounds and also carried out some studies in kinetics. Stewart was celebrated for his *Recent Advances* books,[125,126] each of which went through several editions between about 1908 and 1948. Under the pseudonym J. J. Connington, he also wrote around twenty novels, mainly detective stories, in the period 1923 to 1947.

J. T. Hewitt (1868–1954)[127] had an extremely varied chemical career. At a very early stage therein, well before World War I, he carried out various studies that may be regarded as falling within physical organic chemistry. These included work on fluorescence of anthracene derivatives, in which he explained his findings in terms of oscillation of bonds between equivalent structures, apparently a forerunner of the concept of resonance. E. H. Farmer (1890–1952)[128] became a research student in 1919 under J. F. Thorpe and later C. K. Ingold at Imperial College, having previously taught in secondary schools and served in the army in World War I. He was on the lecturing staff of Imperial College from 1924 to 1938. During this period, much of his research work had physical organic aspects. Problems in obtaining reproducible results in studies of additions of unsymmetrical reagents to olefins led to the abandoning of this work without publication. With the benefit of hindsight we can see that the problems were due to the peroxide effect as discovered later by Kharasch (Section 5.6). In 1938, Farmer became senior chemist of the newly formed British Rubber Producers' Research Association.

H. McCombie (1880–1962),[129] after experience of chemistry in various circumstances, came to Cambridge in 1919 as a Fellow of King's College. Among his students during his tenure of a pre-World War I appointment at Birmingham had been H. A. Scarborough (1891–1969),[130] who rejoined him at Cambridge in 1920. In the 1920s and 1930s they jointly, with various junior collaborators, carried out studies on the kinetics of ester saponification and of quaternary salt formation. They also studied the aromatic substitution of derivatives of biphenyl. The work was of interest in connection with the development of the electronic theory of organic reactions. J. B. Shoesmith (1896–1931)[131] made various contributions to physical organic chemistry in his tragically short career.

5.4 Physical Organic Chemistry in Continental Europe

It is sometimes supposed that chemists of continental Europe contributed very little to the early stages of the development of physical organic chemistry. This is by no means true, although it is the case that significant contributions have often been neglected and tend now to have disappeared from sight. The traditional view is that continental European organic chemists were obsessed with the synthesis of new compounds and investigating the structures of natural products, and had no time for theories of reaction mechanisms or of the influence of structure upon reactivity. This view is particularly applied to German organic chemists, but it is not altogether true even for them. Physical chemists in continental Europe tended to be interested in theories of solutions, thermodynamics, the phase rule, colloids, electrochemistry, etc., but there were those who were also interested in reaction kinetics, and this led to some concern for the kinetics and mechanisms of organic reactions in solution. There were also some organic chemists who were sometimes prepared to adopt the techniques of physical chemistry for the study of organic reactions.

Relatively little has been written specifically on the contribution of continental European chemists to the development of physical organic chemistry, but material relevant to a bibliography can be found. No claim to completeness is made for what follows, but it may serve to indicate that the continental European contribution to the development of physical organic chemistry is worth exploring.

5.4.1 Germany

A. R. Hantzsch (1857–1935) was not merely one of the earliest German chemists to show substantial interest in topics we would regard as essentially physical organic chemistry, but one of the earliest chemists of any nationality to have this outlook.[132,133] After holding junior posts in Leipzig, he was Professor in Zürich, Würzburg and finally Leipzig (1903–1927). As a young man Hantzsch was influenced by both organic and physical chemistry traditions. Up to 1890, he worked in synthetic organic chemistry, but thereafter his work increasingly involved the application of physico-chemical methods to organic problems. The techniques included cryoscopic, electrical conductivity, and visible/UV spectroscopic measurements. His studies of pseudo-acids and -bases (his terms) and the cryoscopy of organic compounds in sulphuric acid (1907 onwards) were particularly notable. Similar cryoscopic work was done at about the same time by the Italian Giuseppe Oddo (1865–1954).[134]

It seems natural to mention Paul Walden (1863–1957),[135] a Baltic German, because of his discovery of inversion of configuration on substitution at an asymmetric centre and because of the importance of this Walden inversion in the development of physical organic chemistry. He also measured the dissociation constants of many organic acids by conductivity measurements and studied the conductivities of potential electrolytes, such as trityl chloride, in non-aqueous solvents like sulphur dioxide. However, his attitude to physico-chemical aspects of organic chemistry appears to have been somewhat ambivalent. According to Tarbell,[136] Walden regarded such matters as very much on the periphery of organic chemistry. Tarbell's view is based on a careful analysis of Walden's continuation of Carl Graebe's *History of Organic Chemistry* for the period from 1880 to about 1935. This article also outlines Walden's rather turbulent career, in which he was Professor in succession at Riga, Rostock, and Tübingen.

Georg Bredig (1868–1944)[137,138] was essentially a physical chemist, who worked in catalysis, reaction kinetics, and electrochemistry. For physical organic chemistry his studies of the catalysed decomposition of diazoacetic ester, of reactions in concentrated sulphuric acid, and of the catalysis of the benzoin condensation by cyanide ion are of interest. The last-mentioned work was done in 1904, when Bredig confirmed, by detailed kinetic studies, a mechanism previously suggested by Lapworth. Bredig was professor in Karlsruhe from 1911 to 1933, retiring shortly after the Nazi party came to power. He left Germany for the USA in 1940.

The list of publications in the obituary of the organic chemist Otto Dimroth (1872–1940)[139] has eleven papers that are classified as *physikalisch-organische Chemie*. They are mainly about tautomerism, intramolecular rearrangements, and solvent effects on tautomeric equilibria. One paper,[140] published in 1933, deals with relationships between reaction velocities and oxidation–reduction potentials for quinone systems and it is evident that Dimroth was a pioneer in developing linear free-energy relationships. He was Professor in Würzburg from 1918 to 1940.

There are both scientific[141] and personal[142,143] biographical articles for Hans Meerwein (1879–1965). He studied chemistry in Bonn and held junior positions there until 1923, when he became Professor at Königsberg. In 1928 he moved to be head of the Chemical Institute in Marburg,[144] with which he was associated for the rest of his life. His work in physical organic chemistry (*c.*1922) began in connection with the rearrangements of derivatives of camphor, for which he made kinetic studies of solvent effects on the reactions and showed that the rearrangements are preceded by ionization. At that time, this was a revolutionary suggestion, and the concept of a carbonium ion (Meerwein used the term cryptoion) was not readily acceptable to many chemists.

Fritz Arndt (1885–1969) is one of several organic chemists who in various ways appear to have anticipated the theory of resonance. In 1924, he interpreted the behaviour of pyrone ring systems in terms of *zwischenstufen*, the actual state of the molecule being intermediate between two structures differing in the arrangement of single and double bonds. This story was told in detail 35 years later in an article by Campaigne,[145] containing many personal communications from Arndt himself. A full obituary notice is available,[146] which includes a list of publications and details of Arndt's somewhat unusual career. This was interrupted both by World War I and his leaving Germany in 1933. From 1934 to 1955 he was a Professor at the University of Istanbul.[147,148]

It is appropriate to mention two of the successors of the above German chemists who in many ways typify the modern situation in which synthetic organic chemists tend also to be physical organic chemists, in that the elucidation of reaction mechanisms, particularly of new reactions, is a matter of central importance for them. Rudolf Criegee (1902–1975)[149] was a pupil of Meerwein at Marburg, but most of his career from 1937 onwards was spent at the Technical University of Karlsruhe. Rolf Huisgen (b. 1920) is of a yet later generation; he was Professor of Organic Chemistry at the University of Munich from 1952 to 1988. He has worked on a great variety of organic reactions, but is best known for his studies of cyclo-additions. Huisgen is the only German chemist who has contributed a volume in the American Chemical Society series of autobiographies 'Profiles, Pathways and Dreams', edited by Jeffrey Seeman. The title he chose for his book well expresses his attitude to working in organic chemistry: *The Adventure Playground of Mechanisms and Novel Reactions*.[150] He contributed an article on the development of physical organic chemistry in Germany to a volume commemorating the 50th anniversary of the *Gesellschaft Deutscher Chemiker*.[151]

Finally, in this section on German chemists, we must mention the brothers Hückel, who in very different ways have exerted a considerable influence on the development of physical organic chemistry. The elder brother, Walter (1895–1973), was an organic chemist who worked on many different topics, often having a physical organic aspect: stereochemistry of decalin, the Walden inversion, many aspects of mechanism and reactivity, dipole moments, *etc*. In 1930 he wrote *Theoretische Grundlagen der organischen Chemie*, a book that proved to be very influential, and reached its 4th edition by 1950.[152] Over the years he was associated with several universities, including Göttingen, Freiberg, Greifswald, Breslau, and finally Tübingen. There is a very full obituary notice in two parts dealing with personal biography[153] and scientific work[154] and also a short biography to 1950.[155] In his retirement, Walter Hückel wrote a long critical-historical study of the development of the non-classical ion hypothesis.[156] The younger brother, Erich (1896–1980), was the Hückel of the Debye–Hückel theory of strong electrolytes. He was by training a physicist, but in his research interests over the years he moved easily from experimental and theoretical physics to physical chemistry and chemical physics. His main contribution to physical organic chemistry was his application of quantum mechanics to benzene and other aromatic molecules (Hückel molecular orbital theory), work which stemmed from an association with Heisenberg and Hund on the theory of the double bond. After much wandering in his early years he became Professor of Theoretical Physics at Marburg in 1937, where he spent the rest of his life. Much material about Erich Hückel is available. He wrote an autobiography[157] in 1975 and there is the transcript of an editorial interview in 1970.[158] There are also brief biographies from 1950[159] and 1965,[160] the latter being by his wife. A commemorative article for the centenary of his birth has appeared.[161]

5.4.2 Russia

The association of the name of N. A. Menshutkin (1842–1907) with the reaction between tertiary amines and alkyl halides to form quaternary ammonium salts, though he was not the discoverer of it, is a tribute to the enormous amount of work he did on the kinetics of this reaction. Nowadays the term 'Menshutkin Reaction' is often extended to reactions of primary or secondary amines (examples of which were also studied by Menshutkin) and to reactions with any alkylating agent. Most organic chemists know of his study (1890) of the rate of the reaction between triethylamine and ethyl iodide in a series of solvents, and his finding that the rates varied by a factor of about 760 between the fastest and slowest solvents. Few chemists are aware, however, of the extent and variety of Menshutkin's work on the kinetics of organic reactions in solution, which was mainly carried out between 1876 and 1907. It was very much work before its time; he could give little interpretation of his findings.

His early education in science was at the University of St. Petersburg. Later he studied under Strecker (Tübingen), Wurtz (Paris), and Kolbe (Marburg). After returning to St. Petersburg, he had a rather turbulent academic career, but from 1869 he held various professorships. For English speakers, the best source of information about his life is the obituary notice in the *Journal of the Chemical Society* by Sir William Tilden.[162] There is one in German by his son,[163] who also wrote a long biography of his father, which was published in the year following N. A. Menshutkin's death.[164] There is an extensive literature of later articles in Russian, often written to commemorate various anniversaries of events in Menshutkin's life. There is also at least one book about him.[165] For an account of the innumerable studies and extensive literature of the Menshutkin reactions during the past century, see the massive review (182 pages, around 800 references) by Abboud and colleagues.[166]

During his researches into the history of the theory of the relationship between colour and constitution of organic compounds,[167,168] Dähne has examined the contributions of W. A. Ismailsky (1885–1973), whose career (mainly in Moscow) spanned both the closing years of pre-revolutionary Russia and the Soviet period.[169,170] Ismailsky appears to have been one of those who anticipated the theory of resonance in connection with the structures of aromatic molecules. This was in his thesis at the Technical University of Dresden in 1913, where he had worked under the direction of Walter Koenig.

Finally, we will mention a paper[171] in English that deals with historical aspects of the Markovnikov Rule for the orientation of addition of an unsymmetrical reagent to C=C. The origins of this rule can hardly be deemed to be part of physical organic chemistry, but the rule has of course been of great significance in the development of various aspects of the subject.

5.4.3 The Netherlands

In a way, the kinetics and mechanism aspect of physical organic chemistry was born in the Netherlands. The connection between kinetics and mechanism was first made by J. H. van't Hoff (1852–1911),[172] through his formulation of kinetics in terms of molecularity. This approach was different from that adopted by Ostwald, who formulated kinetics in terms of order of reaction. With the benefit of hindsight it may seem extraordinary to us that most early workers in chemical kinetics failed to realize that the subject had any bearing on how the molecules of reactants came together and rearranged to form products, but this is in fact the case. Van't Hoff's book *Études de Dynamique Chimique* of 1884 made the connection between kinetics and mechanism quite clear.[173] The second edition of the book was translated into English and published in 1896.[174] This contains accounts of several kinetic studies we would regard as belonging to physical organic chemistry. For instance, rate coefficients are tabulated for the acidic and the alkaline hydrolysis of several series of esters, and the effect of various structural features of the reactants is pointed out. It was work before its time, because no clear explanations could be given. Van't Hoff's early education in science was at Delft and Leiden. Later he studied under Kekulé in Bonn and Wurtz in Paris. He was Professor in the University of Amsterdam, 1878–1896, and then Professor in Berlin.

A. F. Holleman (1859–1953)[175] played a key role in the development of our knowledge and understanding of aromatic substitution. He was the first to realize the importance of measuring rates of substitution and proportions of isomers formed. Much of his work came before the development of electronic theories of organic reactivity. In 1910, he published a book entitled *Die direkte Einführung von Substituenten in den Benzolkern*, which contained a systematic description of all the then known facts about aromatic substitution reactions.[176,177] He had studied under von Bayer in Munich, but later he acted as an assistant to van't Hoff, and the latter's influence certainly turned him in the direction of applying physico-chemical techniques to organic chemistry. He was Professor first at Groningen and then from 1905 to 1924 at Amsterdam.

Another Dutch organic chemist whose work had physical organic aspects was Jacob Böeseken (1868–1949),[178] Professor at the Technical University of Delft from 1907 to 1938. Emeritus Professor Wiendelt Drenth of Utrecht provided some notes on 'The development of physical organic chemistry (until 1960) in a Dutch perspective'.[179] He considers that Böeseken's studies on the increase in the acidity of boric acid by certain dihydroxy compounds (in particular of cyclohexane) involved formulating the fundamentals of conformational analysis many years before this became a recognized topic. Drenth suggests, however, that, in the main, physical organic chemistry did not start to flourish in the Netherlands until after World War II. He mentions in particular the work then of the Royal Dutch Shell Laboratory in Amsterdam and the important roles of E. Havinga (1909–1988)[180] and L. J. Oosterhoff (1907–1974)[181] at Leyden, B. M. Wepster (1920-1992)[182,183] at Delft, F. L. J. Sixma (1923–1963)[184] at Amsterdam, and P. H. Hermans (1898–1979)[185] of the industrial company Algemene Kunstzijde Unie (AKU, now part of AKZO Nobel). Hermans wrote a textbook in 1952, which later appeared in English translation.[186,187]

Drenth has since expanded these notes into a full-blown book (in Dutch) and has extended his coverage to 1994.[188] Only just over a quarter of the book is devoted to the period up to 1970, while there is an exhaustive treatment of more modern times, with respect to the individual workers and institutions and the fields in which they have worked. Drenth has also unearthed biographical material on S. C. J. Olivier (1879–1961), who carried out pioneering work in the kinetics and mechanisms of organic reactions in solution from about 1912 to 1938.[189,190] For many years he was assisted by G. Berger (1892–1942). This work was mentioned in the monographs of Hammett,[191] Waters,[192] and Watson.[193]

5.4.4   Scandinavia

The number of chemists in Scandanavian countries who were involved in the development of physical organic chemistry was small, but their contribution was highly significant. The Norwegian Odd Hassell (1897–1981) will be dealt with in Section 5.7 in connection with conformational analysis. The name of the Dane, Johannes Nicolaus Brønsted (1879–1947) is well known to all physical organic chemists and indeed to most chemists, because in 1923 he proposed definitions of acids and bases in terms of proton transfer. He no doubt thought of himself as a physical chemist, but his work on acid-base catalysis and the relationship between catalytic activity and acid or base strength is one of the foundations of physical organic chemistry. It provided the earliest examples of what became known as linear free-energy relationships, and the study of proton-transfer reactions has become a most important topic within physical organic chemistry. Unfortunately for the English-speaking reader there is little written about Brønsted in English,[194] although there is a biography in Danish.[195] He was Professor of Physical Chemistry at the University of Copenhagen from 1908 until his death in 1947. Among his pupils was the British chemist, R. P. Bell.

5.4.5 France

The book by Nye contains a substantial account of 'The Paris School of Theoretical Organic Chemistry, 1880–1930'; in fact, the account continues well beyond 1930.[15] A main theme of this account is the extremely conservative nature of most French organic chemists during the key period of the 1930s and 1940s. This showed itself in scepticism regarding the development of electronic theories of organic chemistry and a desire to keep French organic chemistry isolated from what were described as Anglo-Saxon influences. There were notable exceptions, such as Charles Prévost (1899-1983)[196] and Albert Kirrmann (1900–1974),[197] but even they sought to develop a distinctive French approach, with peculiar ideas, terminology and symbolism. This was almost entirely ignored outside France. Prévost and Kirrmann were students together in the École Normale Supérieure in Paris after World War I. Having held various university appointments in the provinces, they were together in Paris during the later stages of their careers. Kirrmann was director of the chemical laboratory of the École from 1955 to 1970. During this time, work in physical organic chemistry along international lines began to develop at the École under various younger researchers. Such a development also occurred at around the same time at the University of Paris, where Prévost was Professor. J.-E. Dubois (b. 1920) was particularly influential; he had spent periods of study at University College, London and Columbia University, New York.

Marc Tiffeneau (1873–1945)[198] exerted an important influence in making French chemists ultimately more aware of and receptive of the changes in organic chemistry outside France during the 1930s and 1940s. Throughout his career, he combined work in organic chemistry with work in biological sciences, particularly pharmacology. In his work in organic chemistry, he avoided the conservative outlook of most French organic chemists between World Wars I and II and this led to his creating a school that fully accepted the international developments in organic chemistry. Among the leading members of this group was the Ukrainian émigré Bianca Tchoubar (1910–1990), whose obituary notice makes interesting reading.[199] Tchoubar wrote one of the first accounts in the French language of the electronic theories of organic chemistry and of reaction mechanisms as they had become by the 1950s.[200] This book is in stark contrast to one published by the Pullmans eight years earlier.[201] Its title declares it to be about electronic theories of organic chemistry; it makes one passing reference to Ingold. It is in fact largely about the application of quantum mechanics, such as it was at that time, to organic chemistry. Another early account in French of electronic theories of organic chemistry was by Marc Julia (b. 1920),[202] who succeeded Kirrmann at the École Normale Supérieure in 1970.

5.5   Physical Organic Chemistry in the USA

It must be mentioned at the outset that the *Essays on the History of Organic Chemistry in the United States* by the Tarbells, husband and wife, contain much information on the development of physical organic chemistry in the USA,[203] properly set in context of the development of physical organic chemistry generally. They have been drawn on heavily here. Furthermore, as in Section 5.3, Saltzman's article[204] on the development of physical organic chemistry in the United States and the United Kingdom, pointing out various parallels and contrasts, may be taken to provide a framework for the presentation of this material. In Saltzman's view, development of theoretical organic chemistry in the USA before World War I was dominated by attempts to develop electronic concepts of valence and to apply them to organic structures and reactions. These attempts were in the main derived from J. J. Thomson's proposal in 1903 that the chemical bond was the result of the transfer of a single electron from one atom to another. The theories that attempted to apply this model to the bonds formed by carbon became highly elaborate in the effort to account for the properties of organic compounds and were naturally doomed ultimately to fail. The group of prominent American chemists involved in this enterprise persisted far too long in using this erroneous model. Its ultimate collapse around 1915 led to a decline in interest in theoretical organic chemistry among American chemists for about a decade. These matters are dealt with in some detail in three articles by Saltzman on 'J. J. Thomson and the modern revival of dualism',[205] 'The bonds of conformity-W. A. Noyes and the initial failure of the Lewis Theory in America',[206] and 'Benzene and the triumph of the octet theory'.[207] In Saltzman's opinion, this situation led to American chemists being slow off the mark in the 1920s, when electronic theories of organic chemistry were being developed in the UK, and kinetics and mechanism became recognized as an integral part of organic chemistry (Sections 5.2 and 5.3).

However, as with most broad generalizations, exceptions can be found in various aspects of the American situation. Thus there were chemists in the USA even before World War I with interests in kinetics and mechanism contemporary with those of Lapworth, Lowry, Orton, and others in the UK; indeed, some were even at work in the 19th century. R. B. Warder (1848–1905) of Cincinnati was the first to make satisfactory kinetic measurements on the alkaline hydrolysis of an ester, in 1881.[208] In 1898, Emmet Reid (1872–1973) of The Johns Hopkins University studied the kinetics of the acidic and the alkaline hydrolysis of substituted benzamides and later (1910) correctly identified the normal mechanism of ester hydrolysis as acyl-oxygen fission by analogy with his findings of the behaviour of certain thioesters and thiols. A personal account of all this work may be found in his fascinating autobiography, published in his 100th year.[209] J. F. Norris (1871–1940)[210] discovered in 1907 that t-butyl alcohol reacted with HCl to form the chloride far more rapidly than primary or secondary alcohols and suggested tentatively that a t-butyl carbonium ion might be an intermediate. He thus extended the carbonium ion idea already proposed by J. Stieglitz (1867–1937) in 1899.[211] The latter carried out much kinetic and mechanistic work before World War I. S. F. Acree (1875–1957), also of Johns Hopkins, pursued many studies of the kinetics of various organic reactions from 1904 to 1914.[212] In this period there was also the work of C. G. Derrick (1883-1980) in developing a scale of polarity of atoms and groups from their effects on the ionization constants of carboxylic acids;[213] the application of thermodynamics by Arthur Michael;[214–216] the work by E. C. Franklin (1862–1937)[217,218] and Charles Kraus (1875–1967)[219] on liquid ammonia solutions; and studies of reactive intermediates by J. U. Nef (1862–1915)[220] and Moses Gomberg (1866–1947).

According to Saltzman:

> In the years after 1919 a slow and at times very hesitant development of physical organic chemistry occurred in the US …The person who should have been the catalyst for the development of physical organic chemistry in the US is James Bryant Conant (1893–1978).[221]

For his Ph.D. at Harvard, Conant studied a physical chemistry problem under T. W. Richards (1868-1928) and an organic problem under E. P. Kohler (1865–1938).[222] He obtained his doctorate in 1916, and after war work joined the Harvard faculty in 1919. For the next 14 years he was an effective member of the chemistry department, but in 1933 he was chosen to be President of Harvard, and his chemistry career ended. In those 14 years, he had carried out much research of a physical organic nature. There is plenty of material on Conant. Towards the end of his life he wrote an autobiography.[223] His role in the development of physical organic chemistry has been assessed by Saltzman.[224,225] There are biographical memoirs from both the National Academy of Sciences[226] and the Royal Society of London.[227]

Among Conant's students at Harvard was Paul D. Bartlett (1907–1997), who remained at Harvard and continued research in physical organic chemistry, thus making it one of the foremost centres for the field on the world stage. In 1974 a Bartlett Symposium on physical organic chemistry was held at Fort Worth, Texas, for which those who had been his students prepared a 'Group Autobiography' covering the period 1934–1974.[228] Memoirs were contributed by some 150 people and there were 14 scientific papers on the work done by the Bartlett Group during the period under review. An oral history interview with Bartlett conducted by Leon Gortler in 1978 is on deposit.[229] The graduate lecture course on 'Theoretical Organic Chemistry' given by Bartlett at Harvard in 1938 has been described.[230] Bartlett's impact on physical organic chemistry has been discussed by J. D. Roberts[231] and an obituary notice has appeared.[232]

Another of Conant's students was Frank Westheimer (b. 1912), who, after postdoctoral work with Hammett at Columbia, held a post at the University of Chicago (1936–1954) and then returned to Harvard. Westheimer worked in several areas of physical organic chemistry and engaged in other chemistry-based activities, as revealed in an interview conducted in 1995 by István Hargittai.[233] For much of his career, Westheimer was essentially a physical organic chemist working in biochemistry and he has himself written reflectively on the discovery of the mechanisms of enzyme action over the period 1947–1963[234] and on the application of physical organic chemistry to biochemical problems.[235] Westheimer has also contributed, as *Tetrahedron Perspective* Number 4, an article on 'Coincidences, decarboxylation, and electrostatic effects', which, he writes, '…allows me to review some of my past'.[236]

After being a student under Conant, George Wheland (1907–1982) did post-doctoral work at the California Institute of Technology (CalTech), Oxford, and University College, London and then joined the faculty of the University of Chicago in 1937. He thus became part of the physical organic school that included Kharasch, Westheimer, and Mayo and, as students, Walling and H. C. Brown. Wheland wrote an influential monograph on *The Theory of Resonance*,[237] first published in 1944.

J. F. Norris, already mentioned above in connection with his pre-World War I work, was in the later stages of his career at MIT when Conant was prominent at Harvard. In the 1920s, he made some very competent kinetic studies and published a series of papers entitled 'The reactivity of atoms and groups in organic compounds'.

L. P. Hammett (1894–1987) was a junior contemporary of Conant at Harvard and was also much influenced by E. P. Kohler, who had interests in both physical and organic chemistry. As mentioned in Section 5.1, Hammett is often credited with inventing, or at least bringing into common usage, the term physical organic chemistry. Hammett worked for his Ph.D. at Columbia University, New York and his entire career from 1920 was spent in that university. Around 1966, Hammett reflected on his career in an address given on his receiving the James Flack Norris award in physical organic chemistry. This was subsequently published as 'Physical organic chemistry in retrospect', in which we may read of the influence of Kohler and others in shaping the general direction of Hammett's research interests.[238] Hammett regarded himself essentially as a physical chemist interested in the behaviour of organic compounds and in the article one may see how that this resulted in the Hammett acidity function and the Hammett equation. Hammett's death in 1987 and the centenary in 1994 of his birth stimulated writing about his life and work. The present contributor gave a Hammett Memorial Lecture at the 4th International Conference on Correlation Analysis in Organic Chemistry held at Poznań in 1988.[239] A brief biographical lecture about Hammett was given at the 12th International Conference on Physical Organic Chemistry, held in Padua in 1994.[240] The present contributor has also written about Hammett himself, the Hammett equation, and other matters related to linear free-energy relationships in a semi-popular vein.[241] Previously, the Golden Jubilee of the Hammett equation had been commemorated in an article in a German chemical magazine.[242]

Hammett's 90th birthday was anticipated in 1983 with a joint symposium of the Divisions of the History of Chemistry and of Organic Chemistry at the Fall Meeting of the ACS in Washington DC. An announcement of the symposium with a provisional list of speakers is available,[243] but Hammett's health unfortunately did not permit him to attend. Versions of some of the talks were ultimately published in various journals. These included a talk by Leon Gortler on 'The physical organic community in the United States, 1925–1950: an emerging network'.[244] Summaries of all the talks are in the volume of abstracts for the meeting in question. In a further contribution to the historical literature regarding Louis Hammett and the eponymous equation, Shorter has written on its prehistory,[245] with particular attention to the parallel researches of Hammett and of G. N. Burkhardt in Manchester in the early 1930s.

A few years later a similar symposium was organized to commemorate the centenary of the birth of Frank C. Whitmore (1887–1947). He was a slightly senior contemporary at Harvard of Conant and Hammett, and like the former did a double-barrelled Ph.D. under the supervision of T. W. Richards and E. P. Kohler. The latter was probably responsible for Whitmore's lifelong interest in organic reaction mechanisms. His main career was spent at Northwestern University (1920–1929) and at Pennsylvania State College (1929–1947). Much of his work was in organometallic chemistry, but from 1928 he developed an interest in reactions of aliphatic compounds which result in intramolecular rearrangements. In this connection he was led to postulate intermediate carbonium ions, although the continued resistance to this idea led him to describe the species as involving an open sextet. Saltzman has written a general article about Whitmore's life and work,[246] as well as an article on Whitmore's work on intramolecular rearrangements.[247] Mosher and Tidwell[248] have written on 'Frank C. Whitmore and steric hindrance: a duo of centennials' and Traynham has edited 'Personal recollections of Frank C. Whitmore' by Charles D. Hurd.[249]

In California, physical organic chemistry developed mainly on three campuses: CalTech, University of California at Los Angeles (UCLA), and Berkeley. Howard Lucas (1885–1963)[250] was the initiator of physical organic chemistry at the predecessor institution of CalTech, Throop College of Technology. Lucas went there in 1913 and remained at Throop-CalTech for the remainder of his career. His part in the development of physical organic chemistry in the USA tends to be undervalued. He made several significant contributions in the 1920s and 1930s.[203,251] For instance, in the 1920s he worked on the application of the Lewis theory of the electron-pair bond to organic chemistry, thus in parallel to Robinson and Ingold in the UK. In the 1930s, Lucas originated the first satisfactory explanations of neighbouring group effects. He used the ideas of resonance in his research and in the writing of a textbook, no doubt being much influenced by his CalTech colleague Linus Pauling (Section 5.8). Physical chemistry under A. A. Noyes was the dominant force at CalTech and conditions were not favourable for Lucas' influence to persist. He had two very able students: W. G. Young in the 1920s and Saul Winstein in the 1930s, but their influence is to be found in the establishment of the physical organic chemistry school at UCLA.

Young (1907–1981) moved to UCLA in the early 1930s. Winstein (1912–1969) was an undergraduate at UCLA and, after doing his Ph.D. with Lucas and post-doctoral work with Bartlett at Harvard, was on the faculty at UCLA from 1941 until his untimely death in 1969. Winstein worked on numerous topics in physical organic chemistry, but is most celebrated for the various studies which he deemed to have established the role of non-classical ions. This generated a famous controversy, in which H. C. Brown was the principal exponent of alternative interpretations. We list literature specifically relevant to this controversy below. There is a considerable amount of biographical material for Winstein. Young and Cram[252] wrote a formal obituary notice and Streitwieser[253] wrote a long article that includes a complete list of Winstein's publications. Bartlett[254] has also written on the scientific work of Saul Winstein.

Two of the physical organic chemists associated with UCLA have contributed autobiographies in the ACS series. D. J. Cram (b. 1919)[255] went to UCLA after working for his Ph.D. with Fieser at Harvard, and the rest of his career was spent there. A Nobel Prize winner for 1987, Cram died in 2001.[256] J. D. Roberts (b. 1918)[257] was an undergraduate, graduate student, and post-doctoral fellow at UCLA. After further post-doctoral work at Harvard, Roberts became a faculty member at MIT, but in 1953 returned to California, going to CalTech to succeed Howard Lucas. He has written on the beginnings of physical organic chemistry in the United States,[258,259] and has himself been interviewed for the *Chemical Intelligencer*.[260]

Herbert C. Brown (1912–2004)[261] was born in London, but the Brown family moved to the USA in 1914. He was both an undergraduate and a graduate student at Chicago in the 1930s, and therefore came under the influence of the Chicago school of physical organic chemistry. He worked with Kharasch as a post-doctoral fellow. His Ph.D. work and later post-doctoral work with H. I. Schlesinger were devoted to the study of boron compounds, and this had a profound influence throughout Brown's career. His Nobel Prize in 1979 was awarded for his contributions to synthetic organic chemistry connected with the reactions of organoboranes. His work on physical organic chemistry, which was carried on in parallel with that on organoboranes, was on steric effects and the extension of the Hammett equation to electrophilic aromatic substitution. As already mentioned, he was involved in the famous controversy with Winstein over the existence or otherwise of non-classical carbonium ions. Brown taught at Purdue University from 1947 to 1978 and remained active in science well into retirement.

There is a fair amount of literature about Brown that is relevant to this report. Brown's George Fisher Baker Lectures at Cornell University in 1969 were on boranes in organic chemistry, but were effectively a personal and scientific autobiography.[262] In 1977 Brown published his own account of the non-classical ion problem, with comments by Paul Schleyer.[263] Bartlett published a collection of the principal papers in the controversy, mainly 1949–1964, with commentary.[264] The purpose of a 70-page *Review* in 1966 by G. D. Sargent was declared to be an attempt 'to re-evaluate the viability of the concept of bridged, non-classical ions in light of the criticisms which have been levelled against it'.[265] Arnett and colleagues,[266] with the benefit of distance in time from the controversy, published in 1985 a historical perspective on the non-classical ion disputes, as part of an article on carbocations. Brown was interviewed by István Hargittai[267] and B. P. Coppola has 'looked under the surface of a chemical article' in the case of a contribution to the norbornyl cation problem made in 1961 by R. G. Lawton, then a graduate student at the University of Wisconsin, Madison.[268] S. J. Weininger has surveyed the non-classical cation controversy under the title: 'What's in a Name? From Designation to Denunciation', with copious references.[269] He concentrates on 'the relationship of alternate theoretical formalisms to differing representations of molecules, and the conflicts that arise when new forms of representation are introduced.'

We must now devote some attention to physical organic chemistry at the University of California, Berkeley. It cannot really be claimed that G. N. Lewis (1875–1946) was a physical organic chemist. He did, however, provide important background for the development of physical organic chemistry in his work on the electronic theory of valency,[270] and his broadening of the definition of acids and bases, both of these contributions being made about 1923. The influence and leadership of Lewis were paramount in building up the international reputation of the chemistry department at Berkeley, which he served from 1912 for the rest of his career. He certainly encouraged the establishment of physical organic chemistry as one of the disciplines there. Melvin Calvin has written about the influence of Lewis on physical organic chemists at Berkeley.[271] Much of this influence was by way of encouragement to G. E. K. Branch (1886–1954).[272] Branch was a British West Indian, who initially studied medicine at Edinburgh and then changed to studying chemistry under F. G. Donnan at Liverpool. After some post-graduate study at Liverpool, he left for Berkeley in 1912, and spent the rest of his career there. He made extensive studies of the effect of structure on the strengths of organic acids and participated in the application of the Lewis electron-pair bond theory in organic chemistry. After many years he wrote *The Theory of Organic Chemistry*[273] in collaboration with Calvin, which was published in 1941. Calvin[274] described their collaboration and also included as an appendix an essay on Lewis by Branch, originally written in 1951. The writing of Branch and Calvin is also described in Calvin's autobiography.[275] Calvin and G. T. Seaborg also wrote on 'The College of Chemistry in the G. N. Lewis era: 1912–1946'[276] and Seaborg has contributed 'Gilbert Newton Lewis–some personal recollections of a chemical giant'.[277]

Calvin (1911–1997),[278] after graduate education at the University of Minnesota and post-doctoral work with Polanyi at Manchester, was appointed to the faculty at Berkeley in 1937 and spent the rest of his career there. From purely physical organic work, he moved into bio-organic chemistry. He was awarded the Nobel Prize in 1961 for his work on the path of carbon in photosynthesis. Among the other physical organic chemists associated with Berkeley is Andrew Streitwieser (b. 1927), whose autobiography has appeared in the ACS series.[279]

D. Stanley Tarbell (1913–1999) and Ann Tracy (Hoar) Tarbell (1916–1998), the authors of the invaluable *Essays on the History of Organic Chemistry in the United States, 1875–1955,* have died. A memorial tribute has appeared.[280]

Several physical organic chemists started their careers in Europe and then for various reasons moved to the USA For example, Alan R. Katritzky (b. 1928) was an undergraduate and post-graduate at Oxford, held some appointments at Cambridge, was a founder Professor at the University of East Anglia, and then moved to the University of Florida, Gainesville, in 1980. For an issue of *Heterocycles* dedicated to him for his 65th birthday, he wrote a summary of the 'Katritzky Research Group: scientific results (1954–1993)'.[281] Most of his work has a physical organic as well as synthetic/structural flavour. He has also written on 'Highlights from 50 years of heterocyclic chemistry'.[282] G. A. Olah (b. 1927) left Hungary during the troubles there in 1956. He was awarded the Nobel Prize in 1994 for his work on carbocations.[283] He has been interviewed by G. B. and L. M. Kauffman.[284]

M. J. S. Dewar (1918–1997) began his chemical career in Oxford, where he was associated with Robinson and under whose influence he wrote *The Electronic Theory of Organic Chemistry.*[285] He spent 1945–1951 at the Courtauld laboratory at Maidenhead and then became professor and head of department at Queen Mary College, London. In 1959 he moved to the USA, first to Chicago until 1963, then to the University of Texas at Austin (1963–1990), and finally to active retirement in Gainesville, Florida. Initially, Dewar carried out experimental research, but at an early stage he was much attracted to the application of quantum mechanics to organic chemistry and most of his career was devoted to this, particularly to the development and application of so-called semiempirical treatments. This is reflected in the choice of title for his contribution to the ACS series: *A Semiempirical Life.*[286] Not long before he died, Dewar published some comments on *A Semiempirical Life* in the *Chemical Intelligencer,*[287] in which he made what some considered to be intemperate remarks about certain deceased chemists. These provoked a reply from Laidler, Leffek, and Shorter;[288] see also a review by Davenport of Leffek's biography of Ingold.[289] Obituary notices of Dewar are now available.[290–292] Edward has compared and contrasted the careers of Dewar and D. H. R. Barton, who is considered in Section 5.7.[293]

The American physical organic chemist E. R. Alexander (1920–1950) had a tragically short life, which ended in the crash of a small aircraft. P. R. Jones has written a short account of his contributions to physical organic chemistry, which were both numerous and distinguished, considering the brevity of his career.[294] His publications included a book on ionic organic reactions.[295]

Oral history interviews of various physical organic chemists are now fully available for consultation in the Othmer Library of the Chemical Heritage Foundation in Philadelphia.[296] They include: F. R. Mayo, M. S. Newman, C. C. Price, J. D. Roberts, A. Streitwieser, C. Walling, and F. H. Westheimer.

H. H. Jaffé (1919–1989)[297] was professor of chemistry at the University of Cincinnati. His research interests were quantum chemistry, the basicity of weak bases, spectroscopy, excited state chemistry, and the Hammett equation, of which he wrote a comprehensive review article in 1953.[298] At one time, according to *Science Citation Index*, this was the most cited article in the chemical literature.

R. W. Taft (1922–1996) was professor of chemistry at Pennsylvania State College and then for thirty years at the University of California, Irvine. He did distinguished work in several fields of physical organic chemistry, *e.g.* structure–reactivity relationships, gas-phase reactivity of organic compounds, and the correlation analysis of solvent effects.[299,300]

The autobiography of Paul von R. Schleyer (b. 1930) is scheduled to appear in the ACS series.[301] He has been interviewed.[302] He was associated for a long period with Princeton, but moved to Erlangen in Germany some years ago.

5.6 Free Radical Chemistry

The Tarbells' book contains a good account of the discovery of the triphenylmethyl radical by Moses Gomberg (1866–1947). Gomberg was a child immigrant to the USA from Russia, who worked his way through the University of Michigan, Ann Arbor, to a doctorate in chemistry in 1894. After a year of study with Victor Meyer at Heidelberg, he returned to Ann Arbor and started work that resulted in the serendipitous discovery of triphenylmethyl in 1900. An obituary notice[303] and a biographical memoir[304] for Gomberg are available. See also a review article[305] by Gomberg written in 1924.

The compound formed by the dimerization of triphenylmethyl was erroneously believed for almost seventy years to be hexaphenylethane and was only properly identified as an isomer of quinoid structure by spectroscopic studies in 1968.[306] McBride has examined in detail how the error persisted for such a long time.[307] The centenary of Gomberg's discovery in 1900 of the triphenylmethyl radical has stimulated some historical studies. Tidwell has written on 'The free radical century: Gomberg and beyond.'[308] In addition to mentioning the main participants in the development of free radical chemistry during the first half of the century, he introduces several less important people whose contributions in the late 1920s and early 1930s tend to be overlooked. Tidwell also draws attention to the discussion of the Faraday Society in 1933 on 'Free Radicals'[309] in which various pioneers, including Gomberg, participated.

Shortly before Lennart Eberson died in 2000, he examined the archives of the Royal Swedish Academy of Sciences to try to understand why Gomberg was not awarded a Nobel Prize.[310] The topics dealt with by Tidwell and by Eberson have subsequently been treated again at much greater length. Tidwell's article covers the development of free radical chemistry in the 20th century and cites 399 references.[311] Eberson's investigations of the Nobel Archives are presented in even greater detail.[312] Tidwell has also written an essay on Wilhelm Schlenk (1879–1943) described as 'The Man behind the Flask' who was a pioneer in free radical chemistry.[313]

The Tarbells outline the later development of free radical chemistry, in particular in their Chapter 18, dealing with 1914–1939. The history of free radical chemistry to about 1930 is dealt with by Christine King;[314] Walling covers some of the same ground and continues the story to the 1940s.[315] Both these papers are based on talks given in the Hammett Symposium in 1983 and include some account of free radicals in the gas phase as studied by physical chemists.

The vigorous development of free radical chemistry in the United States began with the work of Morris S. Kharasch (1895–1957) and his group at Chicago in the early 1930s, which was an attempt to resolve confusing and contradictory results for the orientation of the addition of HBr to unsymmetrical olefines. This led to the discovery that the orientation depended on the presence or absence of peroxides. In the former case a free-radical chain reaction was involved, although this was not immediately recognized. The occurrence of this 'peroxide effect' was demonstrated by Kharasch and Mayo in 1933. Progress in the field to 1940 was reviewed by Mayo and Walling.[316] There are two accounts by Mayo of the story of the discovery of the peroxide effect,[317,318] the second being based on a talk given in the Hammett Symposium in 1983. Kharasch was born in the Ukraine, but came to the United States and received his chemical education at the University of Chicago, where he spent most of his career.[319]

Cheves Walling (b. 1916) received his undergraduate training in chemistry at Harvard and joined Kharasch's group in 1937. From 1939 to 1952 he worked in industry, for much of the time being concerned with free radicals as involved in polymerization reactions. He returned to academic life at Columbia University in 1952, and moved to the University of Utah in 1969. His contribution to the ACS series of autobiographies is entitled *Fifty Years of Free Radicals*.[320] Melvin Calvin did much work on free radicals, particularly in connection with photochemistry.[321]

In the 1930s, studies in free radical chemistry in the UK developed in parallel to developments in the United States. This involved D. H. Hey and W. A. Waters. Donald Hey (1904–1987)[322,323] received his chemical training at University College, Swansea and was on the chemistry staff of the University of Manchester (1928–38), and Imperial College (1938–1941). He was Director of the British Schering Research Institute in Manchester from 1941 to 1945, and then professor and later head of department at Kings College, London from 1945 to 1971. From his early days in Manchester, Hey was interested in aromatic substitution and by chance he and a student, W. S. M. Grieve, began to investigate reactions that they found did not conform to the usual rules for the directing effects of substituents in aromatic substitution. Such behaviour was explained in terms of free radical mechanisms. Hey's earliest publications in 1934 brought him into contact with W. A. Waters (1903–1985),[324] who had been a pupil of Lowry at Cambridge (Section 5.3), but by that time was on the chemistry staff of the University of Durham. Waters had suspected for some years that free radicals might participate more widely in organic reactions in solution than was commonly believed at that time. This view arose from studying the literature while writing his book *Physical Aspects of Organic Chemistry,*[325] as well as from experiments of his own. This book contains much valuable material on the early history of free radical chemistry. The contact between Hey and Waters led to their co-authoring in 1937 a review article on some organic reactions involving the occurrence of free radicals in solution.[326] Fifty years later, Walling wrote an article in celebration of the Golden Jubilee of the review by Hey and Waters.[327]

Throughout his time at King's College, Hey's dominant research interest was in free radical aromatic substitution. Waters, particularly after he transferred to Oxford in 1946, specialized in studies of the kinetics and mechanism of oxidations of organic compounds by metal salts, which often involve the intervention of free radical intermediates. Shortly before he died, Waters published his reflections on the development of free radical chemistry.[328] In 1946, he wrote one of the first monographs on the chemistry of free radicals[329] and in 1959 edited the Kharasch memorial volume. A tribute to both Hey and Waters was published by the Chemical Society in 1970 to mark their respective retirements.[330] This collection of articles by some of their former pupils and colleagues was edited by R. O. C. Norman (1932–1993), who had himself been a pupil of Waters at Oxford. Norman was founder Professor of Chemistry at the University of York and made distinguished contributions to free radical chemistry.[331,332]

5.7   Stereochemistry

Several books and articles discuss historical aspects of stereochemistry as related to physical organic chemistry, particularly conformational analysis. The books by the Tarbells and by Brock naturally contain material on stereochemistry in both its classical and modern aspects. Schofield's unpublished volume[333] deals in detail with stereochemical aspects of the S_N1 and S_N2 mechanisms. The history of stereochemistry by O. B. Ramsay[334] in the series 'Nobel Prize Topics in Chemistry' presents a detailed account of the development of many topics in stereochemistry, as well as reprinting key papers by Hassel, Barton, Prelog, and Cornforth, with biographical notes on the authors. There is also a very useful chronology of events and publications in the history of stereochemistry, a glossary of stereochemical terms, and a piece on stereochemical satire.

The Nobel Prize for Chemistry in 1969 was awarded jointly to Odd Hassel (1897–1981) of the University of Oslo and Derek H. R. Barton (1918–1998)[335,336] of Imperial College, London for developing and applying the principles of conformation in chemistry. Hassel's whole career was spent at Oslo[337] and his contribution was the use of physical techniques in the study of the structures of cyclohexane and its derivatives. The work was done from 1930 onwards, but was not well known until after World War II. Barton became acquainted with Hassel's work and in 1949 began to apply the results to interpreting features of the chemistry of steroids, which had hitherto been difficult to understand.

The application of conformational analysis spread rapidly through organic chemistry and biochemistry. Barton contributed to the ACS series of autobiographies.[338] At the time he started the work for which he was awarded the Nobel Prize, he was a lecturer at Imperial College. In 1950 he moved to Birkbeck College, London as Reader and later became Professor there. From 1955 to 1957 he held the Chair of Organic Chemistry at Glasgow, and then returned to Imperial College as Professor. In 1978, he moved to Gif-sur-Yvette in France as Director of the Institut de Chimie des Substances Naturelles. About 1986, he retired and moved to a post as Distinguished Professor at the Texas A & M University.

Much of Barton's work throughout his career illustrates the way in which physical organic chemistry has permeated the whole of organic chemistry.[339] This is also true of the career of Vladimir Prelog (1906–1998),[340] who received the 1975 Nobel Prize in Chemistry jointly with John Cornforth (b. 1917) for their researches in stereochemistry. About 1950, Prelog independently put forward some of the notions that became part of conformational analysis. He too contributed a volume to the ACS series of autobiographies.[341]

A lengthy review article[342] by Jeffrey Seeman on the effect of conformational change on reactivity in organic chemistry contains much historical material. Michelle Beugelmans-Verrier[343] has written (in French) an outline of the history of conformational analysis. Other relevant articles include a survey of a hundred years of stereochemistry by Weyer,[344] with emphasis on classical, but some treatment of physical organic chemical aspects; an account of the concept of strain in organic chemistry by Wiberg;[345] and a survey of stereochemistry since Le Bel and van't Hoff by Eliel.[346] Eliel's own contributions to stereochemistry make his volume in the ACS series of autobiographies relevant to this Section.[347] A large monograph on stereochemistry[348] by Eliel and Wilen contains much material of historical interest. Eliel has been interviewed.[349]

5.8   Quantum Organic Chemistry

Chapter 9 of Nye's book is devoted to 'Quantum chemistry and chemical physics, 1920–1950', with sections on the application of quantum mechanics to molecules in the 1920s, chemists and quantum mechanics in the 1920s and 1930s, and quantum chemistry and chemical physics in the 1930s and 1940s.

It seems useful to list biographical material that is available for some of the pioneers in the application of quantum mechanics to chemistry, particularly organic chemistry, with a few comments. R. S. Mulliken (1896–1986) is best known for his share in the creation of molecular orbital theory. He was associated with the University of Chicago for most of his career,[350,351] and was awarded the Nobel Prize for Chemistry in 1966. Mulliken traced the path to molecular orbital theory in a lecture[352] given in 1970. He wrote an autobiography that was published posthumously.[353]

J. E. Lennard-Jones (1894–1954) is also remembered for his contribution to the development of molecular orbital theory.[354] His Cambridge pupil, C. A. Coulson (1910–1974), carried out much pioneering work in the application of molecular orbital theory to chemistry, particularly organic chemistry.[355] Much of this work was before the advent of sophisticated calculations using computers, but Coulson helped greatly to establish the basis upon which such calculations are performed. He was associated with various universities at different times, but for the last twenty years of his life he held Chairs in Oxford. His book *Valence* (1952) was of great influence; it dealt with both molecular orbital and valence-bond approaches, and included the treatment of conjugated and aromatic molecules.[356] His life and work have been re-examined.[357,358]

The contributions of Erich Hückel to the development of molecular orbital theory have already been mentioned in the subsection on Germany (Section 5.4.1); the development of semi-empirical quantum mechanical treatments in organic chemistry by M. J. S. Dewar has been discussed in Section 5.5. In the early development of the application of quantum mechanics to chemistry, Linus Pauling (1901–1994)[359] was pre-eminent. He was associated with CalTech for most of his career. His work before World War II generated two influential books: the *Introduction to Quantum Mechanics* (with E. Bright Wilson, 1935)[360] and *The Nature of the Chemical Bond* (1939).[361] He favoured the valence-bond treatment and the theory of resonance.

McWeeny has written a tribute to the valence-bond theory pioneers of 1927–1935.[362] Shavitt has outlined the history and evolution of Gaussian basis sets as employed in *ab initio* molecular orbital calculations.[363] Hargittai has interviewed Roald Hoffmann (b. 1937)[364] of Cornell University and Kenichi Fukui (1918–1998)[365] of Kyoto University, who were jointly awarded the Nobel Prize in Chemistry in 1981. Fukui developed the concept of frontier orbitals and recognized the importance of orbital symmetry in chemical reactions, but his work was highly mathematical and its importance was not appreciated until Robert Woodward (1917–1979) and Hoffmann produced their rules for the conservation of orbital symmetry from 1965 onwards.[366]

It should not be overlooked that the quantum theory was one of the components of the development of transition-state theory, which has become so important in connection with organic reaction mechanisms. This topic has been mentioned briefly in connection with the work of Evans and Polanyi (Section 5.3). The history of transition-state theory has been traced by Keith Laidler in an article entitled 'A lifetime of transition-state theory'.[367] The title is appropriate because the author has been closely involved with the topic for the whole of his professional life, beginning in the mid-1930s. He knew personally many of the pioneers; he was a graduate student under Henry Eyring at Princeton and was a co-author with Glasstone and Eyring in 1941 of the first monograph dealing with the topic.[368] Laidler has also discussed the origins of transition-state theory in two earlier publications.[369,370] In his 1996 Dexter Award address, Laidler contrasted the chemical styles of Sidgwick and Eyring.[371]

5.9   Biological Activity of Organic Compounds

Systematic studies of the effects of structure on the biological activities of organic compounds and the analysis of the results are comprised in the term 'Quantitative Structure–Activity Relationships' (QSAR). Many of the treatments employed in the correlation analysis of data in this field closely resemble those used for linear free-energy relationships, *e.g.* the Hammett equation and extensions thereof, and so the study of the biological properties of organic compounds is often regarded as a part of physical organic chemistry. In recent years, some historical study of work in

QSAR has begun. A few papers are known to the present author, but it has not been practicable to carry out a systematic search of the literature.

R. L. Lipnick has contributed several studies, particularly of the work of the pioneer pharmacologist C. E. Overton (1865–1933)[372] on tadpole narcosis at the turn of the century. Lipnick has edited an English translation of Overton's *Studien über die Narkose*, with introductions and appendices.[373] Lipnick has also written articles on Overton,[374] including a re-working of Overton's results by modern methods.[375] He has also been interested in the Soviet pharmacologist, N. V. Lazarev,[376] and the Austrian pioneer H. H. Meyer.[377] Lipnick's scientific papers often contain a substantial historical introduction, with many references.[378,379] Parascandola has written on the controversy over structure–activity relationships in the early 20th century[380] and on early efforts to relate structure and activity.[381] Bynum has contributed 'Chemical structure and pharmacological action: a chapter in the history of 19th century molecular pharmacology'.[382] Balobanov has also written on the same general topic.[383]

5.10   Classical Books and Review Articles in Physical Organic Chemistry

Reference has already been made to a number of books that are now of considerable interest for the history of physical organic chemistry and which are in danger of disappearing from sight in libraries: Hammett,[1] Waters,[52] Watson,[67] Sidgwick,[75,77] Sutton,[96] Dewar,[97] Cohen,[102] Baker,[106,107] Stewart,[125,126] Hückel,[152] van't Hoff,[173,174] Holleman,[176] Hermans,[186] Wheland,[237] Lewis,[270] Branch and Calvin,[273] Alexander,[295] Coulson,[356] Pauling and Wilson,[360] Pauling,[361] and Glasstone, Laidler, and Eyring.[368] To this list we may add the following: Robinson (lectures on electronic theory),[384] Remick (on electronic interpretations),[385] Baker (on electronic theories),[386] Ingold (on structure and mechanism),[387] Newman (on steric effects),[388] Moelwyn-Hughes (on solution kinetics),[389] and Bell (on acid-base catalysis).[390]

The Chemical Society's *Annual Reports on the Progress of Chemistry* contain numerous articles of historical interest. The volumes for 1908, 1909, and 1910 each contain a section of about ten pages devoted to reaction mechanisms.[391–393] The authors include Lapworth. The volumes for 1938 to 1941 contain several interesting articles: Watson (on reaction mechanisms and on the influences of groups upon reactivity,[394] on reaction mechanisms,[395,396] and a long article on physico-organic topics,[397] with several sub-sections dealing with mechanisms of condensation and alkylation reactions, rearrangements, the ortho-effect and steric inhibition of mesomerism; also a sub-section by J. F. J. Dippy on the strengths of organic acids and bases); Smith (on the peroxide effect);[398] Bell (on the role of the solvent in kinetics);[399] and Hey (on free radicals).[400]

The early volumes of the Chemical Society's *Quarterly Reviews* also contain articles of value for the history of physical organic chemistry: Maccoll (on colour and constitution);[401] Bell (on the use of the terms acid and base);[402] Coulson (on molecular orbitals);[403] and Hughes (on steric hindrance).[404] The early volumes of *Chemical Reviews* similarly contain articles of value for the history of physical organic chemistry: Gomberg (on free radicals);[405] Holleman (on factors influencing substitution in benzene);[406] Brønsted (on acid-base catalysis);[407] Ingold (on electronic theories);[408]

Hammett (on quantitative study of very weak bases,[409] reaction rates and indicator acidities,[410] and rate–equilibrium relationships[411]); Hey and Waters (on free radicals);[412] Mayo and Walling (on free radicals);[413] and Dippy (on the strengths of organic acids).[414]

The *Transactions of the Faraday Society* have already been mentioned. Various pre-World War II discussions of the Faraday Society contained a substantial element of physical organic chemistry. In 1928 there was a discussion on 'Homogeneous Catalysis', which included contributions from Böeseken, Brønsted, and Dawson.[415] A discussion on 'Reaction Kinetics' was held in 1937, in which many people previously mentioned took part: Ingold and C. L. Wilson, Bell, M. G. Evans, Hinshelwood, Hughes, Moelwyn-Hughes, Polanyi, Watson, Waters, Hammett, and Eyring.[416] A major and somewhat contentious topic was the very new transition state theory; see also the Chemical Society's *Annual Reports* for 1937.[417] In 1962 a symposium was held to examine progress in transition state theory in the intervening 25 years.[418]

In 1958, two conferences arranged by the Chemical Society dealt with physical organic topics and their proceedings were subsequently published. A symposium at Hull considered 'Steric effects in conjugated systems'.[419] The speakers included several previously mentioned: Wepster, Brown and Dewar, and Coulson. A conference on 'Theoretical organic chemistry' was organized in London as the Kekulé Symposium, to commemorate the centenary of Kekulé's famous paper.[420] Speakers included: Ingold, Hughes, Huisgen, Dewar, J. D. Roberts, Hey, Barton and Pauling. Each symposium had several other contributions that are of interest as indicating the state of particular topics in the late 1950s.

There were far fewer conferences in those days, so it is of interest to note the proceedings of another meeting organized by the Chemical Society at Cork, Ireland, in 1964: 'Organic reaction mechanisms'.[421] Speakers there included Huisgen, Olah, Winstein, and Waters. Also from 1964, it would be appropriate to mention a symposium on 'Linear free energy correlations' held in Durham, North Carolina. A volume of preprints[422] includes papers by Dewar, Hammett, Jaffé, Schleyer and Taft. Others included J. E. Leffler (of Florida State University), who had recently been co-author (with E. Grunwald) of a monograph on 'Rates and Equilibria of Organic Reactions'.[423] It may also be useful to record the publication slightly later of what was claimed to be the first international research monograph in the field of linear free-energy relationships.[424]

Finally, we should mention a review article,[425] some of whose content is already of historical interest and which will be increasingly of such interest as the years go by, since it will effectively summarize the state of the art of an important topic at the time it was written. This is an account by Tsuno and Fujio of the applications of the Yukawa-Tsuno equation (an extended Hammett equation originating in 1959) to carbocationic systems. The article has about 120 pages and more than 200 references.

5.11 Miscellaneous

Bykov's historical sketch of the electron theories of organic chemistry (1965) gives comprehensive coverage, with an emphasis on the application of quantum mechanics.[426] R. E. Kohler has written on the origin of G. N. Lewis' theory of the shared pair bond,[427] on Lewis' views on bond theory,[428] and on the Lewis–Langmuir theory of valence and the chemical community, 1920–1928.[429] There is also a book by Stranges.[430] Criticism of the theory of resonance during the period 1944–1956 has been examined by Kaufman.[431]

Ogata has asked (in Japanese) the question: What kind of learning is physical organic chemistry? Its history, main themes, and research trends are said to be described.[432] Bhatt has written on physical organic chemistry–retrospect and prospect.[433] As *Tetrahedron Perspective* Number 1, a fascinating article by J A. Berson[434] describes some of the historical background of the diene synthesis and the orbital symmetry conservation rules.

The papers of Reichardt on solvents, solvent effects, and solvatochromic dyes often have useful historical introductions.[435,436] Hargittai has interviewed the various discoverers of buckminster fullerene.[437] Marsden and Rae have traced the history of nuclear magnetic resonance in Australia, 1952–1986.[438]

A symposium in print, *Physical Organic Chemistry for the 21st Century*, has appeared under the auspices of IUPAC Commission III.2.[439] Some twenty distinguished authors or groups of authors have tried to foresee the way in which the various parts of the subject may develop in the next decade or so. Tidwell has provided a prologue on the first century of physical organic chemistry and many of the articles reflect on the historical development of the subject.

Acknowledgements

The contributor is very grateful for the assistance of Professor Dr. Christian Reichardt (Marburg) in collecting material for Section 5.4: Germany, for that of Dr. Marie-Françoise Ruasse (Paris) in connection with Section 5.4: France, and for that of Dr. Tom Tidwell (Toronto) in connection with free radical chemistry in Section 5.6. Other colleagues with whom I have been in contact during the writing of this chapter are mentioned in the text.

References

1. L. P. Hammett, *Physical Organic Chemistry*, McGraw-Hill, New York, 1940.
2. W. A. Waters, *Physical Aspects of Organic Chemistry*, Routledge, London, 1935.
3. H. B. Watson, *Modern Theories of Organic Chemistry*, Oxford University Press, London, 1937.
4. J. Fastrez and M.-F. Ruasse, *J. Phys. Org. Chem.*, 1998, **11**, 505.
5. J. H. Brooke, in *Recent Developments in the History of Chemistry*, ed. C. A. Russell, Royal Society of Chemistry, London, 1985, pp. 97–152.
6. A. R. Todd and J. W. Cornforth, *Biog. Mem. Fellows Roy. Soc.*, 1976, **22**, 415.
7. C. W. Shoppee, *Biog. Mem. Fellows Roy. Soc.*, 1972, **18**, 349.
8. *Nat. Prod. Rep.*, 1987, **4**, No. 1: (a) C. A. Russell, 47; (b) M. D. Saltzman, 53; (c) J. Shorter, 61.
9. R. Robinson, *Memoirs of a Minor Prophet: 70 Years of Organic Chemistry*, Elsevier, Amsterdam, 1976, vol. 1.

10. J. Shorter, *Chem. Listy*, 1992, **86**, 318 [Czech translation of ref. 8c].
11. M. D. Saltzman, *Chem. Br.*, 1986, **22**, 543.
12. T. I. Williams, *Robert Robinson: Chemist Extraordinary*, Clarendon Press, Oxford, 1990.
13. M.D. Saltzman, *J. Chem. Educ.*, 1980, **57**, 484.
14. W. H. Brock, *The Fontana History of Chemistry*, HarperCollins, London, 1992.
15. M. J. Nye, *From Chemical Philosophy to Theoretical Chemistry: Dynamics of Matter and Dynamics of Disciplines, 1800-1950*, University of California Press, Berkeley, CA, 1993.
16. G. N. Burkhardt, 'Arthur Lapworth and others', unpublished memoirs. Copies are deposited in the University Libraries at Manchester and Hull, and with the Royal Society in London.
17. K. Schofield, 'The Growth of Physical Organic Chemistry', unpublished volume, 1996. Copies are deposited with the Royal Society in London and with the History of Chemistry Research Group at the Open University in Milton Keynes.
18. J. M. Seddon [Mrs Curtis], 'The Development of Electronic Theory in Organic Chemistry', a thesis submitted for the Final Honour School of Natural Science, Chemistry (Part II), University of Oxford, 1972. The present contributor is greatly indebted to Mrs Curtis for a copy of her thesis. A copy has now been deposited with the Royal Society in London, but there is no copy in any library in Oxford.
19. K. Tokumoto, *Kagakusi Kenkyu*, 1987, **26**, 65.
20. K. Schofield, *Ambix*, 1994, **41**, 87.
21. K. Schofield, 'The Growth of Physical Organic Chemistry', unpublished volume, 1996; ref. 17.
22. A. Maccoll, *Chem. Br.*, 1993, **29**, 880.
23. P. N. Reed, ed., *RSC Historical Group Newsletter*, July 1993 and February 1994. Copies are deposited in the RSC Library.
24. G. K. Roberts, *Br. J. Hist. Sci.*, 1996, **29**, 65.
25. S. Borman, *Chem. Eng. News*, 27 September 1993, 29.
26. *Bull. Hist. Chem.*, 1996, **19**: (a) G. K. Roberts, 2; reprint of ref. 24; (b) D. A. Davenport, 13; (c) T. Benfey, 19; (d) M. D. Saltzman, 25; (e) J. F. Bunnett, 33; (f) D. H. R. Barton, 43; (g) J. D. Roberts, 48; (h) M. J. Nye, 58; (i) F. Basolo, 66; (j) C. A. Bunton, 72; (k) H. J. Shine, 77.
27. M. D. Saltzman and C. L. Wilson, *J. Chem. Educ.*, 1980, **57**, 289.
28. C. A. Bunton, *Pure Appl. Chem.*, 1995, **67**, 667.
29. D. A. Davenport, *CHEMTECH*, 1987, 526.
30. K. T. Leffek, *Sir Christopher Ingold - a Major Prophet of Organic Chemistry*, Nova Lion Press, Victoria B.C., Canada, 1996.
31. I. Eilks and J. Friedrich, *Praxis Chem.*, 1999, **48** (4), 29.
32. M. F. Rayner-Canham and G. W. Rayner-Canham, *Chem. Br.*, 1999, **35** (10), 45.
33. C. K. Ingold, *Biog. Mem. Fellows Roy. Soc.*, 1964, **10**, 147.
34. C. K. Ingold, "Obituary notice: B. J. Flürscheim", *J. Chem. Soc.*, 1956, 1087.
35. J. Shorter, *Nat. Prod. Rep.*, 1987, **4**, 61.
36. J. Shorter, *Ambix*, 2003, **50**, 274.
37. N. K. Adam, 'Obituary notice: D. R. Boyd', *J. Chem. Soc.*, 1956, 2568.
38. M. D. Saltzman, *J. Chem. Educ.*, 1986, **63**, 588.

39. E. H. Rodd, 'H. E. Armstrong', in *British Chemists*, ed. A. Findlay and W. H. Mills, The Chemical Society, London, 1947, p. 58.

40. R. Robinson, 'Arthur Lapworth', in *British Chemists*, ed. A. Findlay and W. H. Mills, The Chemical Society, London, 1947, p. 353.

41. G. N. Burkhardt, 'Obituary notice: Arthur Lapworth', *Nature*, 1941, **147**, 769.

42. C. B. Allsop and W. A. Waters, 'T. M. Lowry', in *British Chemists*, ed. A. Findlay and W. H. Mills, The Chemical Society, London, 1947, p. 402.

43. M. D. Saltzman. *J. Chem. Educ.*, 1972, **49**, 750.

44. J. Shorter, *Nat. Prod. Rep.*, 1987, **4**, 61.

45. G. N. Burkhardt, 'Arthur Lapworth and others', unpublished memoirs; ref. 16.

46. K. Schofield, 'The Growth of Physical Organic Chemistry', unpublished volume, 1996; ref. 17.

47. K. Schofield, *Ambix*, 1995, **42**, 160.

48. R. Robinson, *Memoirs of a Minor Prophet: 70 Years of Organic Chemistry*, Elsevier, Amsterdam, 1976.

49. *Trans. Faraday Soc.*, 1923, **19**, 450.

50. M. D. Saltzman, *Nat. Prod. Rep.*, 1987, **4**, 53.

51. M. D. Saltzman, *Bull. Hist. Chem.*, 1997, **21**, 10.

52. The early editions of W. A. Waters, *Physical Aspects of Organic Chemistry*, Routledge, London, (1st, 1935; 2nd, 1937; 3rd, 1942) contained an 'Introduction' written by Professor Lowry. In the completely rewritten 4th edition of 1950, Lowry's material, as then appropriate, was incorporated in an 'Introduction' written by Waters.

53. M. D. Saltzman, *J. Chem. Educ.*, 1986, **63**, 588.

54. H. K. [Harold King], 'Obituary notice: K. J. P. Orton', *J. Chem. Soc.*, 1930, 1042.

55. F. D.C. [F. D. Chattaway], 'Obituary notice: K. J. P. Orton', *Proc. Roy. Soc. (London) A*, 1930, **129**, xi.

56. G. R. Clemo, *Obit. Not. Fellows Roy. Soc.*, 1944, **4**, 713.

57. G. D. Parkes, 'Obituary notice: F. D. Chattaway', *Nature*, 1944, **153**, 335.

58. H. J. Shine, *Aromatic Rearrangements*, Elsevier, Amsterdam, 1967, pp. 221–230.

59. J. Shorter, *Chem. Br.*, 1995, **31**, 310.

60. J. F. J. Dippy, 'Obituary notice: H. B. Watson', *Chem. Br.*, 1976, **12**, 227.

61. H. B. Watson, 'Obituary notice: A. E. Bradfield', *J. Chem. Soc.*, 1953, 4189.

62. R. E. Corbett, 'Obituary notice: F. G. Soper', *Proc. Roy. Soc. New Zeal.*, 1981–1982, **110**, 33.

63. N. B. Chapman, 'Obituary notice: Brynmor Jones', *Bull. Univ. Hull*, 1989, (117), 8.

64. G. M. Bennett, 'Obituary notice: Gwyn Williams', *J. Chem. Soc.*, 1956, 801.

65. C. K. Ingold, *Biog. Mem. Fellows Roy. Soc.*, 1964, **10**, 147.

66. P. B. D. de la Mare, 'Obituary notice: E. D. Hughes', *Proc. Chem. Soc.*, 1964, 97.

67. H. B. Watson, *Modern Theories of Organic Chemistry*, Oxford University Press, London, 1937; 2nd edn., 1941.

68. M. D. Saltzman, *Bull. Hist. Chem.*, 1994, **15/16**, 37.

69. *Trans. Faraday Soc.*, 1941, **37**, 601.

70. M. C. King, *Ambix*, 1984, **31**, 16.

71. J. Shorter, *J. Chem. Educ.*, 1980, **57**, 411.

72. K. J. Laidler, *Arch. Hist. Exact. Sci.*, 1988, **38**, 197.
73. H. T. Tizard, *Obit. Not. Fellows Roy. Soc.*, 1954, **9**, 237.
74. L. E. Sutton, in *Sidgwick's Organic Chemistry of Nitrogen*, ed. I. T. Millar and H. D. Springall, Clarendon Press, Oxford, 3rd edn., 1966, p.1.
75. N. V. Sidgwick, *The Organic Chemistry of Nitrogen*, Clarendon Press, Oxford, 1910.
76. P. Laszlo, *Ambix*, 2003, **50**, 261.
77. N. V. Sidgwick, *The Electronic Theory of Valency*, Oxford University Press, London, 1927.
78. K. Gavroglu and A. Simoes, *Br. J. Hist. Sci.*, 2002, **35**, 187.
79. C. A. Russell, *Chem. Br.*, 2002, **38** (8), 38.
80. K. J. Laidler, *Arch. Hist. Exact. Sci.*, 1988, **38**, 197.
81. H. W. Thompson, *Biog. Mem. Fellows Roy. Soc.*, 1973, **19**, 375.
82. C. F. Cullis, in *Tutti i Nobel*, Fratelli Fabbri Editori, Milan, 1970, vol. 2.
83. M. Davies, 'Obituary notice: E. A. Moelwyn-Hughes', *Chem. Br.*, 1979, **15**, 397.
84. C. E. H. Bawn, *Biog. Mem. Fellows Roy. Soc.*, 1958, **4**, 193.
85. J. Shorter, *CHEMTECH*, 1985, 252.
86. J. Albery, 'Obituary notice: R. P. Bell', *Chem. Br.*, 1996, **32** (5), 78.
87. B. G. Cox and J. H. Jones, *Biog. Mem. Fellows. Roy. Soc.*, 2001, **47**, 21.
88. J. D. Roberts, *Chem. Intell.*, 2000, **6** (2), 58.
89. J. C. Smith, *The Development of Organic Chemistry at Oxford*, published privately, Oxford, 1975. Part 1 deals with the history to 1930; Part 2 covers the Robinson era, 1930–55, and includes extensive lists of publications. A copy is deposited with the Royal Society in London.
90. J. C. Smith, *The Development of Organic Chemistry at Oxford*, published privately, Oxford, 1975, Part 2, p. 3; ref. 89.
91. E. J. Bowen, *Biog. Mem. Fellows Roy. Soc.*, 1967, **13**, 107.
92. D. Ll. Hammick, 'Obituary notice: T. W. J. Taylor', *J. Chem. Soc.*, 1954, 767.
93. R. G. Denning, 'Obituary notice: L. E. Sutton', *Chem. Br.*, 1993, **29**, 625.
94. D. H. Whiffen, *Biog. Mem. Fellows Roy. Soc.*, 1994, **40**, 369.
95. K. Gavroglu and A. Simoes, *Br. J. Hist. Sci.*, 2002, **35**, 187.
96. L. E. Sutton, *Tables of Inter-Atomic Distances in Molecules and Ions*, Chemical Society, London, 1958.
97. M. J .S. Dewar, *The Electronic Theory of Organic Chemistry*, Clarendon Press, Oxford, 1949.
98. C. W. Shoppee, *Biog. Mem. Fellows Roy. Soc.*, 1972, **18**, 349.
99. K. T. Leffek, *Sir Christopher Ingold - a Major Prophet of Organic Chemistry*, Nova Lion Press, Victoria BC, Canada, 1996.
100. J. Shorter, *Nat. Prod. Rep.*, 1987, **4**, 61.
101. H. S. Raper, 'Obituary notice: J. B. Cohen', *J. Chem. Soc.*, 1935, 1331.
102. J. B. Cohen, *Organic Chemistry for Advanced Students*, Edward Arnold, London, 4th edn., 1923, in 3 Parts.
103. R. Whytlaw-Gray and G. F. Smith, *Obit. Not. Fellows Roy. Soc.*, 1940, **3**, 139.
104. F. Challenger, *J. Roy. Inst. Chem.*, 1953, **77**, 161.
105. E. Rothstein, 'Obituary notice: J. W. Baker', *Chem. Br.*, 1968, **4**, 74.
106. J. W. Baker, *Hyperconjugation*, Clarendon Press, Oxford, 1952.

107. J. W. Baker, *Tautomerism*, Routledge, London, 1934.

108. R. D. Haworth, *Biog. Mem. Fellows Roy. Soc.*, 1959, **5**, 23.

109. K. J. Laidler, *Chem. Intell.*, 1999, **5** (4), 42; 2000, **6** (2), 25.

110. W. J. Albery, *Biog. Mem. Fellows Roy. Soc.*, 1987, **33**, 261.

111. E. E. Turner, *Biog. Mem. Fellows Roy. Soc.*, 1962, **8**, 49.

112. J. Kenyon, 'Obituary notice: R. H. Pickard', *J. Chem. Soc.*, 1950, 2253.

113. T. C. James, 'Obituary notice: J. J. Sudborough', *Proc. Chem. Soc.*, 1964, 68.

114. J. Shorter, *Chem. Soc. Rev.*, 1978, **7**, 1.

115. J. Kendall, 'Obituary notice: James Walker', *J. Chem. Soc.*, 1935, 1347.

116. M. J. ten Hoor, *J. Chem. Educ.*, 1996, **73**, 42.

117. J. Shorter, *Chem. Soc. Rev.*, 1998, **27**, 355; corrected *Chem. Soc. Rev.*, 1999, **28**, 73; and *Chem. Soc. Rev.*, 1999, **28**, 261.

118. G. N. Burkhardt, 'Arthur Lapworth and others', unpublished memoirs; ref. 16.

119. H. W. Melville, *Obit. Not. Fellows Roy. Soc.*, 1953, **8**, 395.

120. E. P. Wigner and R. A. Hodgkin, *Biog. Mem. Fellows Roy. Soc.*, 1977, **23**, 413.

121. G. Palló, *Bull. Hist. Chem.*, 1998, **21**, 39.

122. J. D. Loudon, 'Obituary notice: T. Patterson', *J. Chem. Soc.*, 1949, 1667.

123. S. Smiles, 'Obituary notice: A. W. Stewart', *J. Chem. Soc.*, 1948, 396.

124. G. B. Kauffman, *J. Chem. Educ.*, 1983, **60**, 38.

125. A. W. Stewart, *Recent Advances in Organic Chemistry*, Longmans, Green, and Co., London, 1908, 1911, 1918, 1920, 1927, 1931 (with the addition of Part II by H. Graham), 1936; A. W. Stewart and H. Graham, 1948.

126. A. W. Stewart, *Recent Advances in Physical and Inorganic Chemistry*, Longmans, Green, and Co., London, 1909, 1912, 1919, 1920, 1930; A. W. Stewart and C. L. Wilson, 1944, 1946.

127. E. E. Turner, 'Obituary notice: J. T. Hewitt', *J. Chem. Soc.*, 1955, 4493.

128. G. Gee, *Obit. Not. Fellows Roy. Soc.*, 1953, **8**, 159.

129. F. G. Mann, 'Obituary notice: H. McCombie', *Proc. Chem. Soc.*, 1963, 122.

130. W. A. Waters, 'Obituary notice: H. A. Scarborough', *Chem. Br.*, 1970, **6**, 124.

131. A. L. [A. Lapworth], 'Obituary notice: J. B. Shoesmith', *J. Chem. Soc.*, 1932, 1334.

132. T. S. Moore, 'Hantzsch Memorial Lecture', *J. Chem. Soc.*, 1936, 1051.

133. F. Hein, '[Obituary notice: A. R. Hantzsch]', *Chem. Ber.*, 1941, **74**, IA.

134. For the cryoscopic work of both Hantzsch and Oddo, see G. Scorrano and W. Walter, *J. Chem. Educ.*, 1979, **56**, 728.

135. W. Hückel, '[Obituary notice: Paul Walden]', *Chem. Ber.*, 1958, **91**, XIX.

136. D. S. Tarbell. *J. Chem. Educ.*, 1974, **51**, 7.

137. W. Kuhn, ['Obituary notice: G. Bredig'], *Chem. Ber.*, 1962, **95**, XLVII.

138. K. Fajans, ['Obituary notice: G. Bredig'], *Ber. Bunsenges. Phys. Chem.*, 1968, **72**, 1079.

139. F. Harms, R. Criegee and L. Ebert, *Ber. Deutsch. Chem. Ges.*, 1941, **74** (A), 1.

140. O. Dimroth, *Angew. Chem.*, 1933, **46**, 571.

141. R. Criegee, 'Obituary notice: H. Meerwein', *Angew. Chem. Int. Ed. Engl.*, 1966, **5**, 333.

142. K. Dimroth, 'Obituary notice: H. Meerwein', *Angew. Chem. Int. Ed. Engl.*, 1966, **5**, 338.

143. S. Hunig, *Chem. Ztg.*, 1966, **90**, 301.
144. C. Meinel, *Die Chemie an der Universität Marburg seit Beginn des 19. Jahrhunderts: Ein Betrag zu ihrer Entwicklung als Hochschulfach*, Elwert, Marburg, 1978, pp. 380–401.
145. E. Campaigne, *J. Chem. Educ.*, 1959, **36**, 336.
146. W. Walter and B. Eistert, ['Obituary notice: F. Arndt'], *Chem. Ber.*, 1975, **108**, I.
147. 'Note on a lecture given by Lâle Burk', *Chem. Heritage*, 1995, **13** (1), 34.
148. L. A. Burk, *Bull. Hist. Chem.*, 2003, **28** (1), 42.
149. R. Huisgen and H.-G. Gilde, *J. Chem. Educ.*, 1979, **56**, 369; this is an English translation of a German article: R. Huisgen, *Chem. uns. Zeit*, 1978, **12**, 49.
150. R. Huisgen, *The Adventure Playground of Mechanisms and Novel Reactions*, American Chemical Society, Washington DC, 1994 (in the series 'Profiles, Pathways and Dreams: Autobiographies of Eminent Chemists').
151. R. Huisgen, in *Chemie erlebt - 50 Jahre GDCh*, ed. R. Hoer, GDCh, Frankfurt am Main, 1999, p. 199.
152. W. Hückel, *Theoretische Grundlagen der organischen Chemie*, 2 vols, Akad. Verlag, Leipzig, 1931.
153. R. Neidlein, ['Obituary notice: W. Hückel'], *Chem. Ber.*, 1980, **113**, I.
154. M. Hanack, ['Obituary notice: W. Hückel'], *Chem. Ber.*, 1980, **113**, V.
155. R. E. Oesper, *J. Chem. Educ.*, 1950, **25**, 625.
156. W. Hückel, *Sitzungsber. Heidelb. Akad. Wiss. Math.-Natur Kl.*, 1967–68, 291.
157. E. Hückel, *Ein Gelehrtenleben: Ernst und Satire*, Verlag Chemie, Weinheim, 1975.
158. *Chem. uns. Zeit*, 1970, **4**, 180.
159. R. E. Oesper, *J. Chem. Educ.*, 1950, **25**, 674.
160. A. Hückel, *Nachricht Chem. Tech.*, 1965, **13**, 382.
161. G. Frenking, *Chem. uns. Zeit*, 1997, **30**, 27.
162. W. A. T. [Sir William Tilden], 'Obituary notice: N. A. Menschutkin', *J. Chem. Soc.*, 1908, **93**, 1660.
163. B. Menschutkin, ['Obituary notice: N. A. Menschutkin'], *Ber. Deutsch. Chem. Ges.*, 1907, **40**, 5087.
164. B. N. Menshutkin, *Life and Activity of N. A. Menshutkin*, St. Petersburg, 1908 [in Russian].
165. P. I. Staroselsky and Y. I. Soloviev, *Nicolai Aleksandrovich Menshutkin. Life and Work*, Moscow, 1968 [in Russian].
166. J.-L. M. Abboud, R. Notario, J. Bertrán and M. Solà, *Prog. Phys. Org. Chem.*, 1993, **19**, 1.
167. S. Dähne, *Z. Chem.*, 1970, **10**, 133, 168.
168. S. Dähne, in *Die Allianz von Wissenschaft und Industrie: A.W. Hofmann (1818-1892)*, ed. C. Meinel and H. Scholz, VCH, Weinheim, 1992, p. 257.
169. S. Dähne, *Mitteil. Chem. Ges. Deutsch. Demokrat. Rep.*, 1973, **20**, 257.
170. S. Dähne, *Wiss. Z. Tech. Univ. Dresden*, 1973, **22**, 765.
171. G. Jones, *J. Chem. Educ.*, 1961, **38**, 297.
172. J. Walker, 'van't Hoff Memorial Lecture', *J. Chem. Soc.*, 1913, **103**, 1127.
173. J. H. van't Hoff, *Études de Dynamique Chimique*, F. Muller, Amsterdam, 1884.
174. J. H. van't Hoff, *Studies in Chemical Dynamics*, (trans. T. Ewan), F. Muller, Amsterdam, with Williams and Norgate, London, 1896.

175. J. P. Wibaut, ['Obituary notice: A. F. Holleman'], *Recl. Trav. Chim. Pays-Bas*, 1955, **74**, 1371.

176. A. F. Holleman, *Die direkte Einführung von Substituenten in den Benzolkern*, Veit, Leipzig, 1910.

177. A. F. Holleman, *Chem. Rev.*, 1924, **1**, 187.

178. A. F. Holleman, *Recl. Trav. Chim. Pays-Bas*, 1938, **57**, 489. This is a tribute on the occasion of Böeseken's retirement and is accompanied by a brief tribute from Jocelyn Thorpe.

179. W. Drenth, personal communication, 1995.

180. E. Havinga, *Enjoying Organic Chemistry, 1927–1987*, American Chemical Society, Washington DC, 1990 (in the series 'Profiles, Pathways and Dreams: Autobiographies of Eminent Chemists').

181. E. Havinga, 'Obituary notice: L. J. Oosterhoff', *Chem. Br.*, 1975, **11**, 328.

182. H. van Bekkum and B. van de Graaf, *Chem. Weekbl.*, 1992, **88** (36), 286.

183. J. Shorter, 'Obituary notice: B. M. Wepster', *Newsletter* of the International Group for Correlation Analysis in Chemistry, No. 15, October 1992; deposited in the RSC Library.

184. A brief obituary of Sixma is in the posthumously published F. L. J. Sixma and H. Wynberg, *A Manual of Physical Methods in Organic Chemistry*, Wiley, New York, 1964.

185. D. Heikens and G. Challa, ['Obituary notice: P. H. Hermans'], *Chem. Weekbl.*, 6 April 1979, 2.

186. P. H. Hermans, *Inleiding tot de Theoretische Organische Chemie*, Elsevier, Amsterdam, 1952; *Introduction to Theoretical Organic Chemistry*, Elsevier, Amsterdam, 1954.

187. M. J. ten Hoor has drawn my attention to an historically interesting short section in Hermans's book (p. 355 in the Dutch edition, p. 390 in the English version), beginning with the sentence: 'A firm basis for all subsequent theoretical discussions on benzene was laid in 1913 by Scheffer (outside Holland this most important work remained practically unnoticed)'. Reference is made to a summary by F. E. C. Scheffer and W. F. Brandsma, *Recl. Trav. Chim. Pays-Bas*, 1926, **45**, 522, 531.

188. W. Drenth, *120 Jaar fysisch-organische chemie in Nederland, 1874–1994*, published privately, 2001.

189. H. J. C. Tendeloo, *Chem. Weekbl.*, 1949, **45**, 385 (biography of Olivier).

190. G. B. R. de Graaf, *Chem. Weekbl.*, 1949, **45**, 387 (scientific work of Olivier).

191. L. P. Hammett, *Physical Organic Chemistry*, McGraw-Hill, New York, 1940.

192. W. A. Waters, *Physical Aspects of Organic Chemistry*, Routledge, London, (1st, 1935; 2nd, 1937; 3rd, 1942; 4th 1950).

193. H. B. Watson, *Modern Theories of Organic Chemistry*, Oxford University Press, London, 1937; 2nd edn., 1941.

194. A. S. Veibel, 'Johannes Nicolaus Brønsted'in, *Dictionary of Scientific Biography*, ed. C. C. Gillispie, Scribners, New York, 1970, vol. 2, p. 498.

195. J. A. Christiansen, *Overs. Selsk. Virksomhed*, 1948–49, 57.

196. P. Piganiol, ['Obituary notice: Charles Prévost'], *Ann. Anc. Elèves Ecole Norm. Sup.*, 1985, 46.

197. C. Prévost, ['Obituary notice: Albert Kirrmann], *Bull. Soc. Chim. Fr.*, 1975, 82.

198. J. Lévy, ['Obituary notice: Marc Tiffeneau], *Bull. Soc. Chim. Biol.*, 1946, **3**, 196.

199. M. Charpentier-Morize, ['Obituary notice: Bianca Tchoubar'], *Actual. Chim.*, 1991 (Nov.-Dec), 444.

200. B. Tchoubar, *Les Mécanismes Réactionnels en Chimie Organique*, Dunod, Paris, 1960.

201. B. Pullman and A. Pullman, *Les Théories Electroniques de la Chimie Organique*, Masson, Paris, 1952.

202. M. Julia, *Méchanismes Electroniques en Chimie Organique*, Gauthier-Villars, Paris, 1959.

203. D. S. Tarbell and A. T. Tarbell, *Essays on the History of Organic Chemistry in the United States, 1875-1955*, Folio Publishers, Nashville, TN, 1986.

204. M. D. Saltzman, *J. Chem. Educ.*, 1986, **63**, 588.

205. M. D. Saltzman, *J. Chem. Educ.*, 1973, **50**, 59.

206. M. D. Saltzman, *J. Chem. Educ.*, 1984, **61**, 119.

207. M. D. Saltzman, *J. Chem. Educ.*, 1974, **51**, 498.

208. D. S. Tarbell and A. T. Tarbell, *J. Chem. Educ.*, 1981, **58**, 559.

209. E. E. Reid, *My First One Hundred Years*, Chemical Publishing Co., New York, 1972.

210. J. D. Roberts, *Biog. Mem. Natl. Acad. Sci.*, 1974, **45**, 413.

211. W. A. Noyes, *Biog. Mem. Natl. Acad. Sci.*, 1942, **21**, 275.

212. Ref. 203, p.67.

213. Ref. 203, p.69.

214. L. F. Fieser, *Biog. Mem. Natl. Acad. Sci.*, 1975, **46**, 331.

215. A. B. Costa, *J. Chem. Educ.*, 1971, **48**, 243.

216. D. S. Tarbell, *Chem. Eng. News*, 1976, **54** (15), 110.

217. A. Findlay, 'Obituary notice: E. C. Franklin', *J. Chem. Soc.*, 1938, 583.

218. H. M. Elsey, *Biog. Mem. Natl. Acad. Sci.*, 1991, **60**, 67.

219. R. A. Fuoss, *Biog. Mem. Natl. Acad. Sci.*, 1971, **42**, 119.

220. M. L. Wolfrom, *Biog. Mem. Natl. Acad. Sci.*, 1960, **34**, 204.

221. M. D. Saltzman, *J. Chem. Educ.*, 1986, **63**, 588.

222. J. B. Conant, *Biog. Mem. Natl. Acad. Sci.*, 1952, **27**, 265.

223. J. B. Conant, *My Several Lives. Memoirs of a Social Inventor*, Harper, New York, 1970.

224. M. D. Saltzman, *J. Chem. Educ.*, 1972, **49**, 411.

225. M. D. Saltzman, *Bull. Hist. Chem.*, 2003, **28** (2), 84.

226. P. D. Bartlett, *Biog. Mem. Natl. Acad. Sci.*, 1983, **54**, 91.

227. G. B. Kistiakowsky and F. H. Westheimer, *Biog. Mem. Fellows Roy. Soc.*, 1979, **25**, 209.

228. J. M. McBride, ed., *P. D. and the Bartlett Group at Harvard 1934–1974: A Group Autobiography on the occasion of the Bartlett Symposium on Physical Organic Chemistry*, Fort Worth, TX, 1975.

229. L. Goertler, 'Interview with P. D. Bartlett,' Center for History of Physics, American Institute of Physics, One Physics Ellipse, College Park, MD 20740–3843.

230. L. B. Gortler and M. D. Saltzman, *Bull. Hist. Chem.*, 1998, **21**, 25.

231. J. D. Roberts, *Chem. Intell.*, 1998, **4** (2), 34.
232. J. D. Roberts, 'Obituary notice: P. D. Bartlett', *Chem. Br.*, 1998, **34** (8), 70.
233. I. Hargittai, *Chem. Intell.*, 1996, **2** (2), 4.
234. F. H. Westheimer, *Adv. Phys. Org. Chem.*, 1985, **21**, 1.
235. F. H. Westheimer, *J. Chem. Educ.*, 1986, **63**, 409.
236. F. H. Westheimer, *Tetrahedron*, 1995, **51**, 3.
237. G. W. Wheland, *The Theory of Resonance and its Application to Organic Chemistry*, Wiley, New York, Chapman and Hall, London, 1944.
238. L. P. Hammett, *J. Chem. Educ.*, 1966, **43**, 464.
239. J. Shorter, *Prog. Phys. Org. Chem.*, 1990, **17**, 1.
240. J. Shorter, *Pure Appl. Chem.*, 1995, **67**, 835.
241. J. Shorter, *Chem. Intell.*, 1996, **2** (1), 39.
242. J. Shorter, *Chem. uns. Zeit*, 1985, **19**, 197.
243. *CHOC News*, 1983, **1** (3), 19.
244. L. Gortler, *J. Chem. Educ.*, 1985, **62**, 753.
245. J. Shorter, *Chem. Listy*, 2000, **94**, 210.
246. M. D. Saltzman, *CHEMTECH*, 1987, 484.
247. M. D. Saltzman, *J. Chem. Educ.*, 1977, **54**, 25.
248. H. S. Mosher and T. T. Tidwell, *J. Chem. Educ.*, 1990, **67**, 9.
249. J. G. Traynham, *J. Chem. Educ.*, 1992, **69**, 439.
250. W. G. Young and S. Winstein, *Biog. Mem. Natl. Acad. Sci.*, 1973, **43**, 163.
251. M. D. Saltzman, *J. Chem. Educ.*, 1986, **63**, 588.
252. W. G. Young and D. J. Cram, *Biog. Mem. Natl. Acad. Sci.*, 1973, **43**, 321.
253. A. Streitwieser, *Prog. Phys. Org. Chem*, 1972, **9**, 1.
254. P. D. Bartlett, *J. Am. Chem. Soc.*, 1972, **94**, 2161.
255. D. J. Cram, *From Design to Discovery*, American Chemical Society, Washington DC, 1990 (in the series 'Profiles, Pathways and Dreams: Autobiographies of Eminent Chemists').
256. P. Wright, 'Obituary notice: D. J. Cram', *The Guardian*, 26 June 2001, 16.
257. J. D. Roberts, *The Right Place at the Right Time*, American Chemical Society, Washington DC, 1990 (in the series 'Profiles, Pathways and Dreams: Autobiographies of Eminent Chemists').
258. J. D. Roberts, *Bull. Hist. Chem.*, 1996, **19**, 48.
259. J. D. Roberts, *Chem. Intell.*, 1996, **2** (3), 29.
260. I. Hargittai, *Chem. Intell.*, 1998, **4** (2), 29.
261. P. Wright, 'Obituary Notice: H. C. Brown', *The Guardian*, 23 December 2004, 19.
262. H. C. Brown, *Boranes in Organic Chemistry*, Cornell University Press, Ithaca, NY, 1972.
263. H. C. Brown, *The Nonclassical Ion Problem*, with comments by P. von. R. Schleyer, Plenum, New York, 1977.
264. P. D. Bartlett, *Nonclassical Ions*, Benjamin, New York, 1965.
265. G. D. Sargent, *Quart. Rev.*, 1967, **20**, 301.
266. E. M. Arnett, T. C. Hofelick and G. W. Schriver, *Reactive Intermed.*, 1985, **3**, 189; especially 200–202.
267. I. Hargittai, *Chem. Intell.*, 1997, **3** (2), 4.

268. B. D. Coppola, *Chem. Intell.*, 1998, **4** (2), 40.
269. S. J. Weininger, *Bull. Hist. Chem.*, 2000, **25**, 123.
270. G. N. Lewis, *Valence and the Structure of Atoms and Molecules*, Chemical Catalog Company, New York, 1923; Dover reprint, 1966, with an 'Introduction' by K. S. Pitzer.
271. M. Calvin, *J. Chem. Educ.*, 1984, **61**, 14.
272. G. E. Gibson, M. Calvin, J. H. Hildebrand and J. E. Tippett, *In Memoriam*, University of California, 1954, 27. This is part of a collection of material about Gerald Branch, some of it in typescript, which the present contributor was sent many years ago by B. M. Wepster of Delft. The originals of the material are presumably in University of California archives.
273. G. E. K. Branch and M. Calvin, *Theory of Organic Chemistry*, Prentice-Hall, New York, 1941.
274. M. Calvin, *J. Chem. Educ.*, 1984, **61**, 14.
275. M. Calvin, *Following the Trail of Light*, American Chemical Society, Washington DC, 1992 (in the series 'Profiles, Pathways and Dreams: Autobiographies of Eminent Chemists').
276. M. Calvin and G. T. Seaborg, *J. Chem. Educ.*, 1984, **61**, 11.
277. G. T. Seaborg, *Chem. Intell.*, 1995, **1** (3), 27.
278. G. B. Kauffman and I. Mayo, 'Obituary notice: M. Calvin', *Chem. Intell.*, 1998, **4** (1), 54.
279. A. Streitwieser, *A Lifetime of Synergy with Theory and Experiment*, American Chemical Society, Washington DC, 1997 (in the series 'Profiles, Pathways and Dreams: Autobiographies of Eminent Chemists').
280. G. B. Kauffman, *Bull. Hist. Chem.*, 2000, **25**, 122.
281. A. R. Katritzky, *Heterocycles*, 1994, **37**, 3.
282. A. R. Katritzky, *J. Heterocycl. Chem.*, 1994, **31**, 569.
283. G. A. Olah, *Angew. Chem. Int. Ed. Engl.*, 1995, **34**, 1393.
284. G. B. Kauffman and L. M. Kauffman, *Chem. Intell.*, 1995, **1** (2), 6.
285. M. J .S. Dewar, *The Electronic Theory of Organic Chemistry*, Clarendon Press, Oxford, 1949.
286. M. J. S. Dewar, *A Semiempirical Life*, American Chemical Society, Washington DC, 1992 (in the series 'Profiles, Pathways and Dreams: Autobiographies of Eminent Chemists').
287. M. J .S. Dewar, *Chem. Intell.*, 1997, **3** (1), 34.
288. K. J. Laidler, K. T. Leffek and J. Shorter, *Chem. Intell.*, 1997, **3** (4), 55.
289. D. A. Davenport, *Chem. Intell.*, 1997, **3** (4), 53.
290. N. Bodor, 'Obituary notice: M. J. S. Dewar', *Chem. Intell.*, 1998, **4** (3), 59.
291. N. Bodor and A. Katritzky, 'Obituary notice: M. J. S. Dewar', *Chem. Br.*, 1998, **34** (6), 76.
292. J. N. Murrell, *Biog. Mem. Fellows Roy. Soc.*, 1998, **44**, 127.
293. J. T. Edward, *Chem. Intell.*, 1997, **3** (1), 25.
294. P. R. Jones, *J. Chem. Educ.*, 1987, **64**, 882.
295. E. R. Alexander, *Principles of Ionic Organic Reactions*, Wiley, New York, 1950.
296. *Chem. Heritage*, 1995, **13** (1).

297. M. Charton, 'Obituary notice: H. H. Jaffé', *Newsletter* of International Group for Correlation Analysis in Chemistry, No. 15, October 1992 deposited in RSC Library.

298. H. H. Jaffé, *Chem. Rev.*, 1953, **53**, 191.

299. M. Berthelot and J.-F. Gal, ['An appreciation of Robert Taft'], *Actual. Chim. (Histoire)*, 1997, (3), 28.

300. R. D. Schmidt, 'A Student remembers Robert. W. Taft', *Chem. Intell.*, 1996, **2** (4), 52; and *Newsletter* of International Group for Correlation Analysis in Chemistry, No. 20, December 1996; deposited in RSC Library.

301. P. von R. Schleyer, *From the Ivy League into the Honey Pot*, American Chemical Society, Washington, DC, 1998 (in the series 'Profiles, Pathways and Dreams: Autobiographies of Eminent Chemists').

302. I. Hargittai, *Chem. Intell.*, 1998, **4** (1), 18.

303. C. S. Schoepfle and W. E. Bachmann, *J. Am. Chem. Soc.*, 1947, **67**, 2921.

304. J. C. Bailar, *Biog. Mem. Natl. Acad. Sci.*, 1970, **41**, 141.

305. M. Gomberg, *Chem. Rev.*, 1924, **1**, 91.

306. H. Lankamp, W. Th. Nauta and C. Maclean, *Tetrahedron Lett.*, 1968, 249.

307. J. M. McBride, *Tetrahedron*, 1974, **30**, 2009.

308. T. T. Tidwell, *Chem. Intell.*, 2000, **6** (3), 33.

309. *Trans. Faraday Soc.*, 1934, **30**, 1.

310. L. Eberson, *Chem. Intell.*, 2000, **6** (3), 44.

311. T. T. Tidwell, *Adv. Phys. Org. Chem.*, 2001, **36**, 1.

312. L. Eberson, *Adv. Phys. Org. Chem.*, 2001, **36**, 59.

313. T. T. Tidwell, *Angew. Chem. Int. Ed. Engl.*, 2001, **40**, 331.

314. M. C. King, *CHEMTECH*, 1985, 701.

315. C. Walling, *J. Chem. Educ.*, 1986, **63**, 99.

316. F. R. Mayo and C. Walling, *Chem. Rev.*, 1940, **27**, 351.

317. F. R. Mayo, in *Vistas in Free Radical Chemistry*, ed. W. A. Waters, Pergamon, New York, 1959, p. 139.

318. F. R. Mayo, *J. Chem. Educ.*, 1986, **63**, 97.

319. F. H. Westheimer, *Biog. Mem. Natl. Acad. Sci.*, 1960, **34**, 123.

320. C. Walling, *Fifty Years of Free Radicals*, American Chemical Society, Washington DC, 1995 (in the series 'Profiles, Pathways and Dreams: Autobiographies of Eminent Chemists').

321. M. Calvin, *Following the Trail of Light*, American Chemical Society, Washington DC, 1992 (in the series 'Profiles, Pathways and Dreams: Autobiographies of Eminent Chemists').

322. J. I. G. Cadogan and D. I. Davies, *Biog. Mem. Fellows Roy. Soc.*, 1987, **34**, 293.

323. G. Williams, 'Obituary notice: D. H. Hey', *Chem. Br.*, 1987, **23**, 777.

324. R. O. C. Norman and J. A. Jones, *Biog. Mem. Fellows Roy. Soc.*, 1986, **32**, 597.

325. W. A. Waters, *Physical Aspects of Organic Chemistry*, Routledge, London, (1st edn., 1935; 2nd, 1937; 3rd, 1942; 4th 1950).

326. D. H. Hey and W. A. Waters, *Chem. Rev.*, 1937, **21**, 169.

327. C. Walling, *Chem. Br.*, 1987, **23**, 767.

328. W. A. Waters, *Notes Records Roy. Soc.*, 1984, **39**, 105.

329. W. A. Waters, *Chemistry of Free Radicals*, Clarendon Press, Oxford, 1946.

330. R. O. C. Norman, ed., *Essays on Free Radical Chemistry*, Special Publications No. 24, Chemical Society, London, 1970.

331. B. C. Gilbert, 'Obituary notice: R. O. C. Norman', *Chem. Br.*, 1993, **29**, 815.

332. D. Waddington, *Biog. Mem. Fellows Roy. Soc.*, 1997, **43**, 333.

333. K. Schofield, 'The Growth of Physical Organic Chemistry', unpublished volume, 1996; ref.17.

334. O. B. Ramsay, *Stereochemistry*, Heyden, London, 1981.

335. S. Ley and R. M. Myers, *Biog. Mem. Fellows Roy. Soc.*, 2002, **48**, 1.

336. C. Rees, Obituary notice; D. H. R. Barton', *Chem. Br.*, 1998, **34** (6), 75.

337. J. Daintith *et al.*, eds, *Biographical Encyclopedia of Scientists*, Institute of Physics Publishing, Bristol and Philadelphia, 1994, p. 394.

338. D. H. R. Barton, *Some Recollections of Gap Jumping*, American Chemical Society, Washington DC, 1991 (in the series 'Profiles, Pathways and Dreams: Autobiographies of Eminent Chemists').

339. D. H. R. Barton, *Reason and Imagination. Reflections on Research in Organic Chemistry*, World Scientific Publishing, Singapore, 1996. (This volume reprints selected papers of the author, with comments. His papers on conformational analysis are included, pp. 45–91.)

340. D. Arigoni, J. D. Dunitz and A. Eschenmoser, *Biog. Mem. Fellows Roy. Soc.*, 2000, **46**, 445.

341. V. Prelog, *My 132 Semesters of Chemistry Studies*, American Chemical Society, Washington DC, 1991 (in the series 'Profiles, Pathways and Dreams: Autobiographies of Eminent Chemists').

342. J. I. Seeman, *Chem. Rev.*, 1983, **83**, 83.

343. M. Beugelmans-Verrier, *Actual. Chim.*, 1993, Nov.-Dec., 87.

344. J. Weyer, *Angew. Chem. Int. Ed. Engl.*, 1974, **13**, 591.

345. K. B. Wiberg, *Angew. Chem. Int. Ed. Engl.*, 1986, **25**, 312.

346. E. L. Eliel, *Chemistry*, 1976, **49** (1), 6.

347. E. L. Eliel, *From Cologne to Chapel Hill*, American Chemical Society, Washington DC, 1990 (in the series 'Profiles, Pathways and Dreams: Autobiographies of Eminent Chemists').

348. E. L. Eliel and S. H. Wilen, *Stereochemistry of Organic Compounds*, Wiley, New York, 1994.

349. I. Hargittai, *Chem. Intell.*, 1998, **4** (4), 4.

350. D. Buckingham, 'Obituary notice: R. S. Mulliken', *Chem. Br.*, 1987, **23**, 777.

351. H. C. Longuet-Higgins, *Biog. Mem. Fellows Roy. Soc.*, 1990, **35**, 329.

352. R. S. Mulliken, *Pure Appl. Chem.*, 1970, **24**, 203.

353. R. S. Mulliken, *Life of a Scientist. An Autobiographical Account of the Development of Molecular Orbital Theory*, ed. B. J. Ransil, Springer-Verlag, Berlin, 1989.

354. N. F. Mott, *Biog. Mem. Fellows Roy. Soc.*, 1955, **1**, 175.

355. S. L. Altmann and E. J. Bowen, *Biog. Mem. Fellows Roy. Soc.*, 1974, **20**, 75.

356. C. A. Coulson, *Valence*, Clarendon Press, Oxford, 1952.

357. A. Simoes and K. Gavroglu, *Hist. Stud. Phys. Biol. Sci.*, 1999, **29**, 363.

358. A. Simoes, *Brit. J. Hist. Sci.*, 2004, **37**, 299.

359. J. D. Dunitz, *Biog. Mem. Fellows Roy. Soc.*, 1996, **42**, 315.

360. L. Pauling and E. B. Wilson, *Introduction to Quantum Mechanics with Applications to Chemistry*, McGraw-Hill, New York, 1935.
361. L. Pauling, *The Nature of the Chemical Bond and the Structure of Molecules and Crystals*, Cornell University Press, Ithaca, 1939.
362. R. McWeeny, *Pure Appl. Chem.*, 1989, **61**, 2087.
363. I. Shavitt, *Isr. J. Chem.*, 1994, **33**, 357.
364. I. Hargittai, *Chem. Intell.*, 1995, **1** (2), 14.
365. T. Yonezawa, 'Obituary Notice: Kenichi Fukui', *Chem. Intell.*, 1999, **5** (1), 38.
366. R. B. Woodward and R. Hoffmann, *Angew. Chem. Int. Ed. Engl.*, 1969, **8**, 781.
367. K. J. Laidler, *Chem. Intell.*, 1998, **4** (3), 39.
368. S. Glasstone, K. J. Laidler and H. Eyring, *The Theory of Rate Processes*, McGraw-Hill, New York, 1941.
369. K. J. Laidler and M. C. King, *J. Phys. Chem.*, 1983, **87**, 2657.
370. K. J. Laidler, *The World of Physical Chemistry*, Oxford University Press, Oxford, 1993, pp. 242–249.
371. K. J. Laidler, *Bull. Hist. Chem.*, 1998, **22**, 1.
372. R. L. Lipnick, 'C. E. Overton', in *Dictionary of Scientific Biography*, ed. C. C. Gillispie, Scribners, New York, 1974, vol. 10, p. 25.
373. C. E. Overton, *Studies of Narcosis*, ed. R. L. Lipnick, Chapman and Hall and Wood Library Museum of Anesthesiology, London and Chicago, 1991, being an English translation (with introductions and appendices) of Overton's *Studien über Narkose*, Fischer, Jena, 1901.
374. R. L. Lipnick, *Trends Pharmacol. Sci.*, 1986, **7**, 161.
375. R. L. Lipnick, in *Quantitative Structure-Activity Relationships in Drug Design*, ed. J. L. Fauchère, Alan R. Liss, New York, 1989, p. 421.
376. R. L. Lipnick and V. A. Filov, *Trends Pharmacol. Sci.*, 1992, **13**, 56.
377. R. L. Lipnick, *Trends Pharmacol. Sci.*, 1989, **10**, 265.
378. R. L. Lipnick, in *Fundamentals of Aquatic Toxicology*, ed. G. R. Rand, Taylor and Francis, London, 2nd edn., 1995, 609.
379. R. L. Lipnick, *Environ. Toxicol. Chem.*, 1989, **8**, 1.
380. J. Parascandola, *Pharmacy Hist.*, 1974, **16**, 54.
381. J. Parascandola, *Trends Pharmacol. Sci.*, 1980, **1**, 417.
382. W. F. Bynum, *Bull. Hist. Med.*, 1970, **44**, 518.
383. A. E. Balobanov, Deposited Doc., 1983, *VINITI* 1639–84, 167-178 [in Russian]; avail. *VINITI.*, see *Chem. Abstr.*, 1985, **102**, 148253.
384. R. Robinson, *Outline of an Electrochemical (Electronic) Theory of the Course of Organic Reactions*, Institute of Chemistry, London, 1932.
385. A. E. Remick, *Electronic Interpretations of Organic Chemistry*, Wiley, New York, 1st edn., 1943, 2nd edn., 1949.
386. J. W. Baker, *Electronic Theories of Organic Chemistry*, Clarendon Press, Oxford, 1958.
387. C. K. Ingold, *Structure and Mechanism in Organic Chemistry*, Bell, London, 1st edn., 1953, 2nd edn., 1969.
388. M. S. Newman, ed., *Steric Effects in Organic Chemistry*, Wiley, New York, 1956.
389. E. A. Moelwyn-Hughes, *Kinetics of Reactions in Solution*, Clarendon Press, Oxford, 1st edn., 1933; 2nd edn., 1947.

390. R. P. Bell, *Acid-Base Catalysis*, Clarendon Press, Oxford, 1941.

391. C. H. Desch and G. T. Morgan, *Annu. Rep. Prog. Chem.*, *1908*, 1909, **5**, 74.

392. C. H. Desch and A. Lapworth, *Annu. Rep. Prog. Chem.*, *1909*, 1910, **6**, 68.

393. C. H. Desch and A. Lapworth, *Annu. Rep. Prog. Chem.*, *1910*, 1911, **7**, 60.

394. H. B. Watson, *Annu. Rep. Prog. Chem.*, *1938*, 1939, **35**, 208, 236.

395. H. B. Watson, *Annu. Rep. Prog. Chem.*, *1939*, 1940, **36**, 191.

396. H. B. Watson, *Annu. Rep. Prog. Chem.*, *1940*, 1941, **37**, 229.

397 H. B. Watson, *Annu. Rep. Prog. Chem.*, *1941*, 1942, **38**, 115.

398. J. C. Smith, *Annu. Rep. Prog. Chem.*, *1939*, 1940, **36**, 219.

399. R. P. Bell, *Annu. Rep. Prog. Chem.*, *1939*, 1940, **36**, 82.

400. D. H. Hey, *Annu. Rep. Prog. Chem.*, *1940*, 1941, **37**, 250.

401. A. Maccoll, *Quart. Rev.*, 1947, **1**, 17.

402. R. P. Bell, *Quart. Rev.*, 1947, **1**, 113.

403. C. A. Coulson, *Quart. Rev.*, 1947, **1**, 145.

404. E. D. Hughes, *Quart. Rev.*, 1948, **2**, 107.

405. M. Gomberg, *Chem. Rev.*, 1924, **1**, 91.

406. A. F. Holleman, *Chem. Rev.*, 1924, **1**, 187.

407. J. N. Brønsted, *Chem. Rev.*, 1928, **5**, 231.

408. C. K. Ingold, *Chem. Rev.*, 1934, **15**, 225.

409. L. P. Hammett, *Chem. Rev.*, 1933, **13**, 61.

410. L. P. Hammett, *Chem. Rev.*, 1935, **16**, 67.

411. L. P. Hammett, *Chem. Rev.*, 1935, **17**, 125.

412. D. H. Hey and W. A. Waters, *Chem. Rev.*, 1937, **21**, 169.

413. F. R. Mayo and C. Walling, *Chem. Rev.*, 1940, **27**, 351.

414. J. F. J. Dippy, *Chem. Rev.*, 1939, **25**, 151.

415. *Trans. Faraday Soc.*, 1928, **24**, 545.

416. *Trans. Faraday Soc.*, 1938, **34**, 1

417. W. F. K. Wynne-Jones, *Annu. Rep. Prog. Chem.*, *1937*, 1938, **34**, 43.

418. *The Transition State*, Chemical Society, London, 1962.

419. G. W. Gray, ed., *Steric Effects in Conjugated Systems*, Butterworth, London, 1958.

420. *Theoretical Organic Chemistry*, Butterworth (for IUPAC), London, 1959.

421. *Organic Reaction Mechanisms*, Special Publication no. 19, Chemical Society, London, 1965.

422. 'Symposium on Linear Free Energy Correlations', preprints of papers, U. S. Army Research Office, Durham, N. C., 1964.

423. J. E. Leffler and E. Grunwald, *Rates and Equilibria of Organic Reactions*, Wiley, New York, 1963.

424. N. B. Chapman and J. Shorter, eds, *Advances in Linear Free Energy Relationships*, Plenum Press, London, 1972.

425. Y. Tsuno and M. Fujio, *Adv. Phys. Org. Chem.*, 1999, **32**, 267.

426. G. V. Bykov, *Chymia*, 1965, **10**, 199.

427. R. E. Kohler, *Br. J. Hist. Sci.*, 1975, **8**, 233.

428. R. E. Kohler, *Hist. Stud. Phys. Sci.*, 1971, **3**, 343.

429. R. E. Kohler, *Hist. Stud. Phys. Sci.*, 1975, **6**, 431.

430. A. N. Stranges, *Electrons and Valence - Development of the Theory, 1900–1925*, Texas A & M University Press, College Station, TX, 1982.

431. J. Kaufman, *Synthesis (Cambridge)*, 1977, **4**, 44.

432. Y. Ogata, *Kagaku (Kyoto)*, 1985, **40**, 324. (See *Chem Abstr.*, 1985, **103**, 21831.)

433. M. V. Bhatt, *Curr. Sci.*, 1982, **51**, 1125.

434. J. A. Berson, *Tetrahedron*, 1992, **48**, 3.

435. C. Reichardt, *Chem. uns. Zeit*, 1981, **15**, 139.

436. C. Reichardt, *Chem. Rev.*, 1994, **94**, 2319.

437. I. Hargittai, *Chem. Intell.*, 1995, **1** (3), 6.

438. K. Marsden and I. D. Rae, *Hist. Recs. Aust. Sci.*, 1991, **8**, 119.

439. T. T. Tidwell, Z. Rappoport and C. L. Perrin, eds., *Pure Appl. Chem.*, 1997, **69**, 211.

CHAPTER 6

Physical Chemistry

THEODORE ARABATZIS AND KOSTAS GAVROGLU

Department of History and Philosophy of Science, University of Athens

6.1 General

Several historical studies, both by professional historians and by scientists, address general aspects of the development of physical chemistry.[1] A dissertation discussing the rise of physical chemistry[2] examines the work of the three 'founders', Wilhelm Friedrich Ostwald (1853–1932), J. H. van't Hoff (1852–1911), and Svante Arrhenius (1859–1927). Labelling them 'the ionists', it treats especially their thermodynamic theory of solutions based upon ionic dissociation in electrolytes. Max Planck's (1858–1947) own theory of dissociation, first published in 1887 and largely ignored, is also discussed. The analysis revolves around four themes: the different (national) research traditions of the ionists and Planck; the particularity of each investigator in the context of these traditions; the reception of the innovations they introduced, which depended on the compatibility of their own styles with the styles of other scientists; the character of the ionists' research which, in contrast to Planck whose work was at the forefront of a specialized field, resulted from their mixing diverse, old traditions and problems with new styles and techniques.

Apart from his very influential textbooks, Partington wrote extensively on the history of chemistry.[3] The fourth volume[4] includes a detailed narrative of the various developments in physical chemistry to the 1920s. Partington's references to the original papers are still an invaluable asset of an otherwise descriptive history.

A recent book on physical chemistry,[5] written by a scientist[6] and aimed primarily at other scientists, contains substantial historical information on the beginnings of physical chemistry and on various topics, such as chemical spectroscopy, electrochemistry, chemical kinetics, colloid and surface chemistry, and quantum chemistry. The book also discusses more general topics, such as the development of the physical sciences and the role of scientific journals in scientific communication. The same author has written a brief account of the development of physical chemistry after 1937,[7] emphasizing the application of quantum theory and the invention of new experimental methods: stopped-flow techniques (1940), nuclear magnetic resonance

(1946), flash photolysis (1950–52), crossed molecular beams (1954), temperature jump (1954), laser spectroscopy (1957), and Mössbauer spectroscopy (1958). He has also written about the history of transition-state theory.[8]

There have also been brief, general surveys of the development of physical chemistry,[9–13] a discussion of its early phase in Canada,[14] and in Japan,[15] and an analysis of physical chemistry in higher education in the inter-war period in Poland,[16] as well as more specific studies on its applications to other areas.[17,18]

A book[19] attempts to present the history of an ambivalent attitude concerning the identity of chemistry during the last two centuries and to articulate the methodological complexity of such a reading. The book addresses directly the question of whether chemistry is reducible to physics, but it also examines this issue by studying the formation of the boundaries of physical chemistry, chemical physics, theoretical chemistry and/or quantum chemistry. The author argues that the legitimacy of theoretical chemistry and the drawing of disciplinary boundaries was the result of a whole network of factors involved in the construction of the *identity* of a discipline. She examines that identity by discussing systematically, in various cases, six issues: the genealogy and historical mythology of heroic origins in the initial period of the formation of each discipline; a core literature defining archetypal language and imagery; the practices and rituals that are codified and performed; a physical homeland, including institutions based on citizenship rights and responsibilities; and finally, external recognition and shared values together with unsolved problems. The author traces these elements in the Paris School and in the London–Manchester School of theoretical organic chemistry, both for the period between 1880 and 1930; in Christopher Ingold's (1893–1970) attempt to integrate physical and organic chemistry; and in the development of quantum chemistry in the USA and Britain until the end of the 1940s.

In a previous article,[20] the same author discussed two approaches to chemistry in the 1920s, a British and a French, which shared the same objective, namely the solution of chemical problems by means of concepts borrowed from physics. The former was based on ions and electric charges and employed visual representations and non-mathematical language; the latter employed the concepts of energy and radiation and aimed at constructing a deductive chemistry. Resistance, from the end of the 19th century to the 1930s, to applying the electronic theory to reactive chemistry in France is the subject of a study.[21] Polanyi's work is, also, examined in two articles.[22,23]

Another study presents the emergence of physical chemistry in the USA,[24] by concentrating on a number of institutions and persons who played a pivotal role: the Massachusetts Institute of Technology and A. A. Noyes' initiatives there; the College of Chemistry at Berkeley and G. N. Lewis' (1875–1946) unquestionable dominance in the formation of its academic culture; Cornell and Wilder Bancroft's insistence on the utility of the phase rule, and his influence through his editorship of the *Journal of Physical Chemistry*; and the California Institute of Technology, where Linus Pauling (1901–1994) had his entire professional career from its start to the end of his life. Of particular interest are the discussion of the 'migration' of physical chemistry from Europe to the USA and the analysis of Ostwald's relationship with the many American students who spent some time at his laboratory.

The professionalization of physical chemists at the end of the 19th century and, more generally, the institutional aspects of the history of physical chemistry have been the subject of several articles.[25-33] Finally, there are several studies on topics that were either directly related to physical chemistry, or implicated with its development.[34-38] Physical oganic chemistry is discussed in detail in Chapter 5.

6.2 Biographies

There have been numerous biographies of chemists or other scientists with considerable contributions to physical chemistry. Some of these works are full-fledged biographies, others are popular accounts, and others are papers with short biographical sketches. Among the more substantial works is a full-length study of Edward Frankland's (1825–1899) early years and his work on organic acids, the nature of flame, his founding of organometallic chemistry, his theory of valency and his coining of the term 'bond.'[39] The same author has, also, written a comprehensive biography of Frankland.[40] Svante Arrhenius and his various contributions, ranging from the ionic theory of solutions to environmental science, have been investigated.[41,42] There is also a biography of Alfred Werner (1866–1919) examining his foundational work in coordination chemistry.[43] Walther Nernst (1864–1941) has been studied in relation to science in Germany[44] and his work in physical chemistry.[45] Giorgio Piccardi (1895–1972) and his work on the chemistry of the sun have been investigated,[46,47] as have Kasimir Fajans (1887–1975), his work in radiochemistry,[48] and his idiosyncratic proposal of quanticule theory.[49] Frederic Soddy's (1877–1956) baffling life[50] and his contributions to radiochemistry have been written about and reprints of some of his fundamental papers have been published.[51] There have been numerous studies of Linus Pauling and his many scientific and political activities.[52-58] A selection of Pauling's papers is also available.[59]

William Prout (1785–1850) and his pioneering attempts to understand the nature of matter[60] have been studied, as has Ladislaus Farkas' (1904–1948) role in the establishment of physical chemistry in Israel.[61] Some protagonists of physical chemistry in The Netherlands, J. H. Van't Hoff[62] and Johannes van Laar (1860–1938),[63] have been investigated, as has Johannes Diderik van der Waals (1837–1923) and his contributions to molecular science.[64] Van't Hoff's decisive role in the development of chemical thermodynamics has also been examined.[65]

There have been studies of Chaim Weizmann (1874–1952) and his many activities, both in chemistry and in politics;[66] Jean Perrin (1870–1942) and his intensive work in Brownian motion, which consolidated his view about the reality of atoms and molecules;[67] and of The Svedberg (1884–1971) and his related work.[68] There have been contributions on various aspects of the works of William Harkins (1873–1951),[69] Morris Loeb (1863–1912),[70] Neville Sidgwick (1873–1952),[71] Viacheslav Vasil'evich Lebedinskii (1888–1956),[72] Robert Havemann (1910–1982),[73] Glenn Seaborg (1912–1999),[74] August Horstmann (1842–1929),[75] Frantisek Wald (1861–1930),[76] Elliot Alexander,[77] and Charles Hurd.[78] A dissertation discusses the role of Josiah Willard Gibbs (1839–1903) in the founding of physical chemistry.[79] There is a brief comparative analysis of Gibbs and Ostwald.[80] Fritz London (1900–1954) and his work on the covalent bond has had substantial treatment.[81] There is a study of Alfonso Cossa (1833–1902) and his work in coordination chemistry[82] and also of Wilder Bancroft's (1867–1953) attempts to found the

Journal of Physical Chemistry,[83] as well as of Charles Bury's (1890–1968) contributions to physical chemistry.[84,85]

6.3   Electrochemistry

Ever since the discovery of the wet battery by A. Volta (1745–1827) there have been attempts to understand chemical change by means of electricity. Discussions about electrochemical processes started gaining momentum at the beginning of the 19th century with the discovery of electrolysis – the decomposition of compounds by an electric current – by W. Nicholson (1753–1827) and A. Carlisle (1768–1840). But it was Humphry Davy (1778–1829) whose investigations led him to suggest that chemical affinity was due to electrical mechanisms. J. J. Berzelius (1779–1848) attempted to classify the elements according to their electrochemical properties and considered the compounds as a union of elements with specific electronegativity and positivity. However, his dualistic theory could not explain the formation of the diatomic molecules of elements. Michael Faraday (1791–1867), H. Helmholtz (1821–1894), J. H. van't Hoff and Svante Arrhenius all played important roles in attempting to systematize within their theoretical schemata the vast data that was being collected about the electrical properties of elements and compounds. The advent of the electron, of course, redefined the whole subject of electrochemistry.

Ostwald's classic book on electrochemistry has been translated into English.[86] In addition to the theory itself, the reader is presented with a host of historical details relating to the 18th and 19th centuries as well as a discussion of Ostwald's empiricist views. Another work, arising from a conference held under the auspices of the American Chemical Society, analyses the major developments and 'technologies of … electrochemistry, electrosynthesis, electroanalytical chemistry, industrial electrochemistry, electrode systems, and pH measurement.'[87]

One of the most intriguing achievements of the 19th century was James Joule's (1818–1889) determination of the mechanical equivalent of heat and, therefore, the first law of thermodynamics. There is a study of the background to this paper, through the analysis of Joule's work in electrochemistry.[88]

The German industrialist Emil Rathenau (1838–1915) recognized the industrial potential of electrochemistry. In 1887 he founded the firm Allgemeine Elektrizitätsgesellschaft (A. E. G.) and in 1893 he started a new company, named Elektrochemische Werke, which was headed by his son, Walther (1867–1922). The two men together led the expansion of these companies into the field of electrochemistry.[89]

At the end of the 19th century, the theory of electrolytic dissociation became an important part of physical chemistry. Wilhelm Ostwald, Svante Arrhenius, and Walther Nernst were among the most vigorous supporters of that theory, which also had some severe critics. The ensuing debate has been discussed in a paper, which analyses the arguments on both sides and shows how the proponents of the theory attempted to resolve its difficulties.[90]

The discovery of the piezoelectric effect (the appearance of electrical charges on different surfaces of crystals under mechanical stress) in 1880 by Pierre Curie (1859–1906) and his brother Jacques (1855–1941) is discussed in a brief paper.[91]

J. J. Thomson (1856–1940) performed extensive work in electrochemistry, atomic structure, and valency and his interest in chemistry led him to the construction of a series of models of the chemical atom.[92]

Several works discuss institutional aspects of electrochemistry in Germany.[93–97] The history of the Kaiser Wilhelm Institute for Physics and Electrochemistry, the role of its first director Fritz Haber[98] (1868–1934), and the war work carried on there have been discussed.[99] There are also papers on the researches related to electrochemistry during the Napoleonic period in Tuscany,[100] on the work of Kohlrausch in electrolytic conductivity,[101] on Humphry Davy's work for the Royal Navy on the protection of a ship's body by electrochemical processes,[102] on aspects of silver-based coulometry,[103] on Eben Horsford's (1818–1893) contributions in measuring electrolytic resistance,[104] on the role of Oliver Wolcott Gibbs (1822–1908) and C. Luckow in the early development of electrogravimetry,[105] on the electrochemical school of Edgar Fahs Smith[106] (1878–1913), and on some aspects of the work in electrochemistry of R. Behrend.[107]

6.4   Thermochemistry

The development of thermochemistry and, especially, its use for the experimental study of chemical affinity has been the subject of a number of works. Among them is an analysis of Julius Thomsen's (1826–1909) early work in thermochemistry (1852–1854) and the incorporation of the newly formulated principle of energy conservation into chemical theory.[108] Thomsen attempted to quantify chemical affinity by specifying the magnitude of chemical forces on the basis of thermochemical measurements. The attempt, however, to develop thermochemistry as a programme for reducing chemistry, *via* the mechanical theory of heat, to Newtonian mechanics became increasingly inadequate and in the 1880s was "overtaken" by the development of chemical thermodynamics. By 1884, van't Hoff and Arrhenius were able to relate the effect of temperature increases on reaction velocities.

There has been a discussion of some aspects of the history of thermochemistry, considered as a precursor of the development of chemical thermodynamics.[109] There is also a study of C. M. Guldberg (1836–1902) and P. Waage's (1833–1900) attempts to formulate an equation for the temperature dependence of the rates of chemical reactions.[110] In 1878, Paul Vieille (1854–1934), who did important work in thermochemistry, invented the calorimetric bomb. In 1884, Vieille suggested that this instrument could be used to determine, in a novel and accurate manner, the heats of combustion of carbon and organic compounds.[111]

There is also a paper on the use of platinum in high temperature gas thermometry during the 19th century[112] and there are brief discussions of some aspects of the early developments of microcalorimetry in France,[113] and the thermal dissociation of water.[114]

6.5   Chemical Statics and Dynamics

Between 1864 and 1879, two Norwegians, C. Guldberg and P. Waage,[115] studied heterogeneous systems containing solids in contact with solutions, demonstrating experimentally that an equilibrium is reached in incomplete reactions. They also found that such an equilibrium could be approached from either direction. The driving force for a substitution was mathematically expressed as being directly proportional to the product of the masses, each raised to some definite power. Equilibrium conditions at a given temperature were expressed in terms of molecular concentration, called 'active mass.' In 1877, an equivalent mathematical derivation was reached through thermodynamic considerations by van't Hoff, who suggested that equilibrium is reached when the velocities of opposing reactions become equal and that this dynamic state is related to the concentrations or active masses. There followed a claim of priority by Guldberg and Waage who had written their original papers in Norwegian.

J. H. van't Hoff's classic text on chemical dynamics has been republished[116] and his work on chemical dynamics,[117] as well as that of Ostwald[118] and Arrhenius,[119] has been briefly discussed. There is a comprehensive narrative of the development of chemical kinetics from about the middle of the 19th century, which reconstructs Guldberg and Waage's work on chemical kinetics and compares their efforts to simultaneous work at Oxford by A. G. V. Harcourt (1834–1919) and W. Esson (1839–1916).[120]

Oxford University's College Laboratories played a very important role in the revival of the sciences at Oxford and the work done there in chemistry, and especially chemical kinetics, comprised a rather significant part of their activities. There is a study that provides information about the number of students, the kinds of institutional changes at Oxford and lists the various courses taught there.[121] There is also a history of science, including chemistry, at Oxford in the interwar years.[122]

The role of the early work on chemical kinetics in the evolution of physical chemistry has been examined with reference to van't Hoff's, Ostwald's, and Harcourt's researches prior to the 1880s.[123] There is also a discussion of chemical kinetics and thermodynamics during the 19th century,[124] and an analysis of the relation of chemical kinetics and physical chemistry up to the early part of the 20th century.[125] Studies have been made of the role of instruments and the specific laboratory locales for chemical kinetics in the interwar years,[126] and the work of H. Eyring[127] and J.-A. Muller[128] in chemical kinetics has been analysed.

The proposal, elaboration, and eventual demise in the late 1920s (after considerable controversy) of the 'Radiation Hypothesis', which was introduced in the first decade of the 20th century to account for chemical reactions that were indirectly caused by radiation, has been discussed.[129] There is a book on the history of radical chemistry[130] and also a book co-authored by one of the participants about the development of free radical chemistry during the half century from about the end of World War II.[131] The Dutch School of Catalysis,[132] P. Sabatier's (1854–1941) role in the discovery of catalysis,[133] and the establishment and development of the Ipatieff Laboratory at Northwestern University[134] have also been presented.

6.6   Valency and Structural Chemistry

Edward Frankland's suggestion, in 1852, that an element always appears to combine with the same number of atoms of any given kind started systematic investigations on valence, one of the most fundamental concepts of chemistry. The development of the notion of valency has been systematically discussed in two older books.[135,136] The history of the electron theory of valence has been reconstructed, from the electrostatic models of Thomson and Abegg to G. N. Lewis' proposal and elaboration of the electron pair as the basis of the chemical bond to the development of quantum mechanics.[137] Two further papers examine the 'chemists' electron' for the same period.[138,139] There is also a discussion of the concept of the chemical bond, with a chapter on its prehistory from antiquity to Kekulé, ending with post-World War II developments.[140] Early developments in quantum chemistry [141] and the differences between the various approaches have been analysed.[142,143] There is a special issue of a journal focusing on historical and philosophical issues concerning quantum and/or theoretical chemistry.[144] The controversy between Linus Pauling and the Soviet Chemists concerning resonance theory[145,146] has also been discussed. The work of C. Coulson,[147] R. Fowler,[148] E. Hückel,[149,150] and Pauling and Wheland[151] in quantum chemistry has been systematically examined. Two papers provide an overview of quantum chemistry.[152,153] Finally, there is a paper on some aspects of valence theory, which could be useful in avoiding confusions in teaching.[154]

Several works give an overview the development of the concept of affinity.[155-158] There is an autobiographical narrative by Robert Mulliken (1896–1986), who contributed so decisively to the development of molecular orbital theory.[159] There are also studies on a number of more specific bonding-mechanisms[160-162] and on the reaction of the chemical community to hydrogen bonding.[163]

There is a systematic study of the origins of structural theory in organic chemistry, where the emphasis is on the period from about 1830 to 1861.[164] The same author has examined Kekulé's attempt to account for valency in terms of the internal structure of polyvalent atoms.[165] Robinson's work on strychnine has also been studied.[166] Examples from structural chemistry have been used for illuminating a number of epistemological issues.[167,168]

Furthermore, we have studies on M. Berthelot's (1897–1907) contributions to the development of the notion of isomerism;[169] the errors of Alfred Werner in his account of spontaneous resolution, mainly due to his neglect of the work of others;[170] Adolphe Wurtz's insistence on atomism and its cultural milieu;[171,172] and the early applications of infra-red spectroscopy to chemistry.[173]

6.7 Stereochemistry

J. B. Biot's (1774–1862) discovery that a number of organic compounds and their solutions could rotate the plane of polarized light suggested that optical activity might not be due only to crystal form, but that it might be an inherent property of compounds not, necessarily, related to their crystalline form. The extensive crystallographic studies of Louis Pasteur (1822–1895) pointed to the asymmetrical character of numerous compounds. However, it was J. H. van't Hoff and J. Le Bel (1847–1930) who proposed independently in 1874 the asymmetrical structure of the carbon atom. They observed that when a carbon atom is attached to four different atoms or atomic groups the four substituents can be arranged in two different ways so that the resulting molecules will be mirror images of one another. Van't Hoff talked of tetrahedral arrangements, while Le Bel argued about systems that permitted two arrangements of different substituents around an asymmetrical carbon atom.

The early history of stereochemistry has been explored.[174] An account of stereochemistry from the mid-19th century to 1960 surveys the proposals of van't Hoff and Le Bel as well as those of later workers in the field, such as Odd Hassel and D. H. R. Barton (1918–1998), and J. W. Cornforth and V. Prelog (1906–1998).[175] G. J. W. Bremer's experiments, which gave the initial experimental support to van't Hoff's stereochemical views, have been re-assessed through an analysis of the correspondence between the two chemists.[176] There is a paper arguing that, in effect, van't Hoff and Le Bel produced two different theories[177] and another one suggesting that John Dalton (1766–1844) was the first stereochemist.[178] There is a brief analysis of the independent contributions of Alfred Werner and William Jackson Pope (1870–1939) to stereochemistry.[179]

Discussions concerning the origins of mirror-image isomers, first discovered in the mid-19th century, are the subject of a study examining the causes of this phenomenon, variously attributed to light, magnetism, heat, vitalism, pure chance and Darwinian evolution.[180] The work of the American chemist Arthur Michael (1853–1945), whose discovery that the multiple bonds of halogens exhibit an axial symmetry led him to a critique of stereochemistry, has also been studied. His views, which were difficult to visualize and did not account for axial symmetry as a process, were not preferred over Johannes Wislicenus' more visualizable approach.[181,182] Finally, Charlotte Thomas' early text-book on stereochemistry has been commented upon.[183] See also Chapters, 4 and 5.

6.8   Solutions

There is a brief description of the development of the physical chemistry of non-aqueous solutions from 1920 to 1985[184] and an analysis of van't Hoff's theory of diluted solutions.[185,186] The role of the chemist Paul Walden (1863–1957) in the development of theories about chemical solutions has been discussed.[187] Amedeo Avogadro's (1776–1856) research on the nature of metal salts has been examined.[188] There is, also, a discussion of Thomas Graham's (1805-1869) work on the diffusion of gases and liquids.[189]

6.9   Colloid Chemistry

In 1861, Graham introduced the term "colloid" to distinguish those solutions he called 'crystalloids,' which could pass through membranes, and those which could not go through these membranes, his 'colloids.'[190] Systematic studies of colloids, especially of the conditions of coagulation of colloids by salts, were undertaken before the end of the 19th century, when it was also noticed that colloids migrate in an electric field – a technique used for their purification. Developments in microscopy made possible the counting of colloid particles and the estimation of their size.

Some aspects of the history of colloid chemistry have been explored. There is a study of Isidor Traube (1860–1943), highlighting his work on molecular volumes of pure liquids and particularly on the apparent molar volumes of dissolved molecules in various solvents. It discusses some of the controversies in which he was involved.[191] There is also a systematic treatment of the developments of colloid chemistry in the USA and Canada from the beginning of the 20th century,[192] and of the specific role of laboratory equipment in colloid chemistry laboratories.[193,194] Finally, there is a brief discussion of Ostwald's work on colloids.[195]

6.10   Coordination Chemistry

Coordination chemistry deals with compounds containing a central atom or ion to which are bonded molecules or ions, whose number usually exceeds the value corresponding to the valence of the central atom. The molecules are called ligands and the bonding consists usually of a covalent bond, formed by the donation of a pair of electrons from the ligand (donor) to the central atom or ion (acceptor). The structure and properties of coordination compounds were explained by the coordination theory of Alfred Werner in 1893.[196] An American Chemical Society Symposium marked the centenary of the founding of coordination chemistry.[197] A book and a paper discuss the early developments of coordination chemistry and especially Werner's contributions.[198,199] There is also a historical narrative of the evolution of coordination chemistry to 1930.[200] Two further studies discuss various aspects of coordination chemistry in the USA[201] and Australia.[202] Finally, the origin and the dissemination of the term 'ligand' in chemistry have been discussed.[203–205]

References

1. A. J. B. Robertson, 'Physical chemistry', in *Recent Developments in the History of Chemistry*, ed. C. A. Russell, Royal Society of Chemistry, London, 1985, pp. 153–176.
2. R. S. Root-Bernstein, 'The Ionists: Founding Physical Chemistry, 1872–1890', Ph.D. dissertation, Princeton University, 1980; Univ. Microfilms order no. 81–01554
3. J. R. Partington, *A History of Chemistry*, 4 vols, Macmillan, London, 1961–1970.
4. J. R. Partington, *A History of Chemistry,* Macmillan, London, 1964, vol. 4, pp. 569–746.
5. K. J. Laidler, *The World of Physical Chemistry*, Oxford University Press, New York, 1993.
6. 'Biographical sketch [of K. J. Laidler]', *Chem. Intell.*, 1997, **3**, 50.
7. K. J. Laidler, 'The Henry Marshall Tory medal address: the second half-century of physical chemistry', *Trans. Roy. Soc. Can.*, 1987, **2**, 181–186.
8. K. J. Laidler, 'A lifetime of transition-state theory', *Chem. Intell.*, 1998, **4**, 39–47.
9. S. W. Benson, '50 years of physical chemistry: a personal account', *Annu. Rev. Phys. Chem.*, 1988, **39**, 1–37.
10. D. Herschbach, 'Fifty years in physical chemistry: homage to mentors, methods, and molecules', *Annu. Rev. Phys. Chem.*, 2000, **51**, 1–40.
11. W. Girnus, '100 Jahre aus der Geschichte der physikalischen Chemie – von Johann Gottschalk Wallerius und Carl Friedrich Wenzel zu Svante Arrhenius

und Wilhelm Ostwald', in *Carl-Wilhelm-Scheele Ehrung 1986*, Akademie der Wissenschaften der D. D. R., Berlin, 1987, pp. 109–123.

12. R. Mierzecki, 'The historical development of chemical concepts', *Chemists and Chemistry*, 12, Kluwer, Dordrecht/Polish Scientific Publishers, Warsaw, 1991.

13. E. B. Wilson, 'One hundred years of physical chemistry', *Am. Sci.*, 1986, **74**, 70–77.

14. M. C. King, *E. W. R. Steacie and Science in Canada*, University of Toronto Press, Toronto, 1989.

15. Y. Kikuchi, 'Redefining academic chemistry: Joji Sakurai and the introduction of physical chemistry into Meiji Japan', *Hist. Scientiarum*, 2000, **9**, 215–256.

16. R. Mierzecki, 'Chemistry in Polish schools of higher education between the World Wars: physical chemistry', *Analecta*, 1995, **4**, 171–221 [in Polish].

17. C. Debru, 'La chimie physique et la fonction hemoglobinique, 1925–1951', *Hist. Phil. Life Sci.*, 1986, **8**, 69–79.

18. L. M. Pritykin, 'The origin and development of the physical chemistry of polymers', *Voprosy Istorii Estestvozn Tekhn [USSR]*, 1991, **1**, 15–25; [in Russian] (English abstract by J. H. South).

19. M. J. Nye, *From Chemical Philosophy to Theoretical Chemistry: Dynamics of Matter and Dynamics of Disciplines, 1800–1950*, University of California Press, Berkeley, CA, 1993.

20. M. J. Nye, 'Chemical explanation and physical dynamics: two research schools at the first Solvay Chemistry Conferences, 1922–1928', *Ann. Sci.*, 1989, **46**, 461–480.

21. M. Charpentier-Morize, 'Résistance a l'introduction en France des théories électroniques de la réactivite chimique: Ses conséquences sur l' évolution de la recherche', *Sci. Techn. Perspect.*, 1993, **25**, 101–110.

22. M. J. Nye, 'Laboratory practice and the physical chemistry of Michael Polanyi,' in *Instruments and Experimentation in the History of Chemistry*, ed. F. L. Holmes and T. H. Levere, MIT Press, Cambridge, MA, 2000, pp. 367–400.

23. M. J. Nye, 'Michael Polanyi's theory of surface adsorption: how premature?', in *Prematurity in Scientific Discovery: On Resistance and Neglect*, ed. E. B. Hook, University of California Press, Berkeley, CA, 2002, pp. 151–163.

24. J. Servos, *Physical Chemistry from Ostwald to Pauling: The Making of a Science in America*, Princeton University Press, Princeton, NJ, 1990.

25. D. K. Barkan, 'A usable past: creating disciplinary space for physical chemistry', in *The Invention of Physical Science . . . Essays in Honor of Erwin N. Hiebert*, ed. M.J. Nye *et al.*, Kluwer Academic, Dordrecht, 1992, pp. 175–202.

26. T. Hapke, *Die 'Zeitschrift für physikalische Chemie:' Hundert Jahre Wechselwirkung zwischen Fachwissenschaft, Kommunikationsmedium und Gesellschaft*, Bautz, Hertzberg, 1990.

27. F. Schmithals, 'Die erste Berufung für physikalische Chemie: "Ein Unterfangen von hochster wissenschaftlicher Bedeutung" ', *NTM: Schriftenr. Gesch.*, 1995, **3**, 227–254.

28. K. Nothnagel, 'Von der *Zeitschrift für Elektrochemie* zum *Journal of Physical Chemistry Chemical Physics*', *Ber. der Bunsen-Gesellsch. für Physik. Chem.*, 1998, **102** (12), 1735–1739.

29. J. W. Stout, 'The *Journal of Chemical Physics:* the first 50 years', *Annu. Rev. Phys. Chem.*, 1986, **37**, 1–23.
30. A. K. Nielsen and H. Kragh, 'An institute for dollars: physical chemistry in Copenhagen between the world wars', *Centaurus*, 1997, **39**, 311–331.
31. G. K. Roberts, 'Physical chemists for industry: the making of the chemist at University College London, 1914–1939', *Centaurus*, 1997, **39**, 291–310.
32. P. F. Barbara, 'A brief history of physical chemistry in the American Chemical Society', *J. Phys. Chem.*, 1996, **100**, 12694–12700.
33. V. Past, 'The emergence of physical chemistry: the contribution of the University of Tartu', in *Estonian Studies in the History and Philosophy of Science*, ed. R. Vihalemm, Kluwer, Dordrecht, 2001, pp. 35–50.
34. A. Assmus, 'The Americanization of molecular physics', *Hist. Stud. Phys. Biol. Sci.*, 1992, **23**, 1–34.
35. M. R. S. Creese, 'British women of the 19th and 20th centuries who contributed to research in the chemical sciences', *Brit. J. Hist. Sci.*, 1991, **24**, 275–305.
36. J. Kroh, ed., *Early Developments in Radiation Chemistry*, Royal Society of Chemistry, Cambridge, 1989.
37. M. Guzman-Casado and A. Parody-Morreale, 'Notes on the early history of the interaction between physical chemistry and biochemistry: the development of physical biochemistry', *J. Chem. Educ.*, 2002, **79**, 327–331.
38. J. Schummer, 'Physical chemistry: neither fish nor fowl?' in *The Autonomy of Chemistry*, ed. P. Janich and N. Psarros, Königshausen & Neumann, Würzburg, 1998, pp. 135–148.
39. C. A. Russell, *Lancastrian Chemist: The Early Years of Sir Edward Frankland*, Open University Press, Milton Keynes, 1986.
40. C. A. Russell, *Edward Frankland: Chemistry, Controversy and Conspiracy in Victorian England,* Cambridge University Press, New York, 1996.
41. E. Crawford, *Arrhenius: From Ionic Theory to the Greenhouse Effect*, Science History Publications, Canton, MA, 1996.
42. G. B. Kauffman, 'Svante August Arrhenius, Swedish pioneer in physical chemistry', *J. Chem. Educ.*, 1988, **65**, 437–438.
43. G. B. Kauffman, *Alfred Werner: Founder of Coordination Chemistry*, Springer Verlag, Berlin and New York, 1966.
44. K. Mendelssohn, *The World of Walther Nernst: The Rise and Fall of German Science, 1864–1941*, University of Pittsburgh Press, Pittsburgh, PA, 1973.
45. D. Barkan, *Walther Nernst and the Transition to Modern Physical Science*, Cambridge University Press, Cambridge, 1999.
46. G. B. Kauffman and L. Belloni, 'Giorgio Piccardi (1895–1972): Italian physical chemist and master of the sun', *J. Chem. Educ.*, 1987, **64**, 205–208.
47. G. B. Kauffman and M. T. Beck, 'Self-deception in science: the curious case of Giorgio Piccardi', *Specul. Sci. Technol.*, 1987, **10** (2), 113–122.
48. R. E. Holmes, 'Kasimir Fajans (1887–1975): the man and his work', *Bull. Hist. Chem.*, 1989, **4**, 15–21.
49. J. Hurwic, 'Reception of Kasimir Fajans's quanticule theory of the chemical bond: a tragedy of a scientist', *J. Chem. Educ.*, 1987, **64** (2), 122–123.

50. L. Merricks, *The World Made New: Frederick Soddy, Science, Politics and Environment*, Oxford University Press, New York, 1996.
51. G. B. Kauffman, ed., *Frederick Soddy (1877–1956): Early Pioneer in Radiochemistry*, Reidel, Dordrecht, 1986.
52. A. Serafini, *Linus Pauling: A Man and his Science*, Foreword by Isaac Asimov, Simon & Schuster, New York, 1989.
53. T. Goertzel and B. Goertzel, *Linus Pauling: A Life in Science and Politics*, with the assistance of M. Goertzel and V. Goertzel, with original drawings by G. Goertzel, Basic Books, New York, 1995.
54. T. Hager, *Force of Nature: The Life of Linus Pauling*, Simon & Schuster, New York, 1995.
55. L. Pauling, *Linus Pauling in his own Words: Selected Writings, Speeches, and Interviews*, ed. Barbara Marinacci, introduction by Linus Pauling, Simon & Schuster, New York, 1995.
56. M. Perutz, 'Linus Pauling, 1901–1994', *Naturwissenschaftl. Rundschau*, 1995, **48**, 452–456
57. 'A tribute to Linus Carl Pauling (1901–1994)', *J. Chem. Educ.*, 1996, **73**, 2–32.
58. C. Mead and T. Hager, eds, *Linus Pauling: Scientist and Peacemaker*, Oregon State University Press, Corvallis, OR, 2001.
59. B. Kamb, ed., *Linus Pauling: Selected Scientific Papers*, World Scientific Series in 20th Century Chemistry, vol. 10, World Scientific, River Edge, NJ, 2001.
60. W. H. Brock, *From Protyle to Proton: William Prout and the Nature of Matter, 1785–1985*, Adam Hilger, Bristol, 1985.
61. M. Chayut, 'From Berlin to Jerusalem: Ladislaus Farkas and the founding of physical chemistry in Israel', *Hist. Stud. Phys. Biol. Sci.*, 1994, **24**, 237–263.
62. H. A. M. Snelders, 'J. H. van't Hoff's research school in Amsterdam (1877–1895)', *Janus*, 1984, **71**, 1–30.
63. H. A. M. Snelders, 'The Dutch physical chemist J. J. van Laar (1860–1938) versus J. H. van't Hoff's "osmotic school" ', *Centaurus*, 1986, **29**, 53–71.
64. A. Ya. Kipnis, B. E. Yavelov and J. S. Rowlinson, *Van der Waals and Molecular Science*, Clarendon Press, Oxford, 1996.
65. W. J. Hornix and S. H. W. M. Mannaerts, *Van't Hoff and the Emergence of Chemical Thermodynamics*, Koninklijke Nederlandse Chemische Vereniging, Leidschendam, 2001.
66. G. B. Kauffman and I. Mayo, 'Chaim Weizmann (1874–1952): chemist, biotechnologist, and statesman', *J. Chem. Educ.*, 1994, **71**, 209–214.
67. M. J. Nye, *Molecular Reality: A Perspective on the Scientific Work of Jean Perrin*, Macdonald, London; Elsevier, New York, 1972.
68. M. Kerker, 'The Svedberg and molecular reality: an autobiographical postscript', *Isis*, 1986, **77**, 278–282.
69. G. B. Kaufmann, 'William Draper Harkins (1873–1951): a controversial and neglected American physical chemist', *J. Chem. Educ.*, 1985, **62**, 758–761.
70. M. D. Saltzman, 'Morris Loeb: Ostwald's first American student and America's first physical chemist', *Bull. Hist. Chem.*, 1998, **22**, 10–15.
71. M. Laing, 'Nevil Vincent Sidgwick, 1873–1952: one of the unsung truly greats', *J. Chem. Educ.*, 1994, **71**, 47–473.

72. G. B. Kauffman, 'Viacheslav Vasil'evich Lebedinskii: centenary of the birth of a rhodium chemist', *Plat. Met. Rev.*, 1988, **32**, 141–147.

73. D. Hoffmann, 'Der Physikochemiker Robert Havemann (1910–1982): Eine deutsche Biographie', in *Naturwissenschaft und Technik in der D. D. R.*, ed. D. Hoffmann and K. Macrakis, Akademie Verlag, Berlin, 1997, pp. 319–336.

74. G. B. Kauffman, 'Transuranium pioneer: Glenn T. Seaborg', *Today's Chem.*, 1991, **4** (3), 18–20, 23, 24, 32.

75. A. Kipnis, *August Friedrich Horstmann und die Physikalische Chemie*, ERS-Verlag, Berlin, 1997.

76. K. Ruthenberg and N. Psarros, 'Frantisek Wald und die phänomenologische Chemie', *Mitt. - Ges. Dtsch. Chem., Fachgruppe Gesch. Chem.*, 1994, **10**, 17–30.

77. P. R. Jones, 'The brief career of a prolific, pioneering physical organic chemist, Elliot Ritchie Alexander, Jr.', *J. Chem. Educ.*, 1987, **64**, 882–883.

78. W. J. Hagan, Jr., 'Charles Hurd and colloid research at Union College, 1923–1959', *J. Chem. Educ.*, 1988, **65**, 191–193.

79. G. K. Khatib, 'Change of Phase: The transformation of nineteenth-century thermodynamics. Josiah Willard Gibbs (1873–1878)', Ph.D. dissertation, Cornell University, 1992; Univ. Microfilms order no. 92-36104.

80. R. J. Deltete and D. L. Thorsell, 'Josiah Willard Gibbs and Wilhelm Ostwald: a contrast in scientific style', *J. Chem. Educ.*, 1996, **73**, 289–295.

81. K. Gavroglu, *Fritz London, A Scientific Biography*, Cambridge University Press, Cambridge, 1995.

82. G. B. Kauffman and E. Molayem, 'Alfonso Cossa (1833–1902), a self-taught Italian chemist', *Ambix*, 1990, **37**, 20–34.

83. J. W. Servos, 'A disciplinary program that failed: Wilder D. Bancroft and the *Journal of Physical Chemistry*, 1896–1933', *Isis*, 1982, **73**, 207–232.

84. M. Davies, 'Charles Rugeley Bury and his contributions to physical chemistry', *Arch. Hist. Exact Sci.*, 1986, **36**, 75–90.

85. M. Davies, 'C. R. Bury: his contributions to physical chemistry', *J. Chem. Educ.*, 1986, **63**, 741–743.

86. W. Ostwald, *Electrochemistry: History and Theory*, 2 vols, trans. N. P. Date, Amerind Publishing Co., New Delhi, 1980; orig. German edition, 1896.

87. J. T. Stock and M. V. Orna, eds, *Electrochemistry, Past and Present*, American Chemical Society, Washington, DC, 1989 (preface).

88. W. H. Cropper, 'James Joule's work in electrochemistry and the emergence of the first law of thermodynamics', *Hist. Stud. Phys. Biol. Sci.*, 1988, **19**, 1–15.

89. U. Mader, 'Emil Rathenau und Die Elektrochemischen Werke (1893)', *J. für Wirtschaftsgesch.*, 1990, **4**, 191–227.

90. R. Maiocchi, 'Difficult beginnings: the theory of electrolytic dissociation in the 19th century', *Nuncius*, 1993, **8**, 121–167 [in Italian].

91. R. B. Seymour and G. B. Kauffman, 'Piezoelectric polymers: direct converters of work to electricity', *J. Chem. Educ.*, 1990, **67**, 763–765.

92. S. B. Sinclair, 'J. J. Thomson and the chemical atom: from ether vortex to atomic decay', *Ambix*, 1987, **34**, 89–116.

93. R. Piosik, 'Die Glaselektrode und ihr Miterfinder Zygmunt Klemensiewicz', *Mitt. - Ges. Dtsch. Chem., Fachgruppe Gesch. Chem*, 1993, **8**, 50–60.

94. L. Dunsch, *Geschichte der Elektrochemie: Ein Abriss*, VEB Deutscher Verlag für Grundstoffindustrie, Leipzig, 1985.

95. M. Rasch, 'Zur Situation der Elektrochemie an der TH Berlin-Charlottenburg vor dem 1. Weltkrieg', *Humanismus Techn.*, 1987, **31**, 40–62.

96. M. Rasch, 'Zur Institutionalisierung der Elektrochemie in Deutschland', *Ber. Wissenschaft.: Organ der Ges. für Wissenschaft.*, 1988, **11**, 42–44.

97. B. Sorms, 'Zu den Anfangen der Deutschen Elektrochemischen Gesellschaft (1894)', *Dresdender Beiträge Geschichte Technikwissenschaft.*, 1987, **14**, 31–37.

98. J. T. Stock, 'Fritz Haber (1868–1934) and the electroreduction of nitrobenzene', *J. Chem. Educ.*, 1988, **65**, 337–338.

99. D. J. Stoltzenberg, 'Zur Geschichte des Kaiser-Wilhelm-Instituts für Physikalische Chemie und Elektrochemie: Zur Geplanten Veränderung des Instituts in eine Forschungs- und Entwicklungsstatte des Heeres für den Gaskampf und Gasschutz auch in Friedenszeiten 1916 und nach 1933', *Ber. Wissenschaft.*, 1991, **14**, 15–23.

100. F. Abbri, 'Il misterioso 'spiritus salis': Le ricerche di elettrochimica nella Toscana napoleonica', *Nuncius*, 1987, **2**, 55–88.

101. D. Cahan, 'Kohlrausch and electrolytic conductivity: instruments, institutes, and scientific innovation', *Osiris, 2d Ser.*, 1989, **5**, 167–85.

102. F. A. J. L. James, 'Davy in the Dockyard: Humphry Davy, the Royal Society and the electro-chemical protection of the copper sheeting of His Majesty's Ships in the mid 1820s', *Physis*, 1992, **29**, 205–225.

103. J. T. Stock, 'A century and a half of silver-based coulometry', *J. Chem. Educ.*, 1992, **69**, 949–952.

104. J. T. Stock, 'Eben Horsford (1818–1893) and the measurement of electrolytic resistance', *J. Chem. Educ.*, 1988, **65**, 700–701.

105. J. T. Stock, 'The genesis of electrogravimetry', *Bull. Hist. Chem.*, 1990, **7**, 17–19.

106. L. M. Robinson, The electrochemical school of Edgar Fahs Smith, 1878–1913', Ph.D. dissertation, University of Pennsylvania, 1986; Univ. Microfilms order no. 87-03261.

107. J. T. Stock, 'Robert Behrend's foray into electrochemistry', *J. Chem. Educ.*, 1992, **69**, 197–199.

108. H. Kragh, 'Julius Thomsen and classical thermochemistry', *Brit. J. Hist. Sci.*, 1984, **17**, 255–272.

109. L. Médard and H. Tachoire, *Histoire de la thermochimie: Prélude a la Thermodynamique Chimique*, preface by Alain Horeau, Univ. de Provence, Aix-en-Provence, 1994.

110. P. Øhrstrøm, 'Guldberg and Waage on the influence of temperature on the rates of chemical reactions', *Centaurus*, 1985, **28**, 277–287.

111. L. Médard, 'L' oeuvre scientifique de Paul Vieille (1854–1934)', *Rev. Hist. Sci.*, 1994, **47**, 381–404.

112. I. E. Cottington, 'High temperature gas thermometry and the platinum metals: Some aspects of 19th-century developments', *Plat. Met. Rev.*, 1987, **31**, 196–207.

113. H. Tachoire, 'Louis Menard, Edouard Calvet et les premiers développements de la microcalorimetrie', in *Sciences et techniques en France méridionale*, Comite des Travaux Historiques et Scientifiques, Paris, 1992, 186–194.

114. H. H. G. Jellinek, 'The thermal dissociation of water: the forgotten literature', *J. Chem. Educ.*, 1986, **63**, 1029–1037.

115. R. Bye, 'Peter Waage: Store nordiske kjemikere, V', *Kjemi*, 1998, **9**, 11–13 [in Norwegian].

116. J. H. van't Hoff, *Studien zur chemischen Dynamik (Études de dynamique chimique, 1884)*, trans. into German with an introduction by L. Dunsch, Geest & Portig, Leipzig, 1985.

117. J. Van Houten, 'Nobel centennial essay: a century of chemical dynamics traced through the Nobel Prizes. 1901: J. van't Hoff', *J. Chem. Educ.*, 2001, **78**, 1570.

118. J. Van Houten, 'Nobel centennial essays: a century of chemical dynamics traced through the Nobel Prizes. 1909: Wilhelm Ostwald', *J. Chem. Educ.*, 2002, **79** (2), 146.

119. J. Van Houten, 'Nobel centennial essays: a century of chemical dynamics traced through the Nobel Prizes. 1903: Svante Arrhenius', *J. Chem. Educ.*, 2002, **79** (1), 21.

120. M. C. King, 'Experiments with time: progress and problems in the development of chemical kinetics', *Ambix*, 1981, **29**, 70–82.

121. K. J. Laidler, 'Chemical kinetics and the Oxford College laboratories', *Arch. Hist. Exact Sci.*, 1988, **38**, 197–283.

122. J. B. Morrell, *Science at Oxford, 1914–1939: Transforming an Arts University*, Clarendon Press, Oxford, 1997.

123. K. J. Laidler, 'Chemical Kinetics and the origins of physical chemistry', *Arch. Hist. Exact Sci.*, 1985, **32**, 43–75.

124. J. Berger, ‚Grenzgänge zwischen Physik und Chemie: Thermodynamik und Chemische Kinetik – kein Happy – End im 19. Jahrhundert', *Centaurus*, 1999, **41**, 253–279.

125. V. A. Kritsman, 'Chemical kinetics as part of physical chemistry in the 19th century and at the beginning of the 20th century: analysis of the origin and development of phenomenological kinetics', *NTM: Schriftenr. Gesch.*, 1996, **4**, 19–30.

126. J. L. Ramsey, 'On refusing to be an epistemologically black box: instruments in chemical kinetics during the 1920s and '30s', *Stud. Hist. Phil. Sci.*, 1992, **23**, 283–304.

127. P. Antoniotti, 'H. Eyring e i primi studi quantistici della reattivà chimica,' in *Atti del III Convegno Nazionale di Storia e Fondamenti della Chimica*, Brenner, Cosenza, 1991, 265–277.

128. S. J. Formosinho, 'Jean-Auguste Muller, um precursor das relações termodinâmicas e extra-termodinâmicas em cinética química', *Rev. Univ. Coimbra*, 1989, **35**, 137–153.

129. M. C. King and K. J. Laidler, 'Chemical kinetics and the radiation hypothesis', *Arch. Hist. Exact Sci.*, 1984, **30**, 45–86.

130. C. Ruchardt, *Radikale: Eine chemische Theorie in historischer Sicht*, Sitzungsberichte der Heidelberger Akademie der Wissenschaften, Mathematisch-naturwissenschaftliche Klasse, 1992, **4**, Springer-Verlag, Berlin, 1992.

131. D. Barton (in collaboration with S. I. Parekh), *Half a Century of Free Radical Chemistry*, Cambridge University Press, Cambridge, 1993.

132. J. J. F. Scholten, ed., *A Short History of the Dutch School of Catalysis*, Royal Netherlands Chemical Society, The Hague, 1994.

133. F. Gallais, 'Pierre Sabatier et la découverte de la catalyse', *Comptes Rend. Acad. Sci.*, 1988, **5**, 235–238.

134. H. Pines, *Genesis and Evolution of the Ipatieff Catalytic Laboratory at Northwestern University, 1930–1970*, Department of Chemistry, Northwestern University, Evanston, IL, 1992.

135. C. A. Russell, *The History of Valency*, Humanities Press, New York, 1971.

136. W. G. Palmer, *A History of the Concept of Valency to 1930*, Cambridge University Press, Cambridge, 1965.

137. A. N. Stranges, *Electrons and Valence: Development of the Theory, 1900–1925*, Texas A&M University Press, College Station, TX, 1982.

138. T. Arabatzis and K. Gavroglu, 'The chemists' electron', *Eur. J. Phys.*, 1997, **18**, 150–163.

139. K. Gavroglu, 'The physicists' electron and its appropriation by the chemists', in *Histories of the Electron: The Birth of Microphysics*, ed. J. Z. Buchwald and A. Warwick, MIT Press, Cambridge, MA, 2001, pp. 363–400.

140. B. Vidal, *La Liaison Chimique: Le Concept et son Histoire*, Vrin, Paris, 1989.

141. C. J. Ballhausen, 'Quantum mechanics and chemical bonding in inorganic complexes. 1: Static concepts of bonding; dynamic concepts of valency. 2: Valency and inorganic metal complexes. 3: The spread of ideas', *J. Chem. Educ.*, 1979, **56**, 215–218, 294–297, 357–361.

142. A. I. Simoes, 'Converging trajectories, diverging traditions: chemical bond, valence, quantum mechanics and chemistry, 1927–1937', Ph.D. dissertation, University of Maryland, College Park, 1993; Univ. Microfilms order no. 93-27498.

143. K. Gavroglu and A. I. Simoes, 'The Americans, the Germans and the beginnings of quantum chemistry: the confluence of diverging traditions', *Hist. Stud. Phys. Biol. Sci.*, 1994, **25**, 1–63.

144. *Stud. Hist. Phil. Mod. Phys.*, 2000, **31** (4), 429–469.

145. A. A. Pechenkin, 'The 1949–1951 anti-resonance campaign in Soviet Science', *Llull: Boletin Sociedad Española Hist. Ciencias*, 1995, **18** (34), 135–158.

146. S. Kara-Murza, 'Siguiendo la receta de Orwell: comentario al artículo de A. A. Pechenkin [Following Orwell's trail: comment on A. A. Pechenkin's paper]', *Llull: Boletin Sociedad Española Hist. Ciencias*, 1995, **18** (34), 159–166.

147. A. Simoes and K. Gavroglu, 'Quantum chemistry *qua* applied mathematics: the contributions of Charles Alfred Coulson (1910–1974)', *Hist. Stud. Phys. Biol. Sci.*, 1999, **29**, 363–406.

148. K. Gavroglu and A. Simoes, 'Preparing the ground for quantum chemistry in Great Britain: the work of the physicist R. H. Fowler and the chemist N. V. Sidgwick', *Brit. J. Hist. Sci.*, 2002, **35**, 187–212.

149. A. Karachalios, 'Erich Hückel (1896–1980): Von der Physik zur Quantenchemie,' Ph.D. dissertation, Johannes Gutenberg-Universität Mainz, 2003; forthcoming in English translation from Springer-Verlag.

150. H. Kragh, 'Before quantum chemistry: Erich Hückel and the physics-chemistry interface', *Centaurus*, 2001, **43**, 1–16.

151. B. S. Park, 'Chemical translators: Pauling, Wheland and their strategies for teaching the theory of resonance', *Brit. J. Hist. Sci.*, 1999, **32**, 21–46.

152. B. S. Park, 'The hyperbola of quantum chemistry: the changing practice and identity of a scientific discipline in the early years of electronic digital computers, 1945–1965', *Ann. Sci.* 2003, **60**, 219–247.

153. A. Simoes, 'Chemical physics and quantum chemistry in the Twentieth Century,' in *The Cambridge History of Science*, vol. 5, *The Modern Physical and Mathematical Sciences*, ed. M. J. Nye, Cambridge University Press, Cambridge, 2003, pp. 394–412.

154. D. Zavaleta, 'Paradigms and plastic facts in the history of valence', *J. Chem. Educ.*, 1988, **65**, 677–680.

155. M. Goupil, *Du flou au clair? Histoire de l'affinite chimique, de Cardan a Prigogine*, Comite des Travaux Historiques et Scientifiques, Paris, 1991.

156. M. G. Kim, 'Practice and representation: investigative programs of chemical affinity in the 19th century', Ph.D. dissertation, University of California Los Angeles, 1990; Univ. Microfilms order no. 91-11300.

157. U. Klein, *Verbindung und Affinität: Die Grundlegung der Neuzeitlichen Chemie an der Wende vom 17. zum 18. Jahrhundert*, Science Networks: Historical Studies, no. 14, Birkhäuser, Basel, 1994.

158. M. G. Kim, *Affinity, that Elusive Dream: A Genealogy of the Chemical Revolution*, MIT Press Cambridge, MA, 2003.

159. R. S. Mulliken, *Life of a Scientist. An Autobiographical Account of the Development of Molecular Orbital Theory, with an Introductory Memoir by Friedrich Hund*, ed. B. J. Ransil, Springer-Verlag, Berlin, 1989.

160. W. B. Jensen, 'The historical development of the van Arkel bond-type triangle', *Bull. Hist. Chem.*, 1992–93, **13–14**, 47–59.

161. L. Kuhnert and U. Niedersen, eds, *Selbstorganisation chemischer Strukturen: Arbeiten von Friedlieb Ferdinand Runge, Raphael Eduard Liesegang, Boris Pavlovich Belousov und Anatol Markovich Zhabotinsky*, Geest & Portig, Leipzig, 1987.

162. J. Hurwic, 'Reception of Kasimir Fajans' quanticule theory of the chemical bond: a tragedy of a scientist', *J. Chem. Educ.*, 1987, **64**, 122–123.

163. D. Quane, 'The reception of hydrogen bonding by the chemical community, 1920–1937', *Bull. Hist. Chem.*, 1990, **7**, 3–13.

164. A. J. Rocke, 'Origins of the structural theory in organic chemistry', Ph.D. dissertation, The University of Wisconsin-Madison, 1975; Univ. Microfilms order no. 76-02, 504.

165. A. J. Rocke, 'Subatomic speculations and the origin of structure theory', *Ambix*, 1983, **30**, 1–18.

166. L. B. Slater, 'Woodward, Robinson and strychnine: chemical structure and chemists' challenge', *Ambix*, 2001, **48**, 161–189.

167. A. M. Diamond, Jr., 'The polywater episode and the appraisal of theories', in *Scrutinizing Science: Empirical Studies of Scientific Change*, ed. A. Donovan *et al.*, Kluwer Academic, Dordrecht, 1988, pp. 181–198.

168. S. H. Mauskopf, 'Molecular geometry in 19th-century France: shifts in guiding assumptions', in *Scrutinizing Science: Empirical Studies of Scientific*

Change, ed. A. Donovan et al., Kluwer Academic, Dordrecht, 1988, pp. 125–144.

169. G. Ciancia, 'Marcelin Berthelot et le concept d'isomerie (1860–1865)', *Arch. Int. Hist. Sci.*, 1986, **36**, 54–83.

170. G. B. Kauffman and I. Bernal, 'Alfred Werner's awareness of spontaneous resolutions and the meaning of hemihedral faces in optically active crystals', *Structural Chem.*, 1993, **4** (2), 131–138.

171. A. Carneiro, 'Adolphe Wurtz and the atomism controversy', *Ambix*, 1993, **40**, 75–95.

172. A. J. Rocke, *Nationalizing Science: Adolphe Wurtz and the Battle for French Chemistry*, MIT Press, Cambridge, MA, 2000.

173. C. S. Nunziante and E. Torracca, 'Early applications of infra-red spectroscopy to chemistry', *Ambix*, 1988, **35**, 39–47.

174. P. J. Ramberg, *Chemical Structure, Spatial Arrangement: The Early History of Stereochemistry 1874–1914*, Ashgate, Aldershot, 2002.

175. O. B. Ramsay, *Stereochemistry*, foreword by D. H. R. Barton, preface by J. W. van Spronsen, Heyden & Son, London, 1981.

176. E. Fischmann, 'A reconstruction of the first experiments in stereochemistry: letters from van't Hoff to Bremer in a new chronological sequence', *Janus*, 1985, **72**, 131–156.

177. R. B. Grossman, 'Van't Hoff, Le Bel, and the development of stereochemistry: a reassessment', *J. Chem. Educ.*, 1989, **66**, 30–33.

178. D. H. Rouvray, 'John Dalton: the world's first stereochemist', *Endeavour*, 1995, **19**, 52–57.

179. G. B. Kauffman and I. Bernal, 'Overlooked opportunites in stereochemistry, part 2: the neglected connection between Werner's metal-ammines and Pope's organic onium compounds', *J. Chem. Educ.*, 1989, **66**, 293–300.

180. P. Palladino, 'Stereochemistry and the nature of life: mechanist, vitalist, and evolutionary perspectives', *Isis*, 1990, **81**, 44–67.

181. P. J. Ramberg, 'Arthur Michael's critique of stereochemistry, 1887–1899', *Hist. Stud. Phys. Biol. Sci.*, 1995, **26**, 89–138.

182. P. J. Ramberg, 'Johannes Wislicenus, atomism, and the philosophy of chemistry', *Bull. Hist. Chem.*, 1994, **15/16**, 45–54.

183. M. R. S. Creese and T. M. Creese, 'Charlotte Roberts and her textbook on stereochemistry', *Bull. Hist. Chem.*, 1994, **15/16**, 31–36.

184. Iu. Ia. Fialkov, 'The development of physical chemistry for nonaqueous solutions: periodization and fundamental trends', *Voprosy Istorii Estestvoznaniia Tekh. [USSR]*, 1986, **4**, 86–92; [in Russian, English abstract by A. J. Evans].

185. H. A. M. Snelders, 'J. H. van't Hoff's theory of solid solutions (1890)', *Tijdschrift Geschiedenis Geneeskunde, Natuurwetensch., Wiskunde Techn.*, 1985, **8** (2), 49–57 [in Dutch].

186. H. A. M. Snelders, 'J. H. van't Hoff's theorie van de verdunde oplossing en (1886)' (Van't Hoff's theory of diluted solutions), *Tijdschrift Geschiedenis Geneeskunde, Natuurwetensch., Wiskunde Techn.*, 1987, **10**, 2–19.

187. Iu. Ia. Fialkov and A. N. Zhitomirski, 'Paul Walden's role in the formation and development of the physical chemistry of nonaqueous solutions', *Latvijas PSR*

Zinatnu Akad. Vestis [USSR], 1987, **9**, 65–70; [in Russian, English abstract by R. Vilums].

188. M. Ciardi, 'Theory without experimentation: Amedeo Avogadro's research on the nature of metal salts', *Nuncius*, 1992, **7**, 160–193 [in Italian].

189. B. Pourprix and R. Locqueneux, 'Thomas Graham (1805–1869) et la diffusion, gazeuse et liquide: Une contribution au débat sur la structure de la matière', *Fundamenta Sci.*, 1985, **6**, 179–207.

190. T. Tachibana, 'A historical note on the colloid researches of Thomas Graham', *Kagakushi*, 1995, **22**, 1–14 [in Japanese].

191. J. T. Edsall, 'Isidor Traube: physical chemist, biochemist, colloid chemist and controversialist', *Proc. Am. Phil. Soc.*, 1985, **129**, 371–406.

192. A. G. Ede, 'Colloid chemistry in North America, 1900–1935: the neglected dimension', Ph.D. dissertation, University of Toronto, 1993; Univ. Microfilms order no. DANN 82946.

193. A. G. Ede, 'When is a tool not a tool? Understanding the role of laboratory equipment in the early colloidal chemistry laboratory', *Ambix*, 1993, **40**, 11–24.

194. A. G. Ede, 'Colloids and quantification: the ultracentrifuge and its transformation of colloid chemistry', *Ambix*, 1996, **43**, 32–45.

195. K. Suhnel, '80 Jahre Kolloidchemie: Leben und Werk Wolfgang Ostwalds', *NTM: Schriftenr. Gesch.*, 1989, **26** (2), 31–45.

196. G. B. Kauffman, *Alfred Werner: Founder of Coordination Chemistry*, Springer Verlag, Berlin and New York, 1966.

197. G. B. Kauffman, ed., 'Coordination chemistry: a century of progress', American Chemical Society Symposium Series, American Chemical Society, Washington DC, 1994, no. 565.

198. G. B. Kauffman, *Inorganic Coordination Compounds*, Heyden, London, 1981.

199. T. Yasui, T. Ama and G. B. Kauffman, 'The resolution of a completely inorganic coordination compound', *J. Chem. Educ.*, 1989, **66**, 1045–1048.

200. G. B. Kauffman, 'General historical survey to 1930', in *Comprehensive Coordination Chemistry*, ed. G. Wilkinson, Pergamon Press, Oxford, 1987, Vol. 1, pp.1–20.

201. G. B. Kauffman, G. S. Girolami and D. H. Busch, 'John C. Bailar, Jr. (1904–1991): father of coordination chemistry in the United States', *Coord. Chem. Rev.*, 1993, **128** (1–2), 1–48.

202. D. P. Mellor, 'The development of coordination chemistry in Australia', *Rec. Aust. Acad. Sci.*, 1975, **3**, 29–40.

203. W. H. Brock, 'The origin and dissemination of the term 'ligand' in chemistry', *Ambix*, 1981, **28**, 171–183.

204. W. H. Brock, 'Searching the literature to learn how the term 'ligand' became part of the English language', *J. Chem. Inf. Comp. Sci.*, 1982, **22**, 125–129.

205. G. B. Kauffman, 'Complex terminology', *Chem. Br.*, 1993, **29**, 867–868.

CHAPTER 7

Analytical Chemistry

JOHN A. HUDSON

Formerly Department of Chemistry, Anglia Polytechnic University

7.1 General Works

Those interested in the history of analytical chemistry owe an enormous debt of gratitude to Ferenc Szabadvary, whose comprehensive history of the subject first
appeared in Hungarian in 1960, and in English translation in 1966. The book is still
of considerable value, especially for the history of classical methods. It had been out
of print for many years, and therefore the appearance of a reprint in 1992 was a welcome event.[1a] Translations have also appeared in Russian and Japanese.[1b,c] A serial-
ized account describing the development of analytical chemistry from the latter part
of the eighteenth century until the mid-twentieth century has as been published in
German.[2] A general schematic overview of the historical evolution of analytical
chemistry has been given by J. A. Perez-Bustamente.[3] The same author has emphasized the strong link between the discovery of the elements and analytical chemistry;
a new analytical technique or an improvement in analytical methodology having on
several occasions resulted in the discovery of new elements.[4]

An article by Laitinen describes the emergence of analytical chemistry as an
independent scientific discipline.[5] Laitinen has also studied the history of analytical chemistry in the USA.[6] After referring to some of the individuals who were
responsible for advancing the discipline in the United States in the nineteenth century, such as S. M. Babcock, E. F. Smith, H. W. Wiley and E. Hart, he then describes
how, after the outbreak of World War I, American chemistry began to break away
from European dominance, as several academic centres noted for excellence in analytical chemistry emerged. The remainder of the article, while clearly having an
American focus, serves as a brief general history of analytical chemistry in the
twentieth century. It highlights the impact of World War II research programmes on
analytical chemistry, the development of the wide variety of instrumental methods
currently in use, and the impact of microcomputers on chemical analysis. Many
analytical techniques, especially those developed during the twentieth century, can
be made sufficiently sensitive to detect and measure constituents present only in

trace amounts. An article on the history of trace analysis refers to a wide variety of analytical methods.[7]

Several publications consider the development of analytical chemistry in specific countries or geographical regions: Austria,[8] Canada,[9] Spain,[10] Poland,[11] Japan,[12] Romania,[13] Estonia,[14] Spain in the 20th century,[15] and Denmark up to 1900.[16] An extensive series of articles has dealt with the development of analytical chemistry in various institutes and regions of the former Soviet Union.[17] A paper by E. Homburg discusses the rise of analytical chemistry in Germany over the period 1780–1860, arguing that a very significant precursor to the creation of a distinct chemical profession in that country was the development of analytical chemistry with its associated methods and laboratory skills. Homburg thereby challenges the traditional view that the rise of the scientifically trained professional chemist was largely due to the introduction of Lavoisier's antiphlogistic chemistry.[18]

The histories of some journals and societies devoted to analytical chemistry have been investigated. In the United States, the *Journal of Analytical Chemistry* was founded by E. Hart in 1887, and was merged with the *Journal of the American Chemical Society* in 1893. Another specialist journal was started in 1929 (the *Analytical Edition* of *Industrial and Engineering Chemistry*); in 1948 the name of this publication was changed to *Analytical Chemistry*.[19]

The Russian publication *Zhurnal Analiticheskoi Khimii*, which now appears in English as *Journal of Analytical Chemistry*, celebrated its 50th anniversary in 1996. Articles have appeared describing the *Zhurnal*'s creation and subsequent history,[20] and the development of analytical chemistry in Russia during the *Zhurnal*'s first half century.[21] The role of A. P. Vinogradov, the first editor of the *Zhurnal*, in the development of Russian analytical chemistry has been appraised.[22] The *Zhurnal* also celebrated the 50th anniversary of the Vernadsky Institute by reviewing its contributions to analytical chemistry.[23]

In 1974 the Society for Analytical Chemistry celebrated the centenary of its foundation; it would soon become part of the newly formed Royal Society of Chemistry, continuing within that Society as its Analytical Division. In 1999 a history of the first 25 years of the new Division was published. The book provides an important source for those interested in the recent history of analytical chemistry in the United Kingdom.[24] A collection of articles was published on the occasion of the centenary of the establishment in the United States of the Association of Official Analytical Chemists in 1984. The Association was originally devoted to agricultural chemistry, and was named accordingly.[25]

7.2 Classical Methods

Genuine chemical methods were preceded by techniques such as the touchstone,[26] and also by the direct investigation of the material or object by the use of the senses. A material could be felt, it could be tasted, its colour could be observed, it could be smelt, and the noise it could be induced to make could be listened to. The use of such methods of 'organoleptic' analysis in the work of the Georgian King, Vakhtang VI (1675–1737) have been described.[27] Probably the first demand for a quantitative analytical method arose from a desire to estimate the purity of samples of gold. While the touchstone is a non-chemical technique, the necessity to analyse for gold led to the development of the fire assay, which involved cupellation, and is probably the oldest technique of quantitative chemical analysis.[28,29] The ancient fire assay method is still used in tandem with modern instrumental techniques in the quantitative analysis of precious metals.[30] An article describing highlights in the history of quantitation in chemistry up to the early 19th century commences with the use of the touchstone, and then looks at the work of Boyle, Black and Dalton.[31]

One of the spurs to the development of qualitative analytical procedures during the eighteenth century was the need to analyse mineral waters, and an account has appeared of some of the attempts made by English physicians of the period.[32] As well as mineral waters, a succession of analyses was performed on water from the Dead Sea.[33] Improved procedures for water analysis were published by Bergman in 1778, and it was Bergman who also introduced new methods into quantitative gravimetric analysis. Previous methods had attempted to isolate the pure metal or other substance being estimated, but Bergman precipitated the substance being sought in an insoluble compound of known composition.[34] About a century after Bergman, organic reagents became available for the selective precipitation of some metals. Among the best known were α-nitroso-β-naphthol for cobalt and dimethylglyoxime for nickel. The history of organic reagents in various areas of analytical chemistry has been reviewed.[35]

Two of the most important tools in the early days of analysis by chemical means were the blowpipe and the precision balance. It has often been thought that the introduction of these tools was associated with the early modern development of mining and metallurgy, but it has been demonstrated that they were previously associated with alchemy.[36] The blowpipe was first employed routinely in the qualitative analysis of solids by Cronstedt in the eighteenth century, and its use was greatly extended through the work of Bergman, Gahn, and Berzelius. Several elements were discovered with the aid of the blowpipe. A version of the blowpipe that employed an oxyhydrogen flame was used by Edward Daniel Clarke, who in 1816 claimed to have decomposed alkaline earth oxides and isolated metals such as barium with its aid. Clarke's experimental work and the reception of his claims at the Royal Institution, where Davy had decomposed the materials electrolytically, have been discussed.[37]

Blowpipe apparatus was continuously improved up till the 1860s, when the method began to be replaced by spectral analysis. Blowpipe analysis is now a lost art, and an article outlines the procedures used and contains descriptions (with photographs) of some historic blowpipe kits employed for mineralogical analyses in the field.[38] Many examples of blowpipes have survived.[39–41] A biographical study focusing on the analytical work of one of Bergman's students, Johan Gadolin, has appeared.[42] The history of the assaying of iron ore in America over a 300-year period has been reviewed.[43]

In 1767 William Lewis published a titrimetric procedure to determine the purity of potash, consignments of which were beginning to be imported from the American colonies. His method involved measuring the titrant by weight, employing an indicator, and, most significantly, standardizing the acid against supposedly pure potash.[44] The analytical chemistry of ozone has been the focus of study. Schonbein first introduced the reactions of ozone with neutral potassium iodide and with indigo as qualitative tests. The potassium iodide reaction was subsequently investigated in more detail by Andrews and Tait in the late 1850s, and it is still employed quantitatively, as is the indigo reaction.[45]

An article on Marsh's test for arsenic describes how, in 1861, C. L. Bloxam realized that the chemical generation of arsine could be replaced by an electrolytic reduction. This method avoided the problems caused by the presence of arsenic impurity in the zinc and hydrochloric acid used in the original procedure. By the early 20th century it was possible to prepare reagents of sufficient purity for the original Marsh method to be used.[46] C. A. Krauch performed important work on the purity of reagents, which led to the first manufacture in 1888 of analytical grade reagents by Merck.[47,48]

The need of early chemical manufacturers for rapid methods of quantitative analysis spurred the development of titrimetric methods. C. J. Geoffroy is generally credited with the first description in 1729 of a titrimetric procedure, and subsequently important contributions were made by Lewis, Descroizilles, Gay-Lussac and Ostwald.[49] Bunsen developed several analytical procedures, including iodometric titrations and methods for the determination of industrial gases.[50] One of the names closely associated with the development of titrimetry is K. F. Mohr; his life and work have been reviewed.[51] The most significant twentieth century contribution to the range of volumetric procedures was that of G. Schwarzenbach in 1945, when he introduced complexometric titrations to estimate metals using aminopolycarboxylic acids (especially EDTA) as the reagent.[52] The first non-aqueous titrations were performed by H. Schiff in 1872 when he estimated amines dissolved in benzene by titration with a solution of an aldehyde in the same solvent. The subsequent development of non-aqueous titration methods has been described.[53]

In addition to his work on titrimetry, Gay-Lussac made important contributions in other areas of analytical chemistry. He studied the reactions of hydrogen sulphide in different media, thus laying the foundations for the classification of metals into different analytical groups, and in conjunction with Thenard, he developed the first truly practicable method for the determination of carbon and hydrogen in organic compounds.[54]

The history of acid-alkali indicators from the time of Robert Boyle to the present has been reviewed in an article that also discusses test papers and biological stains.[55]

A general review of the history of classical methods of quantitative analysis has appeared in a paper that focuses on the introduction of gravimetry and titrimetry into the United States.[56] One of the most prolific analytical chemists of the 20th century was I. M. Kolthoff, who had published 933 papers by the time of his 90th birthday in 1984. He worked extensively in volumetric and gravimetric analysis, and also made important contributions in electroanalytical chemistry.[57] Kolthoff died in 1993 in his 100th year. Published posthumously, a revised version of a paper, which Kolthoff had delivered at the centennial meeting of the American Chemical Society in 1976, discusses progress in analytical chemistry and the teaching of the subject in the first quarter of the twentieth century in the United States.[58]

7.3 Organic Analysis and Microanalysis

The first attempts to perform elementary analyses on organic compounds by combustion were performed by Lavoisier using various oils. The method was subsequently refined, principally by the introduction of a solid oxidant, by Gay-Lussac and Thenard, and by Berzelius. In 1820 William Prout designed a combustion apparatus which utilized a spirit lamp rather than the charcoal furnace employed by Berzelius, and, in 1827, he followed this with a completely different piece of apparatus. He reverted to combustion in oxygen and measured the volume of oxygen consumed.[59] However, four years later Liebig introduced his method which once again used a solid oxidant and weighed the carbon dioxide produced after absorption by caustic potash in the 'Kaliapparat'. Because of its relative speed and simplicity, this became the standard method. Around the same time Dumas determined the nitrogen content of organic compounds. Later in the nineteenth century, methods were introduced for the determination of halogens, sulphur, phosphorus, and of oxygen directly.[60–62] A comparative study has been published of the work of Liebig, Dumas and Berzelius over the period 1811–1837. This study focuses especially on the evolution of the Kaliapparat.[63] Liebig's apparatus has been reconstructed.[64]

Johan Kjeldahl first described his method for the determination of nitrogen in proteinaceous materials in 1883, when he was director of the Chemical Laboratory at the Carlsberg Foundation. At that time there was a pressing need for a more convenient and reliable procedure. Wanklyn's albuminoid ammonia method gave low and erratic results. Kjeldahl developed a succession of modifications that resulted in a technique very different from Wanklyn's – one which is still substantially the same today.[65,66]

An early worker in the field of microanalysis was F. Emich, but the introduction of micromethods was largely due to the work of F. Pregl in the early years of the twentieth century. Pregl scaled down the methods of Liebig and Dumas and developed reliable procedures that spread world-wide. He was honoured with the Nobel Prize in 1923, the first time the award had been made for achievements in analytical chemistry. The automation of C, H, N analyses was introduced in the early 1960s.[67]

In the early years of the 20th century, Otto Folin and Donald van Slyke made important contributions to clinical chemistry, especially in devising analytical methods for small volumes of body fluids. Neither had a medical degree, but they demonstrated that chemists could make important contributions to advances in medical diagnosis and the treatment of disease.[68] The period 1948–1960 has been described as a golden age in clinical chemistry, characterized by the introduction of many new instrumental analytical techniques. The period also saw a further improvement in the status of the clinical chemist and biochemist with the formation of professional bodies and the inauguration of new journals.[69] The author of these two articles has also produced a book outlining the development of clinical chemistry over four centuries.[70] A paper on the analysis of sugar describes the development of analytical methods and the steps taken towards their international standardization in the period 1869–1897.[71]

The name of Fritz Feigl will always be associated with spot tests. An article has traced his life and work, from his early years in Europe and his important association with Friedrich Emich, to his later years in Brazil.[72] The microscope has been an important instrument in the analytical laboratory since the early nineteenth century. An article has traced the history of chemical microscopy by surveying the important books that have appeared on the subject since 1827.[73]

7.4 Electroanalytical Methods

The history of electrochemistry received a welcome boost in 1988 when a symposium entitled 'Electrochemistry Past and Present' was held at the third Chemical Congress of North America at Toronto. The published proceedings contain papers describing the historical development of many electroanalytical techniques.[74] Kolthoff has given an account of the state of electroanalytical chemistry prior to World War II based on personal experience.[75]

Electrolytic deposition was used as a qualitative analytical technique in the early years of current electricity, but it was not until 1864 that quantitative electrochemical analysis commenced with the development of electrogravimetry by Wolcott Gibbs.[76,77] Electrolytic techniques of analysis were greatly refined by Edgar Fahs Smith at the University of Pennsylvania, who introduced the rotating anode and double-cup mercury cathode. Smith's book on electrochemical analysis ran to six editions.[78]

The most commonly performed electroanalytical procedure is pH measurement. The pH concept was introduced by Sørensen in 1909; the glass electrode itself evolved from the observations of M. Cremer in 1906 and of Haber and Z. K. Klemensiewicz in 1909.[79] Another account of the invention of the glass electrode focuses on the less well-known Klemensiewicz, and provides a brief outline of his life.[80] The high resistance of the glass electrode prevented its use in the potentiometric determination of pH for many years. Values of pH were estimated using indicators,[81] or potentiometrically using the hydrogen electrode. A new approach was introduced by Biilmann in 1920, who saturated the solution with quinhydrone, and then measured the potential of the resulting system using a bright platinum electrode.[82] When suitable electronics had been developed, the glass electrode could be employed in potentiometric pH determinations, and the pH meter was born.[83,84] The glass electrode sensitive to hydrogen ions was eventually joined by electrodes consisting of glass membranes of different compositions and sensitive to other cations, but the really rapid development in ion-selective electrodes came with the introduction of non-glass membranes.[85,86] The early commercialization of ion-selective electrodes has been described.[87]

One of the exciting developments associated with ion-selective electrodes has been the fabrication of microelectrodes capable of monitoring an intracellular ion concentration. The history of these developments from the mid-1950s has been reviewed.[88] A symposium held in 1996 was devoted to the history of ion-selective electrodes. One paper discussed their development and commercialization,[89] another described how the 1970s was the decade in which they really became established,[90] a third outlined their industrial applications,[91] and a fourth traced the evolution of blood chemistry analyses using them.[92] The first attempts to construct biochemical sensors by immobilizing enzymes on electrodes date from the 1960s.[93]

Polarography has been an extremely valuable electroanalytical method, although its importance has declined in recent years. The life and work of Jaroslav Heyrovsky, the inventor of the technique, was reviewed in the Toronto symposium,[94] and also on the occasion of the centenary of his birth.[95,96] The 75th anniversary of the discovery of polarography occurred in 1997. On that occasion the son of the inventor gave an account of the studies on the dropping mercury electrode and on the polarization curves of solid electrodes that predated the development of polarography proper.[97] The dropping electrode was in use for many years before the invention of polarography, and its history has been reviewed.[98] The development of polarographic instrumentation has been discussed,[99] and the rise and fall of the manufacture of commercial polarographs has been charted.[100] Polarography in turn spawned pulse and square-wave techniques.[101–103]

Polarography was first applied to the analysis of proteins in the 1930s and to the analysis of nucleic acids in the 1950s.[104] Oxygen concentration is measured by employing a special electrode for its reduction. The most important advance in the design of such electrodes was made by L. C. Clark in 1956.[105,106] Anodic stripping voltammetry was discovered fortuitously during experiments to isolate fission products and nuclides produced by neutron bombardment in the period immediately following World War II.[107]

Conductance measurements have long been applied in analytical situations, and an article has traced their history back to 1776. In that year, Cavendish compared the conductivities of solutions by comparing the shocks he received when equally charged Leyden jars were discharged through his body *via* tubes containing the solutions.[108] It was Friedrich Kohlrausch who over the period 1869–1880 devised a relatively simple method of measuring the conductance of solutions.[109] In 1895, Kohlrausch was the first to describe the changes in conductance that occurred when solutions of acids and bases are reacted, but he seems not to have used the phenomenon for analytical purposes. The first conductimetric titrations were performed a few years later by Ostwald.[110] Potentiometric titrations were first performed by Behrend in 1893, in a rare foray outside his normal field of organic chemistry.[111] Kolthoff worked on both conductimetric and potentiometric titrations, and published influential books on each.[112]

Not all techniques that at one time showed promise eventually achieved widespread application. An example is coulometric titrimetry, whose history has been discussed.[113] Reviews have also appeared of electroanalytical chemistry in molten salts[114] and of the early years of the relatively new technique of spectroelectrochemistry.[115] The development of electrochemical instrumentation from its origins, through the electronic age to the computer age has been discussed.[116]

7.5   Chromatography and Electrophoresis

Chromatography has had a punctuated development. M. S. Tswett is rightly regarded as the true inventor of chromatography, but he was preceded by several others whose work bears some relation to subsequent chromatographic techniques. Among these are F. F. Runge, F. Goppelsroeder, D. T. Day and C. Engler. Tswett started his research on chlorophylls, in which he utilized his adsorption columns around 1903, and his two classic papers were published in 1906.[117] There then followed a dormant period in which the method was very little used. This lasted until 1930, when E. Lederer applied the technique to the analysis of the yellow components of egg yolk. Since then the growth has been explosive, and a wide range of chromatography techniques are now in everyday use. By the close of the 20th century, chromatography in all its forms had become the most widely used separation method in chemistry and biochemistry. The evolution of liquid chromatography from the days of its precursors has been reviewed.[118–120] While few would argue with awarding the accolade of the leading *separation* technique of the twentieth century to chromatography, other methodologies might vie for being designated the leading *analytical* technique of the century. An article has pointed out that mass spectroscopy and nuclear magnetic resonance have strong claims, while infrared and atomic absorption spectroscopy trail behind somewhat.[121]

Two brief overviews of the history of the whole field of chromatography have been published.[122,123] In 1999 the magazine *LC-GC* instituted a regular column entitled *Milestones of Chromatography*. It is edited by Leslie S. Ettre, who has had a distinguished career as a chromatographer and has also made important contributions to the history of the technique. The articles contain information of interest both to current practitioners of chromatography and to historians of chemistry. The first focused on the 1949 *Faraday Society Discussion*, which set the scene for subsequent rapid developments in chromatography.[124] Other articles have dealt with the development of gas chromatographic instruments at Shimadzu,[125] preparative liquid chromatography and the Manhattan Project,[126] the development of accessories to adapt gas chromatographs for special tasks,[127] two important symposia held in the early days of HPLC,[128] the activities of A. A. Zhukhovitskii, a Russian pioneer of gas chromatography,[129] the evolution of capillary columns for gas chromatography,[130] the birth of partition chromatography,[131] and the development of thin-layer chromatography.[132]

Runge spotted coloured solutions on to filter paper, and he recognized that a separation of the dissolved components was occurring. He also treated some of his papers with a second reagent, thereby anticipating the idea of a spot test. However, in his subsequent work he turned his attention to making his 'self-grown pictures' more aesthetic. His life and work have been described in an article that contains some beautiful colour photographs of his pictures.[133] Goppelsroeder called his technique 'capillary analysis',[134,135] and the case has been made that he should be considered to be the true inventor of chromatography,[136] but Tswett's claim has been reasserted in a detailed examination of his work.[137] Tswett was, however, aware of Goppelsroeder's work, and he eventually realized that in both methods differential adsorption was occurring. Tswett foresaw that a planar version of chromatography was a possibility.[138] Tswett's original separations were achieved on columns of calcium carbonate with benzene as the mobile phase, and his work has been assessed in the light of a recreation of his experimental system.[139]

A biographical article on Tswett includes many previously unpublished photographs of him.[140] A project to publish his complete correspondence with his Swiss friends was initiated in the 1970s by I. M. Hais; letters to E. Claparède were among the earliest to be published.[141] Those to John Briquet, Tswett's supervisor at Geneva, have also been published. The first group consists of the letters Tswett wrote immediately after his move to Russia in 1896, and also contains Briquet's evaluation of Tswett's Ph.D. thesis.[142] The second group covers the period 1896–8. During this time, Tswett spent a few months unsuccessfully attempting to obtain a position at the University of Odessa, and then moved to St Petersburg, where he encountered similar problems. He had apparently been unaware that a foreign degree was not a sufficient qualification to obtain a senior position at a Russian university. Some of the letters written in this period reflect Tswett's frustration with the situation; he states that he does not feel a true Russian.[143] No letters written to Briquet after 1898 have survived, except for two from 1915 and 1917.[144] In 1898 Tswett was attempting to obtain a degree from the University of Yur'ev (present day Tartu, Estonia). His correspondence with N. I. Kuznetsov, professor of botany at the university, has been published.[145] In 1901, Tswett moved to Warsaw, and it was while he was there that he performed his chromatographic separations. Contemporary illustrations have been published of the two academic institutions at which Tswett worked over the period 1901–1915.[146]

Tswett's key papers were published in Russian or German. Four of the most important have now appeared in English translation, along with a biographical sketch of Tswett and a bibliography of around 500 references to earlier work relevant to the development of chromatography. Most of these are from the period 1840–1910.[147]

Tswett's new technique was taken up by a few others, now almost entirely forgotten. The first was Gottfried Kranzlin, who started work only a few weeks after Tswett's account was published in 1906. Kranzlin reported his results in an obscure journal, and his work remained unknown.[148] Chromatography was also used by Charles Dhere and his co-workers from 1911. They improved Tswett's apparatus, verified his results, and demonstrated that chromatography could achieve better purifications than other methods. Dhere also produced the first biography of Tswett.[149] The most important follower of Tswett was Leroy Sheldon Palmer, who was the first to practise chromatography in the United States. In 1922 he published a book on carotenoids which contained an account of chromatography, and this formed part of the chain that resulted in the adoption of the technique by E. Lederer.[150]

Lederer has given a personal account of his early experiments. In 1930 he arrived at Kuhn's laboratory in Heidelberg to commence his postdoctoral work. Kuhn suggested that the yellow lutein pigment of egg yolk was a mixture of zeaxanthine (which had been isolated from yellow corn) and xanthophyll (from green leaves). Lederer had read Palmer's book, and Kuhn possessed a privately produced German translation of the Russian book that Tswett had published. Using adsorption chromatography, Lederer was able to verify Kuhn's suggestion, and he also used the technique to separate α- and β-carotenes. Lederer was the first person to use adsorption chromatography on a preparative scale.[151,152] M. Lederer has discussed some documents relating to the history of capillary zone electrophoresis, which show that the earliest workers to use filter paper as a stabilizing medium in electrophoresis were D. von Klobusitzsky and P. Konig in Sao Paulo in 1937–39.[153] Lederer also relates how a version of chromatography that did not 'take off' was the use of papers impregnated with ion-exchange resins. He has described the history of the attempts, starting in the early 1950s, to make and use such papers, and suggests that the method might yet experience a revival.[154] Another important figure in chromatography in the 1930s was L. Zechmeister, whose book, written in conjunction with L. Cholnoky, contributed greatly to the revival of the technique.[155]

Liquid–liquid partition chromatography was invented by A. J. P. Martin and R. M. L. Synge in 1941. Martin had earlier devised a multistage solvent extraction process for the separation of vitamin E, although this work was not published. He and Synge then had the idea of holding one of the liquid phases on a stationary support in a column.[156] Liquid–liquid partition chromatography was not widely adopted until the paper chromatography variant was introduced in 1944 by Martin in conjunction with R. Consden and A. H. Gordon. The technique was soon applied to the separation of steroids, and played a major role in the development of steroid-based therapeutic agents.[157]

The history of thin-layer chromatography has been the subject of a book.[158] The first separations on thin layers were performed in 1938. A gas was first used as the mobile phase in adsorption chromatography by Erica Cremer in Innsbruck in 1946. Using hydrogen as carrier gas, she and her student, Fritz Prior, successfully separated air and carbon dioxide using charcoal as the adsorbent. A newly-opened branch of the Deutches Museum in Bonn, devoted to post 1945 developments, has a display featuring the work of Cremer and Prior, with a model of their original apparatus.[159]

Another early worker in this field was J. Janak in Czechoslovakia.[160] Before gas chromatography could come into general use, commercial instruments had to become available. This did not happen until gas–liquid partition chromatography had been invented. In their 1941 paper on liquid–liquid partition chromatography, Martin and Synge had pointed out that the mobile phase does not have to be a liquid but can be a vapour. The suggestion went unheeded, and eventually Martin, working with A. T. James, demonstrated the feasibility of GLC. Their first successful separation was of a mixture of aliphatic amines and employed a titration technique to detect the components as they emerged from the column. The first paper was published in 1952, and the technique was rapidly adopted by industrial research laboratories. James has given a personal account of this work.[161]

James and Martin soon developed a gas-density balance as a detector, but the sensitivity of GLC was increased enormously when the argon ionization detector was invented by James Lovelock; it was he who also produced the electron capture detector in conjunction with S. R. Lipsky in 1959.[162] The flame ionization detector originated in the same year.[163] Another important advance in the early days of GLC was the introduction of capillary columns, which were first used by M. J. E. Golay in 1956.[164,165] The development of some of the first commercial instrumentation has been described.[166,167]

In the years that partition chromatography was being developed, classical liquid chromatography had advanced with the introduction of ion-exchange resins as stationary phases.[168] These had been used in the 1940s in the separation of rare earth elements and nuclear fission products[169] and in 1958 the amino acid analyser was introduced. By the 1960s, attempts were being made to introduce instrumentation into liquid chromatography to produce the same kind of reliability and performance that was by now commonplace in GLC. The first paper on high-performance liquid chromatography was published by C. Horvath and Lipsky in 1966.[170] The new technique was initially applicable only to organic species. Its extension to inorganic ions in the early 1970s in the form of ion chromatography has been described.[171] Two articles have charted the birth and growth of high-performance liquid chromatography.[172,173]

Arne Tiselius announced his electrophoresis apparatus, which was enormous both in size and expense, in 1937. Electrophoresis was designed for the separation and investigation of proteins, and the technique has had its greatest impact in biology and biochemistry.[174] The development of electrophoretic methods after Tiselius has been outlined.[175–177] It was in Tiselius's laboratory that cross-linked dextrans were first prepared. In 1959, their use as the stationary phase in the chromatographic separation of substances on the basis of their molecular size was reported by J. Porath and E. Flodin. Pharmacia immediately started to market Sephadex, and gel filtration was rapidly taken up by biochemists.[178] A similar technique, for the separation of polymers using polystyrene gels, was developed by J. C. Moore in 1962. The theoretical basis of these separations, namely differential exclusion from the gel matrix, was verified in 1964 by T. C. Laurent and J. Killander.[179]

7.6 Spectroscopy

7.6.1 Atomic Spectroscopy

Line spectra were first observed by J. von Fraunhofer, D. Brewster, and J. F. W. Herschel in the 1820s.[180] In the ensuing decades a considerable amount of work was done on spectral phenomena prior to the demonstration by Bunsen and Kirchhoff in 1859 that line spectra could be used for qualitative chemical analysis. Accounts have appeared of the development of the spectroscope both prior and post Bunsen and Kirchhoff.[181–183] Significant observations were undoubtedly made prior to 1860 by Stokes, Stewart, Fox Talbot, and others. The priority claims of Stokes, who recorded his ideas in some private letters to William Thomson, have been examined.[184] The work of Bunsen and Kirchhoff did not owe a great deal to that of their predecessors, with the exception of the demonstration by W. Swan in 1856 that the almost omnipresent yellow line that coincided with Fraunhofer's dark solar D line was due to contamination by minute quantities of sodium salts.[185,186] Platinum played an important role in the early development of spectroscopy. The metal was widely used to support the material in the flame, since it did not colour the flame itself. Bunsen ensured the purity of all his samples for spectrum analysis by recrystallization (sometimes up to fourteen times) in platinum vessels, thereby preventing contamination by minute quantities of salts that could be leached from glass vessels.[187] Sharply contrasting views have been expressed about the failure of chemists prior to Bunsen to exploit spectroscopy.[188–190]

After Bunsen had detected and isolated caesium, spectroscopy was taken up with great enthusiasm by William Crookes, and this led to his detection and isolation of thallium in 1861.[191] Crookes' letters to Charles Hanson Greville Williams, who was also working with the spectroscope, and who felt he deserved some of the credit for the discovery of thallium, have been published.[192] The use of spectrochemistry in the search for hitherto unknown chemical elements in Britain over the period 1860–1869 has been described. It was perceived that, like Crookes, a scientist could make his reputation by discovering a new element. This resulted in several claims for the existence of new elements that later proved to be unfounded.[193] Once Kirchhoff had established beyond doubt that the dark Fraunhofer lines were caused by the same element that caused emission lines of identical wavelengths, the way was open for the chemical analysis of the atmosphere of the sun and stars. This was a process which had been declared to be an impossibility by Auguste Comte less than 30 years previously.[194]

Although Bunsen and Kirchhoff showed that both the absorption and emission spectra of an element could be used for its identification, the emission technique remained the only one in use for many years. When atomic spectroscopy was adapted for quantitative analysis, emission was again employed. The plant physiologist H. Lundegårdh inaugurated the modern era of flame emission spectrometry (flame photometry) in the 1920s. He sprayed the nebulized solution into a premixed air–acetylene flame and recorded the intensities of the emission lines photographically.[195,196] Other techniques included spark excitation, which was especially suitable for solid samples in the metallurgical industries. Photographic detection was replaced by photocells or photomultipliers, and direct reading spectrometers provided simultaneous multielement analyses at high speed.

In 1952, Walsh in Australia realized the inherent advantages of atomic absorption spectroscopy over methods based on flame emission for quantitative analysis, and he has given a personal account of the development of the technique.[197] Walsh's death in 1998 resulted in a memorial issue of the journal *Spectrochimica Acta (B)*. As well as a brief biography of Walsh and a list of his publications, this contained 22 papers on all aspects of the history of atomic absorption spectroscopy. Together they constitute a valuable record of the birth of this important technique, the difficulties of bringing satisfactory instruments to market, and the history of the application of the method to quantify metals in a wide variety of materials and environments.[198]

Walsh worked at CSIRO, and an account has been given of the various contributions made to spectroscopy by its Chemical Physics Division over the period 1944–1986, with atomic absorption spectroscopy featuring most prominently.[199] Walsh published his classic paper in 1955, and almost simultaneously the Dutch physicist C. Th. J. Alkemade published an account of an absorption flame photometer using essentially the same principles as the Walsh instrument. Walsh has rightly been given the credit for the development of atomic absorption spectroscopy, as it was he who campaigned for the acceptance of the new technique. Although he discussed the absorption technique, Alkemade's work was mainly concerned with the fundamentals of emission flame photometry.[200] Acceptance of atomic absorption spectroscopy only came with the demonstration of new analytical applications (*e.g.* direct determination of lead in gasoline) and the commercial production of instruments.[201–203] Another stimulus to the eventual acceptance of the technique was its use in a reliable method of determining calcium and magnesium in blood serum.[204] The evolution of graphite furnace atomic absorption spectroscopy has been described.[205]

Atomic absorption remained the technique of choice until relatively recently. However, with the introduction of plasma sources, atomic emission, in the form of inductively coupled plasma spectroscopy, has made a comeback. This development is now receiving historical attention, and was the subject of a symposium held in 1999. Papers discussed atomic emission analysis prior to 1950,[206] the fact that emission techniques developed continuously, even in the period when absorption methods were dominant,[207] and the development of the plasma sources on which the new techniques depend.[208] Also discussed was the powerful hyphenated technique of ICP-MS,[209] and the history of one of the leading manufacturers of atomic emission instruments.[210]

7.6.2   Visible and Ultraviolet Absorption Spectroscopy

The simplest technique of quantitative absorption spectroscopy to estimate a solute concentration is that of colorimetry. In the earliest methods, the colours were matched visually. In 1684–5, Robert Boyle attempted to estimate the iron content of Tunbridge water by visual colour matching, using gall-nut extract as his colour reagent for iron.[211] Simple instruments to assist in the colour matching process (colorimeters) go back to 1827 with a publication of J. J. Houtton de la Billardière.[212,213] The most famous of the early colorimeters was first described by Jules Duboscq in 1868. This utilized the fact that absorbance is proportional not only to concentration but also to pathlength, and the concentrations of two solutions of the same solute could be compared by finding the ratio of the two pathlengths that gave identical shades of colour.[214] The theoretical foundation of quantitative analysis by light absorption is usually attributed to A. Beer. However, Beer's paper of 1852 does not contain an absorption law in which concentration appears as an explicit variable. It was Bunsen and Roscoe in 1857 who first used measurements of absorbance to determine an unknown concentration.[215] Commencing around 1900, Otto Folin devised many colorimetric procedures for the analysis of the constituents of urine for use in clinical laboratories.[216]

The development of photodetectors enabled the human eye to be replaced by a much more sensitive detector of light intensity. The evolution of modern colorimeters and of spectrophotometers capable of operation in both the ultraviolet and visible regions of the spectrum has been discussed.[217,218] The phenomenon of fluorescence was first employed for quantitative analysis in the 1930s, when the first filter fluorimeters were constructed. An article has outlined the development of fluorescence analysis up to 1980.[219] Lasers have now been employed long enough in analytical chemistry for a historical account to be given.[220]

7.6.3   Infrared and Raman Spectroscopy

A comprehensive historical review of the analytical applications of infrared spectroscopy from the first experiments to the introduction of FTIR spectrometers has appeared.[221] The first study of the absorption of infrared radiation by a range of chemical substances was made in 1881 by Abney and Festing, after the former had developed a photographic method of detecting radiation in the near-infrared region. Over the next 25 years a number of other studies were made. This early phase culminated in the work of W. W. Coblentz in the United States, which was published in 1905.[222] It became evident from Coblentz's data that infrared spectra were related to molecular structure, but IR spectroscopy remained principally the province of researchers in university physics departments until World War II.

Chemists only began to use the technique routinely when commercial IR spectrometers had been developed.[223–226] Three wartime research programmes created the initial demand, and provided the impetus for the production of commercial instruments. These were the US synthetic rubber programme,[227] the British project to identify hydrocarbons in fuels from enemy aircraft,[228] and the joint British–US penicillin programme. The mineral oil mull technique for obtaining the IR spectrum of a powdered solid was first employed in 1942.[229] The availability from the late 1950s of low cost IR spectrophotometers designed for the bench chemist removed the technique from the realm of the specialist.[230] The adoption of infrared spectroscopy by chemists has been considered as a case-study of technological innovation in science.[231] Today, dispersive IR spectrophotometers are being replaced by FTIR machines, and the evolution of these instruments has been described.[232,233] The development of commercial instrumentation over a 50 year period in both dispersive and FT infrared, and in Raman spectroscopy, has been described.[234] The centenary of the birth of C. V. Raman occurred in 1988, and several publications have appeared describing his life and work, and in particular his discovery of the Raman effect and the evolution of Raman spectroscopy.[235–239]

7.6.4 Mass Spectroscopy and Nuclear Magnetic Resonance

Prior to World War II, the mass spectrometer was a research tool. The few instruments in use were built in research laboratories and were used in the determination of the relative abundances of isotopes or in the study of the dissociation and ionization of small molecules after electron impact. The mass spectrometer demonstrated its application in analytical situations during the war. Over 100 instruments were constructed to monitor the process stream in the UF_6 diffusion plant built for the Manhattan Project.[240] At the same time, mass spectrometers were first employed by the petroleum industry to analyse complex hydrocarbon mixtures. This stimulated research on fragmentation and rearrangement of organic species in the mass spectrometer,[241] and as a result mass spectroscopy become a powerful technique of structure determination. Accounts have appeared by some of those who worked with the first commercial mass spectrometers shortly after World War II.[242–244] The quadrupole mass spectrometer originated in work performed in the 1950s, and from the late 1960s these instruments have been used as detectors on gas chromatographs.[245]

The fact that certain nuclei possess spin and a magnetic moment was confirmed in the early 1930s. The first observation of nuclear magnetic resonance was made by I. I. Rabi in 1939 using molecular beams. The first successful demonstration of NMR in bulk matter was published in 1946 by two independent groups led by F. Bloch and E. Purcell. By 1950 it was realized that variations in the signal observed from the same atom in different environments were due to differing electron densities around the nuclei. This was initially an annoyance to the physicists, but it provided the key to the application of NMR to chemistry. Commercial instruments were available from the mid-1950s, and since then NMR has found very wide application. The history of NMR, from the first observations to the present has been discussed,[246–248] and the development of some of the instrumentation has been described.[249–251] A brief history of the technique has appeared, which focuses on developments in Germany, where one of the pioneers, F. Bloch, received his doctorate as a student of Heisenberg.[252] A short note has discussed the origin of the term *chemical shift*.[253] An account has been given of the history of NMR in Australia.[254] As has happened with IR, there has been a Fourier transform revolution in NMR, and its origins can be traced back to 1966.[255] The early development of NMR at Harvard University within the experimental tradition of molecular spectroscopy has been investigated.[256]

The phenomenon of NMR, initially discovered by physicists and exploited by chemists,[257] was soon applied to biology and medicine. Nuclei such as ^{31}P proved especially useful. These developments were greatly aided by the development of high field, high resolution FT spectrometers in the 1960s. In a different approach, using a magnetic field gradient, each proton responds with its own NMR frequency determined by its position. This is the basis of magnetic resonance imaging, now widely employed in medicine. The first NMR image, of two test tubes of water, was obtained by P. C. Lauterbur in 1973.[258] A symposium held to honour the 65th birthday of Oleg Jardetsky, one of the pioneers of biological NMR, included historical papers by Jardetsky himself and two other early workers in the field, M. Cohn and R. G. Shulman.[259–261]

7.7   Thermal Methods of Analysis

A prerequisite of thermal methods is the attainment and measurement of high temperatures. Thus, of particular importance was the development of the thermocouple into an accurate temperature measuring device by H. Le Chatelier in 1886.[262] The prehistory of thermogravimetry goes back to the relatively crude experiments carried out by Bryan Higgins in 1780 when he studied the weight changes induced by heating chalk and limestone to different temperatures. Two years later, Josiah Wedgwood performed similar experiments on clays. In 1905, after the development of the microbalance, O. Brill found that the alkaline earth carbonates decomposed at different temperatures, and he used this in the analysis of them in admixture.[263] The study of the weight loss of a substance as a function of time when heated at a constant temperature was initiated by J. B. Hannay in 1877. He was able to propose the existence of several new hydrates. However, the material was not weighed directly; the weight loss was deduced from the mass of water expelled.[264]

In 1915, K. Honda in Japan described the first true thermobalance, in which the substance under investigation was weighed continuously while subjected to a programmed temperature rise.[265] Modifications to this apparatus were introduced in the 1920s by Honda's co-workers.[266] A much earlier device, although not strictly a thermobalance, was introduced in 1833 by Talabot for use in the silk industry in Lyons. Silk usually contains around 8% water, and may absorb up to 15% without appearing wet. To avoid fraudulent practice amongst dealers, Talabot invented an apparatus to dry the silk to constant weight in a constant temperature oven.[267] This apparatus was further refined in 1853 and was known as the Talabot–Persoz–Rogeat desiccator.[268,269] It is possibile that one or more of these may have been acquired by the silk conditioning houses established in Manchester and London. An article by Keattch on these establishments has revealed that the consulting analytical chemist F. Crace-Calvert was operating a conditioning house in Manchester from 1858.[270]

The first measurements of temperature as a function of time during a cooling or heating process were made by J. F. E. Rudberg in Sweden in 1829. Other early workers were M. L. Frankenheim (1837) and H. Le Chatelier (1883 and 1887), both of whom seem to have been unaware of the earlier work.[271] Le Chatelier was followed by W. C. Roberts-Austin, who initiated differential thermal analysis in 1899. The development of this technique, from its introduction to the 1970s, has been discussed.[272] Hungarian work in thermal analysis over the period 1950–1990 has been described.[273]

7.8 Analytical Laboratories

When Liebig extended his Chemical Institute in Giessen in 1839, he added a laboratory mainly devoted to analytical work. The Institute is now a museum, and a new guidebook has a modern photograph of the analytical laboratory which nicely complements the famous sketch of the same scene made in 1840.[274] Liebig has been the subject of a major biography.[275]

The Excise Laboratory, the precursor of the Laboratory of the Government Chemist, was set up in London in 1842. At that time tobacco was being mixed with a wide range of substances after importation but prior to sale. This added weight to the tobacco and therefore there was a discrepancy between the amount of tobacco that was sold and the amount on which import duty had been paid. The Excise Laboratory was thus founded to protect the government's revenue, not to safeguard the consumer.[276] George Phillips, Head of the Laboratory (and for a while its only member of staff), was soon asked to analyse other adulterated substances such as pepper, beer, and coffee. By the end of the nineteenth century, the laboratory was performing work on the effect of substances on the environment and the population, both by carrying out analyses and by providing personnel to serve on committees. Topics investigated included lead in pottery glazes, phosphorus in matches, and preservatives and colouring matters in food. The Laboratory's history has been studied.[277–279]

In the early days of the chemical industry, manufacturers were reluctant to set up laboratories, but towards the end of the nineteenth century firms started to build laboratories both for analysis and research.[280,281] Analysis also played a role in the development of the laboratory in the early pharmaceutical industry.[282] The history and development of the analytical and quality control laboratories at the Merck company in Darmstadt has been examined.[283] The changing role of the chemist in the British alkali industry has also been investigated.[284,285]

In 1791 A. F. Fourcroy proposed that laboratories should be established in hospitals so that 'excretions, urine and various discharges of the sick' could be subjected to chemical analysis. Some laboratories were established, but were mainly devoted to research and teaching. Such diagnostic tests as were available were carried out at the bedside until the mid-nineteenth century, after which they began to be performed in purpose-built clinical laboratories. The origin of such laboratories, with special emphasis on German examples, has been described.[286] Early examples were at Berlin, Würzburg, and Vienna.[287] A photographic record of analytical instruments used in clinical laboratories at various periods has been produced.[288] The twentieth century has seen the establishment of large numbers of clinical laboratories, and it is estimated that there are now 100 000 worldwide. Testing for glucose in urine has been an important clinical application of chemical analysis, and the role of analytical chemistry in the conquest of diabetes has been outlined.[289]

Portable laboratories were manufactured from the late eighteenth century onwards, and these cabinets of apparatus and chemicals enabled mineralogists to perform simple analytical operations in the field. However, they were soon acting as a source of both amusement and instruction, and are the forerunners of both modern chemistry sets and home experiment kits. Their history has been reviewed.[290] One of the first manuals of qualitative analysis was published by J. A. Goettling in 1790 to accompany the portable laboratory that he was offering for sale.[291]

7.9   Instrumentation

Broader issues arising from the literature on the history of chemical instruments and apparatus are considered in Chapter 9. In this chapter, instrumentation will be considered in relation to the history of chemical analysis. While classical qualitative analysis relies mainly on powers of observation, quantitative analysis, both classical and modern, implies measurement, and for this some kind of measuring instrument is required. The equal arm two-pan balance was probably in use in Egypt as early as 5000 BC, and the evolution of the chemical balance from its more primitive forbears has been considered briefly.[292] H. R. Jenemann published in 1997, in German–English parallel text form, an illustrated history of the balance from ancient Egypt to the present day.[293] A brief history of the balance in society points out that it was used for millennia before it became a precision instrument and that initially the weights it measured were indicators of value rather than of mass. It early achieved a metaphorical significance, being used to symbolize a comparison of ethical values. Later it became the symbol of justice, and various organizations have used a representation of the balance to demonstrate their sincerity and accuracy.[294] To get a balance to read reasonably quickly some method of damping is necessary. A survey of damping methods has appeared, ranging from the manual restraint of the beam practised by the ancient Egyptians to the fast reacting nullification of eddy signals employed in modern digital balances.[295] A note in the *Journal of Chemical Education* surveys the papers on the history of the balance that appeared in previous issues of that publication.[296]

As modern scientific instruments become ever more sophisticated, concealing their electronics within unexciting black (or grey) boxes, older instruments become all the more fascinating. An early instrument used in the analysis of solutions was the hydrometer, or in Lavoisier's term 'the chemist's balance for fluids'. The various forms of this instrument and some of its applications over the period 1770–1810 have been discussed.[297] While most designs were produced to determine the concentrations of alcoholic solutions, variants were marketed for use with a wide range of other liquids. Among these instruments was the urinometer, and an early version was described by William Prout.[298]

A paper by J. T. Stock refers to a number of historical analytical instruments, including balances, an automatic precipitate washer for gravimetric analysis, and automatic carbon dioxide monitors for flue gases.[299] Stock has also examined some of the apparatus invented by Bunsen, apart from his eponymous burner, some of which, such as the constant level water bath and the water operated suction pump, would have found employment in analytical work.[300] An article on the apparatus of Edward Frankland refers to his method of determining carbon and nitrogen in organic compounds. Frankland used this method on the residues obtained by evaporating samples of water intended for domestic use, and from the results he calculated the previous sewage contamination to which the water had been subjected.[301]

Analytical instruments increased enormously in their complexity during the twentieth century, and this had a fundamental effect on the practice of analytical chemistry.

An analytical instrumentation industry grew to satisfy the new demand, and the industry, in turn, advanced the use of the new instrumentation.[302] Since 1947, the Pittsburgh Conference has provided a meeting point for those involved in developing and acquiring analytical instrumentation, and has thereby assisted both in the spread and further development of new instruments.[303] One author has claimed that the transition to the new analytical chemistry represents a discontinuity no less striking than that from the horse to the locomotive.[304] Another makes a similar claim from an examination of the changes in analytical chemistry over the period 1920–1950 and points out that not only have new scientific instrument manufacturing companies emerged but a new level of capital expenditure is now necessary to practise analytical chemistry. The author maintains that such changes have been widespread throughout the natural sciences, and he claims the changes can be identified as the fourth big scientific revolution.[305]

Automated analysis, which is common in today's laboratories and manufacturing plants, has a long history. An automatic device for the repeated washing of a gravimetric precipitate was described over 100 years ago, as was a method for continuously raising and lowering the gas–liquid interface in the tubes on an Orsat gas analyser.[306] The monitoring of carbon dioxide in flue gases provides an early industrial example of automated analysis. The first patent was taken out in 1893, and the analysis enabled furnaces to be controlled in such a way as to utilize the fuel as economically as possible.[307] An important development in automated analysis occurred in clinical laboratories, where the large number of samples being handled caused L. Skeggs in 1951 to produce a continuous flow system for the analysis of blood samples. This was subsequently marketed as the AutoAnalyser.[308]

A book dealing with instruments that have been developed by a community connected to both science and industry (termed the 'research-technology community') has chapters on Fourier-transform spectrometers and liquid scintillation counters.[309,310] Precision scientific instruments were produced by specialist manufacturers; the growth of the instruments industry in Britain and France between 1870 and 1939 has been the subject of a book.[311] Of the analytical instruments developed in the 20th century, the most widely used has been the pH meter, and its invention and marketing by A. O. Beckman has been described.[312]

7.10 Industrial Analysis

The use of chemical analysis to monitor the quality of the raw materials or finished products of industrial processes goes back a long way.[313,314] Indeed, some techniques owe their development to the need of industry for rapid analytical techniques. However, analytical methods are now often intimately bound up with the production itself, and supply much of the information required for the control and regulation of the process.[315] A good example of a continuous monitoring technique that can be used in process control is that of electrodeless conductivity measurement; its history has been described.[316] A history of early industrial pH measurement and control systems has been given.[317]

A major history of the British chemical industry focuses on the industry's impact on the environment. Although the book is not primarily concerned with chemical analysis, there are several interesting references. Among these is an account of the scandal surrounding the analysis of chemical manures in the latter part of the nineteenth century. The manures were sold on the basis of analytical figures for nitrogen, phosphorus and potash content, and depending on the results an analyst was known to provide he was referred to as being 'high' or 'low'. Concern about the situation led to the publication of official methods of analysis, which the commercial analysts eventually had no choice but to adopt.[318] Accounts have appeared of the history of some individual industrial analyses, including the analysis of fertilizers[319] and animal feeds.[320] The various methods used to measure the fat content of milk over the last 100 years have been reviewed.[321] Polarimetry has found its principal analytical application in sugar refineries, breweries and in the food industry in general. Although polarimetry was developed as a quantitative tool by Biot commencing in 1815, the first relevant observations were made by F. Arago in 1811.[322]

7.11 Analytical Education

One way to study the history of analytical education, and indeed to chart the development of analytical chemistry itself, is to survey the textbooks used and the lecture courses attended by former generations of students. The syllabi of the chemistry lecture courses of P. Shaw (1734), R. Watson (1771), H. Moyes (1784 and 1786) and W. K. Sullivan (1856) have been examined. It emerges that in the eighteenth century analytical chemistry was regarded as an integral part of chemistry, but by the nineteenth century it was treated as a specialism within the overall syllabus.[323] Thomas Thomson established the first course in practical chemistry at a British University in Glasgow in 1818. Much of the experimental work was analytical in nature, and an insight into the course has been provided by the discovery of the laboratory notebook of Walter Crum, a student in the first class of 1818.[324] Qualitative organic analysis was first introduced into the undergraduate curriculum at MIT by Samuel Parsons Mulliken around 1895. It subsequently became an important component of undergraduate chemical education in all universities. Mulliken went on to develop a more extensive scheme of analysis appropriate for professional analysts, while the book published by Shriner and Fuson in 1935 became the standard undergraduate text in the United States.[325]

In the nineteenth century, C. R. Fresenius published influential books on both qualitative and quantitative analysis, and their style and content has been discussed.[326] Attention has been drawn to the role of Fresenius in the rationalization and systematization of chemical analysis.[327] 1992 marked the 150th anniversary of the publication of Fresenius' book on qualitative analysis and the 100th anniversary of appearance of F. P. Treadwell's book. These works have been compared.[328]

The *Textbook of Quantitative Inorganic Analysis* by Kolthoff and E. B. Sandell, published in 1936, did much to establish analytical chemistry as a separate academic discipline in the United States.[329] The dramatic changes in analytical chemistry in the twentieth century, and especially after World War II, have had a major impact on university and college courses. The changing nature of such courses in the United States, both introductory and advanced, has been traced by surveying the contents of various analytical textbooks over a fifty year period.[330] A similar study looks at seven successive editions of the same analytical textbook over the period 1948–1988, thereby charting the rise and fall of a number of instrumental methods. The first edition dealt with techniques such as colorimetry and classical electroanalytical methods, but also included mass spectroscopy. Chromatography did not feature at all, but merited three chapters in the sixth edition. The seventh edition gave special attention to inductively coupled plasma spectroscopy, but optical rotatory dispersion, circular dichroism and polarimetry, all of which had been present in the fifth and sixth editions, had disappeared.[331] A strong case has been made that classical quantitative methods should continue to be taught in the modern era of analytical instrumentation.[332]

In the United Kingdom, analytical chemistry as a separate discipline was relatively neglected in the nineteenth century, although a revival occurred at a few universities in the middle decades of the 20th century.[333] Another author, after surveying developments in recent decades, concludes that the teaching of analytical chemistry remains superficial in many UK chemistry programmes, and suggests how courses might develop in the future.[334]

7.12   The Social Dimension of Analytical Chemistry

A book describing the development of chemistry in Britain over the period 1760–1820 in relation to the contemporary social context makes repeated reference to analytical chemistry. One of the themes developed is that chemistry in general, and analytical chemistry in particular, was seen as a means to social improvement through its applications in agricultural chemistry and mineral analysis. The availability and relative simplicity of much of the apparatus (*e.g.* the portable laboratories referred to earlier) meant that the appropriate analyses could be widely performed. Chemical analysis also held out the prospect of advances in medicine by applying both simple techniques and the latest technology, especially the voltaic pile, to attempts to analyse body fluids.[335]

Consulting analysts played a major role in the Victorian period. A biography of Edward Frankland has emphasized how extensive was his range of consultancy work, much of it analytical in nature. Early in his career, while at Manchester, he was analysing coal and coal gas, fertilizers, water, linseed oil, gypsum, gold mud and potassium chloride, among many other substances. Later in London he continued his analytical consultancy work, and after retirement he performed a huge number of water analyses in his private laboratory. All this work helped Frankland to amass a considerable fortune.[336]

Analytical chemistry, or at least the results of chemical analyses, probably impinge on the public consciousness more than most other aspects of chemistry. A symposium held in 1999 on the interaction between analytical chemistry and the law contained three historical papers. The first concentrated mainly on the British and Irish contexts,[337] the second, by the grandson of C. R. Fresenius, compared the present-day position of the analyst as expert with that obtaining in his grandfather's day,[338] and the third discussed the development of expertise in forensic chemical analysis illustrated by case studies from the Viennese Institute.[339]

Forensic evidence in a court case, or a drugs test on an athlete, is often front page news. A collection of essays on various aspects of the history of forensic science has appeared.[340] An overview of forensic chemistry in nineteenth century Britain describes how in 1820 toxicology and forensic medicine were much more advanced in some European countries. Better training for doctors was brought in, with Thomas Stevenson amongst those helping to establish the profession of Public Analyst.[341] Analytical chemistry sometimes creeps into fictional writing. Balzac's novel *Cesar Birotteau*, published 1833, contains a scene in which the chemist Vauquelin explains to the fictional character Birotteau the results of his experiments on the chemical analysis of human hair. The account is a good popular exposition of the main findings in Vauquelin's paper on the subject published in 1806.[342] Arthur Conan Doyle's fiction was certainly influenced by his medical education and Marsh's test for arsenic is probably just as well known from the account in Dorothy L. Sayers' *Strong Poison* as from any chemistry textbook.[343]

Chemists were analysing mineral waters in the eighteenth century, but in the nineteenth their attention turned to public water supplies after John Simon had demonstrated that some of the epidemics that had swept through industrial cities were water-borne. Bitter disputes arose about what in water might be dangerous. It was thought by many that the presence of organic matter indicated harmful contamination, but there were disagreements as to how organic matter might best be analysed. In 1865, W. A. Miller devised the potassium permanganate process that measured the absorption of oxygen by 'putrefying matter'. In 1867, J. A. Wanklyn introduced a method that distilled and measured the 'albuminoid ammonia'. In 1868, Edward Frankland, who served as London's official water analyst from 1865 to 1876, described his combustion process by which the residue from the water was oxidized with copper oxide and the resulting quantities of carbon dioxide and nitrogen were measured. From the ratio of carbon and nitrogen he calculated an estimate of 'previous sewage contamination'. The higher the estimate, the greater the likelihood of the water containing invisible 'germs'. It was not until the 1890s that the question of water contamination passed, at least for a time, mainly from the realm of chemistry to that of bacteriology.[344–346]

Victorian chemists had more success in measuring some of the components of air pollution. The Alkali Acts, which had the aim of reducing hydrogen chloride emissions from factories employing the Leblanc process, became law in 1863. The first Alkali Inspector was Angus Smith, who undertook a wide range of research projects. His book, containing chemical analyses of air and rain sampled at locations throughout the British Isles, was published in 1872. Not surprisingly, rain sampled in industrial locations was found to be significantly higher in chloride, sulphate and ammonia than that from rural locations.[347] The moral basis of Smith's work has been discussed.[348]

Another cause of concern in the nineteenth century was food adulteration, especially the addition of colouring materials. In 1851 *The Lancet* set up the Analytical Sanitary Commission, whose main members were Arthur Hassall and Henry Letheby. The reports of the Commission showed that food was indeed being contaminated by poisonous additives. These findings led to the first Food and Drugs Act of 1860, and from the 1870s Local Authorities started to appoint public analysts.[349]

Chemical analysis thus provides the foundation on which rests the sub-discipline of environmental chemistry. A discussion on themes in the history of environmental chemistry points out that features frequently encountered today have a long history. The issues discussed are characterized as 'expert disagreement', 'glorious vistas and dire threats', and 'satisfactory authority'.[350] A study of the history of analytical chemistry reveals how chemists have always been seeking to detect and quantify ever smaller amounts of analytes. As a result of the ability to measure lower concentrations, regulatory bodies have progressively lowered the permitted levels of contaminants. After mass spectrometers were linked to gas chromatographs, it became possible to identify and quantify very low levels of pollutants, and eventually the US Environmental Protection Agency specified the technique as a mandated method. As a result GC-MS became widely adopted.[351] James Lovelock makes the point that the unique sensitivity of the electron capture detector to estimate certain types of compound has had a profound impact on green politics.[352] A European Union directive for drinking water quality specifies that the level of any pesticide should not exceed one-tenth of a part per billion, and this has now been incorporated into British law.[353] This regulation applies to any pesticide, irrespective of toxicity, and the wisdom of such a stipulation is currently questioned by many.

Analytical chemistry probably continues to grow at a faster rate than any other branch of the subject, and the number of papers published annually in *The Analyst* has doubled within the last 25 years. Historical studies of analytical chemistry are also appearing in increasing numbers, and many are concerned with techniques that were developed relatively recently. The title of a recent RSC-sponsored lecture for school students on the current direction of analytical chemistry was entitled 'Finding Out More and More about Less and Less'. The present state of the history of analytical chemistry could be described as 'Finding Out More and More about More and More'.

References

1. F. Szabadvary, *History of Analytical Chemistry*, (a) Gordon and Breach, Reading, Berks, 1992; (b) Mir, Moscow, 1984; (c) Uchida Rokakuho Publishing Co., Tokyo, 1988.
2. H. Kelker, *GIT Fachz. Lab.*, 1990, **34**, 25, 773, 882; 1991, **35**, 305, 471, 576, 797; 1992, **36**, 454; 1993, **37**, 328.
3. J. A. Perez-Bustamente, *Fresenius' J. Anal. Chem.*, 1997, **357**, 151.
4. J. A. Perez-Bustamente, *Fresenius' J. Anal. Chem.*, 1997, **357**, 162.
5. H. A. Laitinen, in *Euroanalysis IV; Reviews on Analytical Chemistry*, ed. L. Niinisto, Akadémiai Kiadó, Budapest, 1982, p. 15.
6. H. A. Laitinen, *Talanta*, 1989, **36**, 1.
7. H. A. Laitinen, *J. Res. Natl. Bur. Stand. (U.S.A.)*, 1988, **93**, 175.
8. H. Malissa, *Mikrochim. Acta*, 1991, **2**, 3.
9. W. A. E. McBryde, *Can. Chem. News*, 1990, **42** (7), 21.
10. S. Arribas Jimeno, *Introducción a la História de la Química Analítica en España*, Univ. Oveido, Oveido, 1985.
11. L. Górski, in *Euroanalysis V; Reviews on Analytical Chemistry*, ed. A. Hulanicki, Akadémiai Kiadó, Budapest, 1986, p. 9.
12. S. Fujiwara, *Anal. Sci.*, 1985, **1**, 99.
13. G.-L. Radu and G.-E. Baiulescu, *Fresenius' J. Anal. Chem.*, 1997, **357**, 189.
14. H. Hödrejärv, *Fresenius' J. Anal. Chem.*, 1997, **357**, 191.
15. J. M. C. Pavon, *Quim. Anal.*, 1996, **15**, 99; J.M.C. Pavon and A. Garcia de Torres, *Quim, Anal.*, 1997, **16**, 143.

16. H. T. Nielsen, *Dan. Khemi*, 2000, **81**, 20.
17. E. S. Boichinova, *J. Anal. Chem.*, 1992, **47**, 129; A. N. Tananaeva and Yu. M. Polezhaev, *ibid.*, 133; S. I. Gusev, V. P. Zhvopistev, E. G. Bondareva and I. Petrov, *ibid.*, 138; G. K. Budnikov, V. F. Toropova, L. A. Anisimova and O. Yu. Timofeeva, *ibid.*, 141; A. A. Tumanov, I. A. Gur'ev and N. G. Chernorukov, *ibid.*, 144; N. I. Kokurin and S. A. Aleksandrova, *ibid.*, 147; A. T. Pilipenko, *ibid.*, 151; I. V. Pyatnitskii, *ibid.*, 159; O V. T. Mishchenko, S. V. Bel'tyukova and V. P. Antonovich, *ibid.*, 164; V. T. Velikanova and V. M. Zhukovskii, *ibid.*, 415; M. K. Kozlovskii, S. A. Chernyavskaya, and R. M. Salikhdzhanova, *ibid.*, 418.
18. E. Homburg, *Ambix*, 1999, **46**, 1.
19. G. K. Lee, L. Voress, M. Warner, F. Wach and D. Noble, *Anal. Chem.*, 1994, **66**, 251A.
20. G. M. Kolesov, *J. Anal. Chem.*, 1996, **51**, 19 and 32.
21. N. M. Kuzmin and Yu A. Zolotov, *J. Anal. Chem.*, 1996, **51**, 2.
22. G. M. Varshal and Yu. A. Zolotov, *J. Anal. Chem.*, 1996, **51**, 29.
23. B. F. Myasoedov and M. P. Volynets, *J. Anal. Chem.*, 1997, **52**, 413.
24. J. D. R. Thomas, *A History of the Analytical Division of the Royal Society of Chemistry*, Royal Society of Chemistry, Cambridge, 1999.
25. K. Helrich, *The Great Collaboration: The First 100 Years of the Association of Official Analytical Chemists*, A.O.A.C., Arlington, Virginia, 1984.
26. W. A. Oddy, *Endeavour*, 1986, **10**, 164.
27. N. Bagaturia, A. Dolidze and R. Chagunava, *Bull. Georgian Acad. Sci.*, 2000, **161**, 365.
28. W. A. Oddy, *Gold Bull.*, 1983, **16**, 52.
29. J. O. Nriagu, *J. Chem. Educ.*, 1985, **62**, 668.
30. S. Kallmann, *Anal. Chem.*, 1984, **56**, 1020A.
31. D. T. Burns, *Fresenius' J. Anal. Chem.*, 1990, **337**, 205.
32. N. G. Coley, *Med. Hist.*, 1982, **26**, 123.
33. A. Nissenbaum, *J. Chem. Educ.*, 1986, **63**, 297.
34. W. A. Smeaton, *Endeavour*, 1984, **8**, 71.
35. V. V. Kuznetsov, S. B. Savvin and E. B. Strel'nikova, *J. Anal. Chem.*, 1992, **47**, 99.
36. W. R. Newman, in *Instruments and Experimentation in the History of Chemistry*, ed. F. L. Holmes and T. H. Levere, MIT Press, Cambridge, MA, 2000, p. 34.
37. B. P. Dolan, *Ambix*, 1998, **45**, 137.
38. K. S. Basden, *Microscope*, 1997, **45**, 165.
39. W. B Jensen, in *The History and Preservation of Chemical Instrumentation*, ed. J. T. Stock and M. V. Orna, Reidel, Dordrecht, 1986, p. 123.
40. U. Burchard, *Mineral. Rec.*, 1994, **25**, 251.
41. L. Niinisto, *Frezenius' J. Anal. Chem.*, 1990, **337**, 213.
42. O. Makitie, in *Euroanalysis IV; Reviews on Analytical Chemistry*, ed. L. Niinisto, Akadémiai Kiadó, Budapest, 1982, 23.
43. K. K. Olsen, *Bull. Hist. Chem.*, 1995, **17/18**, 41.
44. F. G. Page, *Bull. Hist. Chem.*, 2001, **26**, 66.
45. D. T. Burns, in *The Chemistry of the Atmosphere - Oxidants and Oxidation in the Earth's Atmosphere*, ed. A. R. Bandy, Royal Society of Chemistry, Cambridge, 1995, 149; D. T. Burns, *Fresenius' J. Anal. Chem.*, 1997, **357**, 189.

46. R. J. Spring, *Talanta*, 1982, **29**, 883.
47. G. Schwedt, *LaborPraxis*, 1989, **13**, 30.
48. I. Possehl, *Pharm. Ztg.*, 1989, **134**, 36.
49. A. Johansson, *Anal. Chim. Acta*, 1988, **206**, 97.
50. J. A. Perez Bustamente, *Quim. Anal.*, 1999, **18**, 217.
51. F. Kober, *Prax. Naturwiss. Chem.*, 1992, **41**, 33.
52. L. M. Venanzi, in *Highlights of Chemistry as Mirrored in Helvetica Chimica Acta*, ed. M. V. Kisakürek and E. Heilbronner, VHCA, Basel, 1994, ch. 2.
53. V. Ya. Zakharans and T. A. Komarova, *J. Anal. Chem.*, 1992, **47**, 93.
54. F. Szabadvary, *Talanta*, 1978, **25**, 611.
55. H. T. Pratt, *Text. Chem. Color.*, 1991, **23**, 25.
56. C. M. Beck, *Anal. Chem.*, 1991, **63**, 993A.
57. H. A. Laitinen and E. J. Meehan, *Anal. Chem.*, 1984, **56**, 248A.
58. I. M. Kolthoff, *Anal. Chem.*, 1994, **66**, 241A.
59. W. H. Brock, *From Protyle to Proton*, Hilger, Bristol, 1985.
60. V. Ya. Zakharans and T. A. Komarova, *J. Anal. Chem.*, 1992, **47**, 96.
61. V. Ya. Zakharans and T. A. Komarova, *The History of Organic Elementary Analysis*, Khimya, Vestn. MGU, 1990.
62. B. Bobranski, *Wiad. Chem.*, 1982, **36**, 425.
63. A. J. Rocke, in *Instruments and Experimentation in the History of Chemistry*, ed. F. L. Holmes and T. H. Levere, MIT Press, Cambridge, MA, 2000, p. 273.
64. M. C. Usselman, A. J. Rocke, C. Reinhart and K. Foulser, *Ann. Sci.*, 2005, **62**, 1.
65. P. Morries, *J. Assoc. Public Anal.*, 1983, **21**, 53.
66. H. A. McKenzie, *Trends Anal. Chem.*, 1994, **13**, 138.
67. D. T. Burns, *Anal. Proc.*, 1993, **30**, 272.
68. L. Rosenfeld, *Bull. Hist. Chem.*, 1999, **24**, 40.
69. L. Rosenfeld, *Clin. Chem.*, 2000, **46**, 1705.
70. L. Rosenfeld, *Four Centuries of Clinical Chemistry*, Gordon and Breach, Amsterdam, 1999.
71. G. Bruhns, *Zuckerindustrie*, 1998, **123**, 351.
72. A. Espinola, *Bull. Hist. Chem.*, 1995, **17/18**, 31.
73. J. G. Delly, *Microscope*, 1999, **47**, 13.
74. J. T. Stock and M. V. Orna, eds, *Electrochemistry, Past and Present*, American Chemical Society, Washington DC, 1989.
75. I. M. Kolthoff, *J. Electrochem. Soc.*, 1982, **129**, 59C.
76. J. T. Stock, in ref. 74, p. 469.
77. J. T. Stock, *Bull. Hist. Chem.*, 1990, **7**, 17.
78. L. M. Robinson, in ref. 74, p. 458.
79. C. E. Moore, B. Jaselskis and A. von Smolinski, in ref. 74, p. 286.
80. R. Piosik, *Mitt. - Ges. Dtsch. Chem.*, *Fachgruppe Gesch. Chem.*, 1993, **8**, 50.
81. B. Fantini and N. Nicolini, *Chim. Ind. (Milan)*, 1990, **72**, 830.
82. J. T. Stock, *J. Chem. Educ.*, 1989, **66**, 910.
83. D. S. Tarbell and A. T. Tarbell, *J. Chem. Educ.*, 1980, **57**, 133.
84. B. Jaselskis, C. E. Moore and A. von Smolinski, in ref. 74, p. 272.
85. J. D. R. Thomas, in ref. 74, p. 303.
86. B. P. Nikol'skii, *J. Anal. Chem.*, 1992, **47**, 91.
87. M. S. Frant, *Analyst (Cambridge U.K.)*, 1994, **119**, 2293.

88. J. A. M. Hinke, *Can. J. Physiol. Pharmacol.*, 1987, **65**, 873.
89. M. S. Frant, *J. Chem. Educ.*, 1997, **74**, 159.
90. J. Ruzicka, *J. Chem. Educ.*, 1997, **74**, 167.
91. T. S. Light, *J. Chem. Educ.*, 1997, **74**, 171.
92. C. C. Young, *J. Chem. Educ.*, 1997, **74**, 177.
93. G. K. Budnikov and E. P. Medyantseva, *J. Anal. Chem.*, 1992, **47**, 50.
94. P. Zuman, in ref. 74, p. 339.
95. L. R. Sherman, *Chem. Br.*, 1990, **26**, 1165.
96. J. Koryta, *Electrochim. Acta*, 1991, **36**, 221.
97. M. Heyrovsky, *Chem. Listy*, 1997, **91**, 1034.
98. M. Heyrovsky, L. Novotny and I. Smoler, in ref. 74, p. 370.
99. R. M.-F. Salikhdzhanova, *J. Anal. Chem.*, 1992, **47**, 91.
100. R. C. Rooney, *Analyst (Cambridge, U.K.)*, 1992, **117**, 1829.
101. J. Osteryoung and C. Wechter, in ref. 74, p. 380.
102. P. Zuman, *Analyst (Cambridge, U.K.)*, 1992, **117**, 1803.
103. G. C. Barker and A. W. Gardner, *Analyst (Cambridge, U.K.)*, 1992, **117**, 1811.
104. E. Palacek, *Bioelectrochem. Bioenerg.*, 1981, **8**, 469.
105. M. V. Orna, in ref. 74, p. 196.
106. L. C. Clark and E. W. Clark, *Int. Anesthesiol. Clin.*, 1987, **25**, 1.
107. L. B. Rogers, in ref. 74, p. 396.
108. J. T. Stock, *Anal. Chem.*, 1984, **56**, 561A.
109. D. Cahan, *Osiris*, 1989, **5**, 167.
110. F. Szabadvary and R. A. Chalmers, *Talanta*, 1983, **30**, 997.
111. J. T. Stock, *J. Chem. Educ.*, 1992, **69**, 197.
112. H. A. Laitinen and E. J. Meehan, *Anal. Chem.*, 1984, **56**, 248A.
113. G. W. Ewing, in ref. 74, p. 402.
114. H. A. Laitinen, in ref. 74, p. 417.
115. W. R. Heinemann and W. B. Jensen, in ref. 74, p. 442.
116. H. Gunasingham, in ref. 74, p. 236.
117. K. I. Sakodynskii, *J. Anal. Chem.*, 1993, **48**, 897.
118. L. S. Ettre, in *High Performance Liquid Chromatography, Advances and Perspectives*, ed. C. S. Horvath, Academic Press, New York, 1980, vol. 1, p. 1.
119. L. S. Ettre, *J. Chromatogr.*, 1990, **535**, 3.
120. R. P. W. Scott, *J. Liq. Chromatogr.*, 1987, **10**, 1547.
121. E. R. T. Adlard, *CAST, Chromatogr. Sep. Technol.*, 2000, **11**, 12.
122. J. C. Touchstone, *J. Liq. Chromatogr.*, 1993, **16**, 1647.
123. L. S. Ettre, *Chromatographia*, 2000, **51**, 7.
124. L. S. Ettre, *LC-GC*, 1999, **17** (6), 524.
125. Y. Nagayanagi, S. Takimoto and H. Saito, *LC-GC*, 1999, **17** (10), 930.
126. L. S. Ettre, *LC-GC*, 1999, **17** (12), 1104.
127. L. S. Ettre, *LC-GC*, 2000, **18** (4), 392.
128. L. S. Ettre and V.R. Meyer, *LC-GC*, 2000, **18** (7), 704.
129. L. S. Ettre and V. G. Berezkin, *LC-GC*, 2000, **18** (11), 1148.
130. L. S. Ettre, *LC-GC*, 2001, **19** (1), 48.
131. L. S. Ettre, *LC-GC*, 2001, **19** (5), 506.
132. L. S. Ettre and H. Kalasz, *LC-GC*, 2001, **19** (7), 712.

133. H. H. Bussemass, G. Harsch and L. S. Ettre, *Chromatographia*, 1994, **38**, 243.
134. H. Newesely, *Chromatographia*, 1990, **30**, 595.
135. G. D'Ascenzo and N. Nicolini, *Fresenius' J. Anal. Chem.*, 1990, **337**, 232.
136. P. Jossgang, *Nature (London)*, 1992, **356**, 100.
137. L. S. Ettre and K. I. Sakodynskii, *Chromatographia*, 1993, **35**, 223, 329.
138. K. I. Sakodynskii, *J. Planar Chromatogr. - Mod. TLC*, 1992, **5**, 210.
139. V. R. Meyer, *Chromatographia*, 1992, **34**, 342.
140. K. I. Sakodynskii, *J. Chromatogr.*, 1981, **220**, 1.
141. I. M. Hais, *J. Chromatogr.*, 1988, **440**, 509.
142. L M. Hais, M. Niang and L. S. Ettre, *Chromatographia*, 1997, **44**, 545.
143. I. M. Hais, M. Niang and L. S. Ettre, *Chromatographia*, 1997, **44**, 651.
144. I. M. Hais, M. Niang and L.S. Ettre, *Chromatographia*, 1997, **44**, 663.
145. L. S. Ettre and Y. Kazakevich, *Chem. Anal. (Warsaw)*, 1998, **43**, 481.
146. L. S. Ettre, *Chromatographia*, 1997, **46**, 444.
147. V. G. Berezkin (compiler), *Chromatographic Adsorption Analysis: Selected Works by Mikhail Semenovich Tswett*, Ellis Horwood, New York, London, 1990.
148. H. H. Bussemass and L. S. Ettre, *Chromatographia*, 1994, **39**, 369.
149. V. R. Meyer and L. S. Ettre, *J. Chromatogr.*, 1992, **600**, 3.
150. L. S. Ettre and R. L. Wixom, *Chromatographia*, 1993, **37**, 659.
151. E. Lederer, *New J. Chem.*, 1988, **12**, 249.
152. M. Lederer, *Chem. Intell.*, 1999, **5** (3), 31.
153. M. Lederer, *Chem. Intell.*, 1998, **4** (3), 55.
154. M. Lederer, *Chem. Intell.*, 1999, **5** (4), 32.
155. L. S. Ettre, *Anal. Chem.*, 1989, **61**, 1315A.
156. L. S. Ettre, *Int. Lab.*, September 1991, 18.
157. A. Zaffaroni, *Steroids*, 1992, **57**, 642.
158. U. Wintermeyer, *Die Wurzeln der Chromatographic: Historischer Abriss von den Anfangen bis zur Dunnschicht-Chromatographie*, GIT Verlag, Darmstadt, 1989.
159. O. Bobleter, *Chromatographia*, 1996, **43**, 444.
160. L. S. Ettre, *LC-GC*, 1990, **8**, 716.
161. A. T. James, *Membranes in Gas Separation and Enrichment*, Spec. Publ.-R. Soc. Chem., No. 62, Royal Society of Chemistry, London, 1986, p. 175.
162. J. E. Lovelock, *LC-GC*, 1990, **8**, 854.
163. I. G. McWilliam, *Chromatographia*, 1983, **17**, 241.
164. L. S. Ettre, *Anal. Chem.*, 1985, **57**, 1419A.
165. L. S. Ettre, *Chromatographia*, 1982, **16**, 18.
166. K. P. Dimick, *LC-GC*, 1990, **8**, 782.
167. C. E. Bennet, in *The History and Preservation of Chemical Instrumentation*, ed. J. T. Stock and M. V. Orna, Reidel, Dordrecht, 1986, p. 79.
168. V. A. Shaposhnik, *J. Anal. Chem.*, 1992, **47**, 110.
169. C. M. Beck, *Anal. Chem.*, 1991, **63**, 993A.
170. L. S. Ettre, *Int. Lab.*, October 1991, 18.
171. H. Small, in *The History and Preservation of Chemical Instrumentation*, ed. J. T. Stock and M. V. Orna, Reidel, Dordrecht, 1986, p. 97.
172. L. R. Snyder, *J. Chem. Educ.*, 1997, **74**, 37.
173. B. L. Karger, *J. Chem. Educ.*, 1997, **74**, 45.

174. L. E. Kay, *Hist. Phil. Life Sci.*, 1988, **10**, 51.
175. O. Verterberg, *J. Chromatogr.*, 1989, **480**, 3.
176. H. Rilke, *Electrophoresis*, 1995, **16**, 1354.
177. L. A. Sherman and J. A. Goodrich, *Chem. Soc. Rev.*, 1985, **14**, 225.
178. J.-C. Janson, *Chromatographia*, 1987, **23**, 361.
179. T. C. Laurent, *J. Chromatogr.*, 1993, **633**, 1.
180. F. A. J. L. James, *Ambix*, 1985, **32**, 53.
181. D. T. Burns, *J. Anal. At. Spectrom.*, 1987, **2**, 343.
182. D. T. Burns, *J. Anal. At. Spectrom.*, 1988, **3**, 285.
183. J. A. Bennett, *The Celebrated Phenomena of Colours*, Whipple Museum, Cambridge, 1984.
184. I. D. Rae, *Ambix*, 1997, **44**, 131.
185. F. A. J. L. James, *Ambix*, 1983, **30**, 30.
186. M. Saillard, *Bull. Union Physic.*, 1983, **77**, 1157.
187. M. A. Sutton, *Plat. Met. Rev.*, 1988, **32**, 28.
188. F. A. J. L. James, *Hist. Sci.*, 1985, **23**, 1.
189. M. A. Sutton, *Hist. Sci.*, 1986, **24**, 425.
190. F. A. J. L. James, *Hist. Sci.*, 1986, **24**, 433.
191. F. A. J. L. James, *Notes Rec. Roy. Soc. Lond.*, 1984, **39**, 65.
192. F. A. J. L. James, *Ambix*, 1981, **28**, 131.
193. F. A. J. L. James, *Br. J. Hist. Sci.*, 1988, **21**, 181.
194. F. A. J. L. James, *Proc. R. Inst. G. B.*, 1986, **58**, 17.
195. B. Nygård and R. Petterson, *Kem. Tidskr.*, 1990, **102**, 53.
196. B. Nygård and R. Petterson, *Fresenius' J. Anal. Chem.*, 1990, **337**, 186.
197. A. Walsh, *Anal. Chem.*, 1991, **63**, 933A.
198. Biography: Anon., *Spectrochim. Acta, B*, 1999, **54**, 1933; List of publications: *ibid.*, 1939; Papers: A. Walsh, *ibid.*, 1943; P. T. Beale, *ibid.*, 1953; J. B. Willis and N. S. Ham, *ibid.*, 1955; J. P. Shelton, *ibid.*, 1961; D. J. David, *ibid.*, 1967; J. B. Willis, *ibid.*, 1971; M. D. Amos, *ibid.*, 1977; C. B. Belcher, *ibid.*, 1983; P. L. Boar, *ibid.*, 1989; J. W. Robinson, *ibid.*, 1993; P. Butler, *ibid.*, 1999; K. Fuwa, *ibid.*, 2005; J. B. Dawson and W. J. Price, *ibid*, 2011; T. S. West, *ibid.*, 2017; M. D. Amos, *ibid.*, 2023; C. Mitchell, *ibid.*, 2041; W. Slavin, *ibid.*, 2051; H. L. Kahn, *ibid.*, 2057; H. J. Sloane, *ibid.*, 2061; B. V. L'vov, *ibid.*, 2063; D. S. Gough, *ibid.*, 2067; L. de Galan, *ibid.*, 2073.
199. J. B. Willis, *Hist. Rec. Aust. Sci.*, 1991, **8**, 151.
200. J. B. Willis, *Spectrochim. Acta, B*, 1988, **43**, 1021.
201. J. W. Robinson, *Anal. Chem.*, 1994, **66**, 472A.
202. S. R. Koirtyohann, *Anal. Chem.*, 1991, **63**, 1024A.
203. W. Slavin, *Anal. Chem.*, 1991, **63**, 1033A.
204. J. B. Willis, *Clin. Chem. (Washington DC)*, 1993, **39**, 155.
205. B. V. Lvov, *Anal. Chem.*, 1991, **63**, 924A.
206. R. E. Jarrell, *J. Chem. Educ.*, 2000, **77**, 573.
207. G. M. Hieftje, *J. Chem. Educ.*, 2000, **77**, 577.
208. S. Greenfield, *J. Chem. Educ.*, 2000, **77**, 584.
209. R. S. Houk, *J. Chem. Educ.*, 2000, **77**, 598.
210. R. F. Jarrell, F. Brech and M. J. Gustafson, *J. Chem. Educ.*, 2000, **77**, 592.

211. D. T. Burns, *Anal. Proc.*, 1986, **23**, 75.
212. A. T. Pilipenko, *J. Anal. Chem.*, 1992, **47**, 70.
213. L.-G. Oltra and C.-M. Verdu, *Bull. Hist. Chem.*, 2001, **26**, 57.
214. J. T. Stock, *J. Chem. Educ.*, 1994, **71**, 967.
215. G. B. Levy, *Int. Lab.*, June 1993, 4.
216. A. B. Costa, *J. Chem. Educ.*, 1982, **59**, 645.
217. R. Altemose, *J. Chem. Educ.*, 1986, **63**, A216.
218. R. Jarnutowski, J. R. Ferraro and D. C. Lankin, *Spectroscopy (Eugene, Oregon)*, 1992, **7** (7), 22.
219. G. Shenk, *Spectroscopy (Eugene, Oregon)*, 1997, **12** (8), 48.
220. F. E. Lytle, *Anal. Chem.*, 2000, **72**, 477A.
221. R. N. Jones, in *Chemical Biological and Industrial Applications of Infrared Spectroscopy*, ed. J. R. Durig, Wiley, New York, 1985, p. 1.
222. S. N. Cesaro and E. Torracca, *Ambix*, 1988, **35**, 39.
223. P. A. Wilks in *The History and Preservation of Chemical Instrumentation*, ed. J. T. Stock and M. V. Orna, Reidel, Dordrecht, 1986, p. 27.
224. P. A. Wilks, *Anal. Chem.*, 1992, **64**, 833A.
225. F. A. Miller, *Anal. Chem.*, 1992, **64**, 824.
226. A. N. Sheppard, *Anal. Chem.*, 1992, **64**, 877A.
227. P. J. T Morris, *The American Synthetic Rubber Research Programme*, University of Pennsylvania Press, Philadelphia, PA, 1989.
228. A. N. Sheppard, *Anal. Chem.*, 1992, **64**, 877A.
229. F. A. Miller, *Appl. Spectrosc.*, 1992, **46**, 1096.
230. P. A. Wilks, in *The History and Preservation of Chemical Instrumentation*, ed. J. T. Stock and M. V. Orna, Reidel, Dordrecht, 1986, p. 27.
231. Y. M. Rabkin, *Isis*, 1987, **78**, 31.
232. P. R. Griffiths, *Anal. Chem.*, 1992, **64**, 868A.
233. P. Connes, *J. Phys. II*, 1992, **2**, 565.
234. J. R. Ferraro, R. Jarnutowski and D. C. Lankin, *Spectroscopy (Eugene, Oregon)*, 1992, **7** (2), 30.
235. G. Venkataraman, *Journey into Light; Life and Science of C. V. Raman*, Indian Academy of Sciences, Bangalore, 1988.
236. A. Jayaraman and A. K. Ramdas, *Phys. Today*, 1988, **41**, 56.
237. S. Ramaseshan, *Indian J. Pure Appl. Phys.*, 1988, **26**, 27.
238. J. C. D. Brand, *Notes Rec. Roy. Soc. Lond.*, 1989, **43**, 1.
239. F. A. Miller and G. B. Kauffman, *J. Chem. Educ.*, 1989, **66**, 795.
240. A. O. Nier, *J. Chem. Educ.*, 1989, **66**, 385.
241. S. Meyerson, *Anal. Chem.*, 1994, **66**, 960A.
242. S. Meyerson, *Org. Mass Spectrom.*, 1986, **21**, 197.
243. A. Quayle, *Org. Mass Spectrom.*, 1987, **22**, 569.
244. J. H. Beynon, *Org. Mass Spectrom.*, 1991, **26**, 353.
245. R. E. Finnigan, *Anal. Chem.*, 1994, **66**, 969A.
246. D. M. Grant and R. K. Harris, eds, *Encyclopedia of Nuclear Magnetic Resonance*, Wiley, Chichester, 1996, vol. 1: *Historical Perspectives*.
247. R. Richards, *Chem. Br.*, 1996, **32** (6), 33.
248. E. D. Becker, *Anal. Chem.*, 1993, **65**, 295A.

249. R. Richards, *Chem. Br.*, 1991, **27**, 243.

250. J. N. Shoolery, *Anal. Chem.*, 1993, **65**, 731A.

251. D. C. Lankin, J. R. Ferraro and R. Jarnutowski, *Spectroscopy (Eugene, Oregon)*, 1992, **7** (8), 18.

252. H. Pfeifer, *Magn. Res. Chem.*, 1999, **37**, 154.

253. S. G. Levine, *J. Chem. Educ.*, 2001, **78**, 133.

254. K. Marsden and I. D. Rae, *Hist. Rec. Aust. Sci.*, 1991, **8**, 119.

255. R. Freeman, *Anal. Chem.*, 1993, **65**, 743A.

256. C. Reinhardt, *Ann. Sci.*, 2004, **61**, 1.

257. F. Feeney, *Educ. Chem.*, 1996, **33**, 96.

258. E. R. Andrew, *Br. Med. Bull.*, 1984, **40**, 115.

259. O. Jardetsky, in *Biological NMR Spectroscopy*, ed. J. L. Markley and S. J. Opella, Oxford University Press, New York, 1997, p. 3.

260. M. Cohn, in *Biological NMR Spectroscopy*, ed. J. L. Markley and S. J. Opella, Oxford University Press, New York, 1997, p. 16.

261. R. G. Shulman, in *Biological NMR Spectroscopy*, ed. J. L. Markley and S. J. Opella, Oxford University Press, New York, 1997, p. 20.

262. R. C. Mackenzie, *Themochim. Acta*, 1984, **73**, 251.

263. C. J. Keattch and D. Dollimore, *J. Therm. Anal.*, 1991, **37**, 2089.

264. C. J. Keattch and D. Dollimore, *J. Therm. Anal.*, 1991, **37**, 2103.

265. C. J. Keattch and D. Dollimore, *J. Therm. Anal.*, 1993, **39**, 97.

266. C. J. Keattch and D. Dollimore, *J. Therm. Anal.*, 1993, **39**, 755.

267. C. J. Keattch and D. Dollimore, *Thermochim. Acta*, 1999, **340–1**, 31.

268. C. Eyraud and P. Rochas, *Thermochim. Acta*, 1989, **152**, 1.

269. C. J. Keattch, *J. Therm. Anal.*, 1995, **44**, 1211.

270. C. J. Keattch, *J. Therm. Anal.*, 1996, **46**, 1501.

271. R. C. Mackenzie, *Isr. J. Chem.*, 1982, **22**, 203.

272. R. C. Mackenzie, *Themochim. Acta*, 1984, **73**, 307.

273. F. Paulik, *J. Therm. Anal.*, 1996, **47**, 659.

274. S. Heilenz, *The Liebig Museum in Giessen*, Universitäts-Buchhandlung, Giessen, 1991.

275. W. H. Brock, *Justus von Liebig: The Chemical Gatekeeper*, Cambridge University Press, Cambridge, 1997.

276. P. W. Hammond, *Chem. Br.*, 1992, **28**, 796.

277. P. W. Hammond and H. Egan, *Weighed in the Balance*, HMSO, London, 1992.

278. P. W. Hammond, *Anal. Proc.*, 1992, **29**, 311.

279. R. D. Worswick, *Analyst (Cambridge, U.K.)*, 1993, **118**, 583.

280. R. Stanley, *Lab. Pract.*, 1987, **36**, 27.

281. J. F. Donnelly, *J. Social Hist.*, 1996, **20**, 779.

282. S. Tomic, *Ann. Sci.*, 2001, **58**, 287.

283. I. Possehl, *Mitt. - Ges. Dtsch. Chem.*, *Fachgruppe Gesch. Chem.*, 1993, **9**, 31.

284. J. F. Donnelly, in *The Chemical Industry in Europe, 1850–1914: Industrial Growth, Pollution and Professionalization*, ed. E. Homburg, A. S. Travis and H. G. Schröter, Kluwer Academic, Dordrecht, 1998, p. 203.

285. J. F. Donnelly, *Technol. Culture*, 1994, **35**, 100.

286. J. Büttner, *Eur. J. Clin. Chem. Clin. Biochem.*, 1992, **30**, 585.

287. J. Büttner, *Mitt-Ges. Dtsch. Chem., Fachgruppe Gesch. Chem.*, 1997, **13**, 23.
288. J. Büttner and C. Habrich, *Roots of Clinical Chemistry*, GIT Verlag, Darmstdt, 1987.
289. H. M. Free and A. H. Free, *Anal. Chem.*, 1984, **56**, 664A.
290. B. Gee, in *The Development of the Laboratory*, ed. F. A. J. L. James, Macmillan, Basingstoke, 1989, p. 37.
291. G. Schwedt, *Dtsch. Apoth. Ztg.*, 1990, **130**, 2781.
292. L. Niinisto, *Frezenius' J. Anal. Chem.*, 1990, **337**, 213.
293. H. R. Jenemann, *Die Waage des Chemikers: The Chemist's Balance*, Dechema and Hans R. Jenemann-Stiftung, Frankfurt am Main, 1997.
294. H. R. Jenemann and E. Robens, *J. Therm. Anal. Calorim.*, 1999, **55**, 339.
295. Th. Gast, H. R. Jenemann and E. Robens, *J. Therm. Anal. Calorim.*, 1999, **55**, 347.
296. K. R. Williams, *J. Chem. Educ.*, 2001, **78**, 434.
297. B. Bensaude-Vincent, in *Instruments and Experimentation in the History of Chemistry*, ed. F. L. Holmes and T. H. Levere, MIT Press, Cambridge, MA, 2000, p. 153.
298. J. Burnett, in *Making Instruments Count*, ed. R. G. W Anderson, J. A. Bennett and W. F. Ryan, Variorum, Aldershot, 1993, p. 242.
299. J. T. Stock, *Bull. Hist. Chem.*, 1994, **15/16**, 1.
300. J. T. Stock, *School Sci. Rev.*, 1998, **80**, 75.
301. C. A. Russell, in *Instruments and Experimentation in the History of Chemistry*, ed. F. L. Holmes and T. H. Levere, MIT Press, Cambridge, MA, 2000, p. 311.
302. J. K. Taylor, in *The History and Preservation of Chemical Instrumentation*, ed. J. T. Stock and M. V. Orna, Reidel, Dordrecht, 1986, p. 1.
303. D. Nelson, *Anal. Chem.*, 1992, **64**, 588A.
304. Y. Zabkin, in *Chemical Sciences in the Modern World*, ed. S. H. Mauskopf, University of Pennsylvania Press, Philadelphia, PA, 1993, p. 25.
305. D. Baird, *Ann. Sci.*, 1993, **50**, 267.
306. J. T. Stock, *Educ. Chem.*, 1983, **20**, 7.
307. J. T. Stock, *Trends Anal. Chem.*, 1983, **2**, 14.
308. W. T. Caraway, in *History of Clinical Chemistry*, ed. J. Büttner, Walter de Gruyter, Berlin, 1983, p. 77.
309. S. F. Johnston, in *Instrumentation: Between Science, State and Industry*, ed. B. Joerges and T. Shinn, Kluwer, Dordrecht, 2001, p. 121.
310. H. J. Rheinberger, in *Instrumentation: Between Science, State and Industry*, ed. B. Joerges and T. Shinn, Kluwer Academic, Dordrecht, 2001, p. 143.
311. M. E. W. Williams, *The Precision Makers: A History of the Instruments Industry in Britain and France 1870–1939*, Routledge, London, 1994.
312. C. E. Moore and B. Jaselskis, *Bull. Hist. Chem.*, 1998, **21**, 32.
313. R. F. Bud and G. K. Roberts, *Science versus Practice: Chemistry in Victorian Britain*, Manchester University Press, Manchester, 1984.
314. J. F. Donnelly, *Br. J. Hist. Sci.*, 1990, **24**, 3.
315. K. H. Koch, *Fresenius' J. Anal. Chem.*, 1990, **337**, 229.
316. T. S. Light, in ref. 74, p. 429.
317. J. T. Stock, *Bull. Hist. Chem.*, 1991, **10**, 31.

318. C. A. Russell, ed., *Chemistry, Society and Environment*, Royal Society of Chemistry, Cambridge, 2000.
319. F. J. Johnson, *J. Assoc. Off. Anal. Chem.*, 1984, **67**, 483.
320. V. C. Midkiff, *J. Assoc. Off. Anal. Chem.*, 1984, **67**, 851.
321. I. Rosenthal and B. Rosen, *J. Chem. Educ.*, 1993, **70**, 480.
322. J. Rosmorduc, *Rev. Hist. Sci.*, 1988, **41**, 25.
323. D. T. Burns, *Fresenius' J. Anal. Chem.*, 1993, **347**, 14.
324. D. Duff, *Chem. Br.*, 1997, **33**, 46.
325. D. L. Adams, *Bull. Hist. Chem.*, 1999, **24**, 16.
326. G. Schwedt, *Fresenius' J. Anal. Chem.*, 1983, **315**, 395.
327. J. A. Perez-Bustamente, *Quim. Anal.*, 1997, **16**, 139.
328. R. Soloniewiez, *Kwart. Hist. Nauki Tech.*, 1992, **37** (4), 141.
329. G. D. Christian, *Anal. Chem.*, 1995, **67**, 532A.
330. R. W. Murray, *Talanta*, 1989, **36**, 11.
331. A. Ratcliffe and H. A. Mottola, *J. Chem. Educ.*, 1991, **68**, 543.
332. C. M. Beck, *Anal. Chem.*, 1991, **63**, 993A.
333. R. Belcher, *Phil. Trans. R. Soc. London, Ser. A*, 1982, **305**, 475.
334. J. F. Alder, *Anal. Proc.*, 1988, **25**, 251.
335. J. Golinski, *Science as Public Culture*, Cambridge University Press, Cambridge, 1992.
336. C. A. Russell, *Edward Frankland, Chemistry, Controversy and Conspiracy in Victorian England*, Cambridge University Press, Cambridge, 1996.
337. D. T. Burns, *Fresenius' J. Anal. Chem.*, 2000, **368**, 544.
338. W. Fresenius, *Fresenius' J. Anal. Chem.*, 2000, **368**, 548.
339. W. Vycudilik, *Fresenius' J. Anal. Chem.*, 2000, **368**, 550.
340. S. M. Gerber and R. Saferstein, eds, *More Chemistry and Crime: From Marsh Arsenic Test to DNA Profile*, American Chemical Society, Washington DC, 1997.
341. N. G. Coley, *Endeavour*, 1998, **22**, 143.
342. T. L. Sourkes, *Ambix*, 1992, **39**, 11.
343. S. M. Gerber, ed., *Chemistry and Crime: From Sherlock Holmes to Today's Courtroom*, American Chemical Society, Washington DC, 1983.
344. C. Hamlin, *Bull. Hist. Med.*, 1982, **56**, 56.
345. B. Luckin, *Pollution and Control*, Adam Hilger, Bristol, 1986.
346. C. Hamlin, *A Science of Impurity*, Adam Hilger, Bristol, 1990.
347. P. Brimblecombe, *The Big Smoke*, Methuen, London, 1987.
348. C. Garwood, *Ann. Sci.*, 2004, **61**, 99.
349. W. A. Campbell, *Chem. Br.*, 1990, **26**. 558.
350. C. Hamlin, in *Chemical Sciences in the Modern World*, ed. S. H. Mauskopf, University of Pennsylvania Press, Philadelphia, PA, 1993, p. 295.
351. R. E. Finnigan, *Anal. Chem.*, 1994, **66**, 969A.
352. J. E. Lovelock, *LC-GC*, 1990, **8**, 854.
353. The Water Supply (Water Quality) Regulations (As Amended), HMSO, London, 1989.

CHAPTER 8

Medical Chemistry and Biochemistry

NOEL G. COLEY

History of Chemistry Research Group, The Open University

8.1 Origins of Medical Chemistry

Chemistry was the handmaid of medicine long before it became established as a scientific discipline in its own right and this association remains important as one of the origins of chemistry.[1] The use of chemical techniques in the preparation and prescription of drugs is as old as medicine itself and there is a long tradition of extracting medicinal substances from plants by solvents and distillation.[2] Until the nineteenth century plant extracts, with a few animal preparations, formed the bulk of the *Materia Medica*, the physician's therapeutic resources. Some were known to be very powerful and have more recently been found to contain significant amounts of active principles. A popular compendium of such medicines and folk remedies, first published in England in 1656, has been re-issued.[3] Some of these ancient remedies contain still unidentified compounds with potentially useful therapeutic properties, and a systematic scientific investigation of their medicinal value remains a worthwhile enterprise in the continuing search for new drugs and medicines. In the ever-changing priorities of modern medicine it is not impossible that promising candidates for investigation may still be found amongst the chemical components of ancient remedies. In this inter-disciplinary, though so far largely neglected, field historians of medicine, classical scholars and modern pharmaceutical researchers can co-operate effectively.[4] Some think that there is a good case for promoting the study of early therapies and the chemical components of such ancient medicines.

In the sixteenth century, Paracelsus added some mineral compounds to the *Materia Medica* and powerful medicines were obtained from the ores of metals like mercury and antimony by alchemical processes. Antimony has a very long history in medicine, but it came into greater prominence at a critical time in the sixteenth century when syphilis, then a new killer disease, was rapidly spreading across Europe.[5] Paracelsus' mercurial preparations were also effective, although their poisonous qualities, like those of antimony, made them highly dangerous in use. Nevertheless, mineral substances increased the importance and value of 'chemical medicines' due to their potency.[6] These metallic compounds, along with others like zinc, also introduced by Paracelsus, have remained useful in medicine.[7]

For Paracelsian physicians, both diagnosis and treatment were dependent on chemistry, while knowledge of the therapeutic properties of drugs and medicines was considered essential. Chemistry embraced pharmacy, as it did other chemical arts, and chemical texts were often written with the applications of chemical preparation and purification processes to medicine and pharmacy in mind. As Paracelsianism infiltrated medical thought it also ensured that human physiology and pathology came to be thought of in chemical terms. The development of Paracelsian alchemy and medical chemistry depended heavily on patronage.[8,9] Based on its proven efficacy in medicine, Paracelsianism became the most important chemical theory of the sixteenth and seventeenth centuries, extending far beyond the confines of medical therapeutics. It has been thoroughly explored and analysed by Allen Debus,[10] Piyo Rattansi[11] and especially Walter Pagel, who saw Paracelsian philosophy as the root of Renaissance medicine.[12,13]

Paracelsus' influence in medicine remained long after the usefulness of his philosophy had declined. Ever since the sixteenth century, compounds of antimony and mercury have figured in the treatment of venereal diseases.[14] In the seventeenth century, the Dutch physician-chemist J. B. van Helmont brought Paracelsianism to new heights, giving rise to a new breed of 'chymical physicians'. Shunned and despised by the Galenists, Paracelsians and Helmontians worked for the reform of medicine and the recognition of the role of chemistry within it. Van Helmont, who recognized the acidity of the gastric juice and introduced the concept of 'gas' as a physical entity, identified important links between chemistry and medicine,[15,16] treating disease as a chemical malfunction of the organs. The approach of the Helmontians was first to identify the nature of the malfunction and then find appropriate chemical medicines with which to correct it.[17] Thus, Paracelsian and Helmontian chemical theories of the treatment of disease began to change the face of medicine.[18] But historians have encountered great difficulties in their efforts to identify the true nature of Paracelsianism. The influence of the Paracelsian corpus extends far beyond chemistry and medicine; recent studies have included reference to its impact on religious thought.[19]

The new chemical medicine was well-known throughout Europe, though accepted with varying degrees of enthusiasm in different European countries. The contrast is well drawn by Debus in his studies of Paracelsianism in England[20] and France,[21] but its influence was also felt in other national contexts[22] and in its effects on the intellectual life of European courts.[23] Paracelsus' chemical work has also received special attention. In the 1950s, Multhauf found the origins of medical chemistry in Paracelsianism[24] and more recently Debus has argued persuasively that it would be undesirable to isolate Paracelsian chemistry from its medical applications. Already by the seventeenth century pharmacy was becoming established as a discipline in some universities.[25]

Chemical theories in medicine were not confined to the diagnosis and treatment of disease, for it was gradually realized that normal physiological functions such as respiration, digestion and the assimilation of food also depended upon chemical processes. This recognition began the study of iatrochemistry,[26] the first step on a long and tortuous path that would lead through physiological chemistry to biochemistry by the late nineteenth century. Following William Harvey's discovery of the circulation of the blood, seventeenth-century physicians and chemists began to investigate the changes occurring in the blood during respiration.[27] In the Royal Society, comparative studies of venous and arterial blood, coupled with experiments on the air, began to throw new light on the action of the air on blood as it passed through the lungs. Thus the physiology of respiration was opened up to investigation, leading to the view that the air contained a component responsible for maintaining life. Several Royal Society members, including Robert Hooke, proposed the theory of 'nitro-aerial particles', which support both respiration and combustion. The theory received its fullest expression in the work of John Mayow. However, these speculations were soon superseded by the phlogiston theory of Becher and Stahl, which dominated theoretical chemistry throughout most of the eighteenth century.

The seventeenth century also saw a revival of interest in the use of mineral waters and spas, long fallen into disuse.[28] Hydrotherapy, using mineral waters both for bathing and drinking as medicines, became popular throughout Europe during the eighteenth century. Besides the therapeutic value of natural mineral waters, the study of their chemical composition made useful contributions to solution analysis and the development of precision in quantitative chemical methods. Another interesting development in the history of medical chemistry concerns investigation of the role of common salt in diet and human physiology.[29–31]

8.2 Contributions from Pharmacy

Pharmacy has an interesting history, not commonly studied by chemists, but worthy of attention for the contributions it has made to experimental techniques.[32] The interdependence of pharmacy and medical chemistry resulted in gains for both. The skill and precision required in purifying medicinal compounds and mixing them accurately in the dispensary benefited the chemist while chemical theory supported the work of the apothecary and pharmacist.[33] The practice of pharmacy developed along scientific lines even from the sixteenth century, when the work of Franz Joel (1508–1579), an early German pharmacist, began to exert an influence on the development of chemistry. For a long time the work of the pharmacist was shrouded in almost as much secrecy as that of the alchemist.[34,35] This was especially true of medicines that were known to contain poisonous substances. For example, the composition of tartar emetic, a useful medicine often found in the home, was kept secret to obscure its antimony content.[36] Important eighteenth-century German chemists like Trommsdorff, Hermbstaedt and Bley found their initial training in pharmacy.[37] The history of pharmacy has many links with the chemistry of natural products, as is shown in a recent bibliography of the subject.[38] Despite the essential role pharmacists played in the history of medicine and chemistry, they were generally regarded as artisans and tradesmen practising a useful art.[39] Until the late eighteenth century, this art, like others that depended on the use of chemistry, was seen as a primary reason for studying chemistry. It was the chemical revolution, followed in the early nineteenth century by developments such as Dalton's atomic theory and the discovery of electrolysis, which fostered the notion of chemistry as a separate discipline, distinguished from such practical arts as pharmacy which were not really part of 'mainstream' chemistry.[40] But, however they were viewed by the chemists, it cannot be denied that pharmacists and apothecaries made important contributions to practical chemical skills and accurate quantitative work.[41]

An important development of recent years is the rise of pharmaceutical medicine as a new discipline.[42] The history of the American pharmaceutical industry has recently been studied,[43–45] as has the industry in Britain and France.[46] There are several histories of particular pharmaceutical companies and individuals[47–50] and a useful annual bibliography of pharmaceutical history, first published in 1952, was revived in 1993 after a long gap.[51] New studies of important individual workers in the field have also appeared, emphasizing the links between the pharmaceutical industry and academic research.[52,53] Concerned with the health and safety of the public, the pharmaceutical industry has always developed under strict controls, as the powerful drugs, which it brings to the market, have sometimes shown serious flaws. The ethics of drug manufacture and use have always figured prominently in the industry.[54] With the growth of the pharmaceutical industry the need for secrecy to prevent piracy increased; one study concerns the problem in France.[55]

8.3 Pharmacology

Pharmacology is as old as medicine itself and tests on the physiological effects of medicines extend in an unbroken sequence from ancient times to the present. Beginning with the work of Paul Erlich,[56] the organic chemical industry in the nineteenth century developed numerous therapeutic compounds, which were introduced into medicine as chemo-therapeutics. No less important for the history of chemistry have been the attempts to relate pharmacological activity to the chemical structures of effective drugs.[57,58] The introduction of new drugs is a very long process, but many new entities are always being developed;[59] in recent years, rapid changes in pharmacology have been needed to keep pace with the introduction of new drugs and treatments.[60] As a necessary part of research into the potential and safety of new drugs,[61] and of older ones put to new uses, pharmacology has a vital part to play. There are also important lessons to be learned from the history of pharmacology.[62–64] Apart from efforts to discover new drugs, one modern consequence of a renewed interest in historical pharmacology is a return to some older well-tried drugs that have suffered an eclipse in recent years. Prominent among these is aspirin.[65]

The history of aspirin, which has received much attention recently,[66] dates from early times when salicylates of plant origin were used to deaden pain.[67] Its value as an anti-thrombotic agent along with other well-known thrombolytics such as heparin and warfarin has been re-examined.[68] Aspirin was patented, manufactured and introduced into medicine a century ago, making it one of the longest serving drugs, with plenty of useful life still left. After a period in which there were doubts, it has made its return as an effective treatment for some forms of heart disease. As a result, there has been increased interest in the drug in recent years. Besides general histories of aspirin,[69–71] there are some shorter accounts covering aspects of its development and uses that show its continuing value.[72,73] Other studies of the history of aspirin place special emphasis on its origins in traditional medicine,[74] or on the life of Felix Hoffmann (1863–1946) who first synthesized it in pure form at Bayer in 1897.[75] Pharmacological studies by Hoffmann and Heinrich Dresser followed.[76] A little-known property of aspirin relates to the history of the drug as a plant regulator (a phytohormone).[77] There is also an account of the discovery, use and action of another early drug, the antibiotic sulphanilamide. The use of hypoglycemic sulphonamides in the treatment of diabetes and their effects on pancreatic beta cells has aroused interest.[78]

Medicinal chemistry is a complex subject that has recently been reviewed in simple terms.[79] The links between pharmacology, clinical medicine and clinical chemistry are also complex.[80] In recent years, reviews have also appeared of the history of clinical chemistry, in several contrasting contexts.[81,82] As it deals with such a variety of complex medical conditions, the claim of clinical chemistry to be regarded as a separate scientific discipline has been discussed[83] and the rise of clinical enzymology as a specialism within clinical chemistry has also been noted.[84] The role of molecular biology in elucidating drug action has been considered with respect to the importance of metabolic pathways as a guide to drug design.[85] The possibilities arising from the ever-increasing knowledge of the molecular structure of the human genome are also being explored with the intention of developing new drugs that will influence the *causes* of diseases rather than merely their effects.[86,87] Among the most powerful drugs at present in use are those that affect the mind and personality.[88] The control of drug use and abuse is essential; it is the latter that causes major problems in many countries.[89]

Much recent literature relates to the discovery of individual specific drugs.[90] Among the alkaloids whose history has been discussed are strychnine[91] and the social problems concerning cocaine use and abuse in Hungary.[92] In this paper, developments from the establishment of its constitution by Willstätter to work since 1980 on addiction and the introduction of new chemical antagonists are discussed. The stereochemistry and structural analysis of the quinoline alkaloids from Pasteur to Prelog has also been reviewed recently.[93] The uses of biosynthesis in preparing certain alkaloids have also been investigated[94] and other specific drugs are discussed in *Chronicles of Drug Discovery*.[95] The introduction of antibiotics[96] and the rise of the Spanish antibiotics industry have been reviewed.[97] The discovery, production and uses of penicillin, the most widely known of the antibiotics, has continued to stimulate interest in Denmark.[98] Other valuable accounts of the history of the medical uses of rare earth compounds[99] and of the search for new drugs have also appeared,[100] including some potentially interesting drugs with specific properties.[101,102] A paper on the development of anti-malarial drugs demonstrates continuing interest in the rise of chemotherapy.[103] A closely related problem is that of insect resistance to chemical products.[104]

8.4 Origins of Biochemistry

Apart from its uses in the dispensary and *Materia Medica*, eighteenth-century medical chemistry was closely linked with animal and vegetable chemistry. In both, there were two main branches, one concerned with the composition of natural substances and the other with the chemistry of natural processes. In animal chemistry the former included the chemical constitution of substances such as flesh, bone, blood, urine, mucus and so on, and the latter included the chemistry of functions like respiration, digestion, assimilation and metabolism, secretion and excretion, the subject matter of early biochemistry.[105] In the eighteenth-century, the most difficult problems faced by the animal chemists resulted from the chemical complexity of natural substances and the vital functions, problems far beyond the capabilities of the chemistry of the time. Animal chemistry required a reliable knowledge of the composition of animal substances, but throughout the eighteenth century the usual method of 'animal analysis' was by destructive distillation. This resulted in the breakdown of complex organic matter into watery and oily fractions called 'spirits', together with some 'air' and a residue of charcoal.[106,107] These products were considered to be the components, or 'proximate principles', of the animal matter and were supposed to have been released by the action of heat. In fact, they were often decomposition products formed during the process itself. The true composition of animal matter therefore remained obscure until the nineteenth century. After about 1810, the situation changed as quantitative techniques of combustion analysis were introduced. These involved carefully weighing the organic sample, mixing it with an excess of an oxidizing agent, commonly copper oxide, and heating the mixture to redness. Methods such as this were introduced by Gay Lussac and Thenard, improved by Berzelius and Prout, and brought to a high degree of reliability by Liebig and his school at Giessen,[108] though Liebig's work on alkaloids was hampered by much uncertainty.[109]

However, if animal analysis was notoriously difficult, the complex chemical changes involved in animal functions raised such apparently insoluble problems that it seemed that special vital forces, unknown to mineral chemistry, must be at work in all living organisms. Most physiologists, who were reluctant to allow that chemistry, concerned with inanimate matter, had anything useful to say about the vital functions, readily accepted vitalism.[110] Many chemists, unable to provide convincing explanations of these complex processes, also felt constrained to accept the notion of a vital force in living matter, which made the chemistry of life quite different from mineral chemistry. The issue of vitalism continued to trouble chemists throughout the nineteenth century.[111] As few physicians paid much attention to chemistry, antipathy to chemical studies continued in medicine until the beginning of the twentieth century.[112,113]

The mechanisms of digestion and respiration occupied the attention of some chemists and most physiologists in the eighteenth century and, as the usual route to chemical knowledge was through a medical education, many who were interested in these topics were physicians. Many were unwilling to employ chemical methods, or to accept the results, but there were some who recognized the utility of chemistry in understanding the vital functions.[114] In 1752, Réaumur described experiments he had made on birds of prey such as the kite in which he showed that the gastric juice was acidic. In 1771, Edward Stevens extended these experiments to animals and concluded that a powerful solvent secreted by the stomach lining was the cause of digestion. Lazaro Spallanzani had shown that the gastric juice could digest food outside the stomach and so undermined the belief that digestion could only occur in the living organ. The medical history of gastric physiology and pathology has been reviewed.[115]

In Sweden, during the eighteenth century, Torbern Bergman (1735–1784) made advances in qualitative and quantitative chemical analysis,[116,117] while Carl Wilhelm Scheele examined many new substances.[118–120] The latter made numerous chemical discoveries, and his identification and analyses of many naturally occurring compounds gave his work special value for the animal chemists. Scheele examined tartaric, citric, malic, mucic, oxalic and lactic acids, all of which he prepared in a pure state and analysed; he also investigated the components of urinary calculi and was the first chemist to recognize uric acid. Scheele's work was a distinct advance on all that had gone before and provided a strong indication to chemists that the known methods of analysis and synthesis used in mineral chemistry could also be applied to natural substances found in living matter.

Joseph Fruton has discussed the origins of biochemistry[121] and major steps in its development are recorded using seminal extracts and authoritative analytical commentaries in a recent documentary study.[122] Fruton has also supplied students of the history of biochemistry with an invaluable bio-bibliographic resource giving extensive references to sources for the work of a very large number of chemists and biochemists.[123,124] There is also another useful bibliographical account of medical and biomedical biography, which although not specifically biochemical is nevertheless highly relevant.[125] Another recent work on modern biochemistry includes a historical account of the development of biochemical concepts, methods and applications.[126]

8.5   Early Physiological Chemistry

Between about 1760 and 1820 the numbers of animal substances whose composition was known with reasonable reliability was largely extended by Bergman and Scheele in Sweden and others, notably A. F. Fourcroy and L. N. Vauquelin in France. However, despite these discoveries substantial progress in animal chemistry was not possible in the eighteenth century until the theoretical foundations of anti-phlogistic chemistry had been firmly established. The chemical revolution that overturned the phlogiston theory and refurbished chemical thought in other ways has always attracted considerable attention from historians of chemistry and much has been written in celebration of the publication of Lavoisier's *Éléments de Chimie* (1789) and his untimely death in 1794 (see Chapter 2). Following the establishment of the oxygen theory, Lavoisier turned his attention to the chemistry of the vital functions, beginning with respiration. He showed that one of the main products of respiration was carbon dioxide and his view that respiration was a form of combustion in the lungs led him to suggest erroneously that this was the *sole* source of animal heat. The experimental work and reasoning by which Lavoisier arrived at his new ideas on animal chemistry have been thoroughly researched by F. L. Holmes.[127] Marco Beretta discusses the importance of the respiration work for Lavoisier's reputation as an experimentalist. He also makes the interesting point that Madame Lavoisier was perhaps the first woman to play a part in such experimental work.[128]

Berzelius became interested in animal chemistry and its applications in medicine from the late eighteenth century. He began to lecture on the subject and in 1806–08 published a two-volume chemical textbook in Swedish based on his lectures. This was never translated into English and only limited parts of it became known in this country, but Berzelius became the recognized leader among animal chemists before Liebig. He insisted on the need to base deductions in physiological chemistry on experimental observations. Alan Rocke has discussed Berzelius' physiological chemistry in a broad contextual study of his work as a chemist of the Enlightenment.[129]

From the late eighteenth century, it had become clear that study of the chemistry of animal secretions and functions in health and disease could greatly improve the reliability of medical diagnosis and treatment.[130] With this in mind, nineteenth-century physicians often studied animal chemistry. The chemical constitution of the blood and its changes in health and disease was a topic of central importance in physiological chemistry and a problem for physicians. Work on this subject in Britain and France from the late eighteenth to the mid-nineteenth centuries has been reviewed.[131] The value of animal chemistry was early recognized in the study of urinary calculi – a disease that was a scourge of the upper classes during the eighteenth century. It was widely studied and there was much literature on it, often linked to the uses of natural mineral waters that were thought to be capable of preventing the formation of bladder stones, or of dissolving stones already formed.[132] The high incidence of bladder stone in the seventeenth and eighteenth centuries ensured that much attention was paid to this disease and to efforts to prevent or cure it.[133] However, the causes of bladder stone remained an unsolved pathological problem until recent times.[134]

Attempts were also made to correlate clinical observations with the results of chemical analyses of the urine and its deposits and to post-mortem observations. No one was more successful in this field than Richard Bright working at Guy's Hospital in London with a team of young physicians and medical students.[135–137] They studied the symptoms of kidney disease by clinical and chemical examinations and related their observations to post-mortem examination of the morbid kidney. Bright, whose work was published in 1825, was one of the first physicians to couple his hospital treatments systematically with chemical and physiological research directed towards precision in diagnosis and treatment.[138]

Besides such applications of chemistry to physiological and pathological problems, there were also many chemists who worked on natural products in the chemical field. Of these Chevreul was among the most important, not only because between 1811 and 1823 he elucidated the structures of animal fats, but also because he introduced the use of melting points as a precise means of determining the purity of the fats. His researches showed that the animal fats are glyceryl esters of high molecular weight that may be compared with inorganic compounds such as salts.[139]

The chemical nature of the brain has long aroused great interest. Much work has been done on its neurological functions, but the chemistry of brain tissues, which seem superficially to resemble the fats, was also investigated from Fourcroy's time throughout the nineteenth century.[140] The most notable nineteenth-century physiological chemist to work on the chemistry of the brain was J. L. W. Thudichum, a London physician of German origin.[141,142] Thudichum investigated various natural substances, including the matter of gall-stones, the pigment urochrome, and the red matter of the blood, which he called haematoporphrin. Thudichum offended some influential members of the scientific establishment by denying the existence of 'protagon', the supposed major constituent of the brain, proposed by O. Liebrich in 1864 and adopted by leading physiological chemists in Britain and Germany. The heat of the controversy resulted in an almost total lack of recognition of Thudichum's work until Otto Rosenheim of King's College, London revived his remarkable discoveries in 1930.

By the middle of the nineteenth century, a great many chemical facts were known about natural substances and functions, but there was as yet no chemical theory of physiological functions. The first serious attempt to provide such a theory had been made by the London physician, William Prout,[143] after a long series of analyses and experiments on respiration, digestion, and the functions of the kidney. In 1824, as he was about to publish a theory of metabolism, Prout discovered that the acid present in the gastric juice is hydrochloric acid and this made him reconsider his proposed theory. Those who continued to think that the acidity of the gastric juice was due to lactic acid, or some other organic acid rejected Prout's discovery, but his observations were soon supported from other quarters. J. G. Children identified hydrochloric acid in cases of dyspepsia, while, in extensive investigations on digestion, F. Tiedemann and L. Gmelin in Heidelberg recognized the presence of hydrochloric acid in the stomach. However, the most startling evidence came from William Beaumont, an American physician who through a strange chance had been able to make physiological tests on the gastric juice in the living human stomach.[144,145] Beaumont had been called to treat a serious gunshot wound sustained by Alexis St Martin, a French Canadian. The wound healed leaving a gastric fistula through which Beaumont was able to conduct a long series of experiments under all conditions.[146,147] Beaumont initiated a field of study that has been pursued ever since.[148]

With renewed confidence bolstered by the supporting evidence accrued since his original discovery, Prout returned to the task of creating his metabolic scheme. He divided foods into three categories: albuminous, oleaginous and saccharine (comparable with the modern categories of proteins, fats and carbohydrates). He identified several stages in the digestion of food as it passed through the organs, and he suggested that chyle and chyme, formed during the process, were intermediate steps in the fabrication of blood. As the latter was considered a vital fluid, this series of changes seemed to exemplify the transformation from inanimate food to living blood, though the precise point at which the transformation occurred was not clear.[149,150] Prout's scheme, published in 1834, was the most comprehensive then available, but it was superseded eight years later by that of Liebig, who based his ideas on Lavoisier, Berzelius, Gmelin and others, including the Dutch animal chemist J. G. Mulder, whose 'proteine' theory was first proposed in a letter to Liebig published in 1838. Mulder had identified an organic radical that he claimed was common to all protein in both animals and plants; it seemed that this common compound would make the transformation of plant foods into muscle tissues easy, and Liebig eagerly adopted Mulder's 'proteine' in his theory. Liebig and Mulder later disagreed about the methods of physiological chemistry,[151] but in 1842 Liebig constructed various chemical routes by which animal proteins could in theory yield heat and degradation products that included uric acid and, ultimately, urea.[152] These compounds were excreted in the urine and Liebig took their quantity to be a measure of the muscle tissues used up by the animal during exertion. He was not a physiologist and seems not to have understood the complexity of metabolic processes. His system, based upon the crude comparison of intake and output, was ultimately doomed to failure. Yet, as it was seductively simple, it gave rise to controversy over the following three decades as physiologists and chemists took sides in efforts to provide the detailed evidence required to support or refute Liebig's theory.

By 1870, it had become clear that the theory was so seriously flawed as to be untenable, but in the meantime much had been accomplished, not least the incipient breakdown of resistance on the part of many physiologists to chemical methods. Instrumental in this was the work done by those like Bidder and Schmidt, Bischoff and Voit, leading to the establishment of a school of physiological chemistry in Munich using quantitative methods based on those of Lavoisier and Boussingault[153] in France and of Liebig in Germany.[154] Although Bischoff and Voit ultimately failed to substantiate Liebig's theories, there were those who sought to apply his ideas in medical research. Henry Bence Jones did so with some degree of success.[155–157] The so-called 'Bence Jones proteins' have later been recognized to be fundamentally important in the auto-immune system.[158] Others too, who did not follow Liebig's methods precisely, were nevertheless encouraged to apply chemical methods in medical research.[159] Thus Liebig's work stimulated the development of experimental physiology in the nineteenth century,[160] but it was equally important in promoting biochemistry. This was especially true of the biochemistry of muscular activity, for Liebig's whole thesis was built on the notion that muscular energy resulted from the oxidative degradation of muscle tissue. While this idea contains a germ of truth, the process is very much more complex than could ever have been imagined by Liebig, and a long series of biochemical researches were needed before the causes and consequences of muscular contraction could be clarified.[161] This is an important study not only in biochemistry, but also in medical chemistry, since the health of the muscular system is essential for the general health of the body. Working in Cambridge, where biochemistry was prominent in the first half of the twentieth century, Dorothy Needham found the stimulus to investigate the experimental history of this aspect of biochemistry up to 1970.[162]

Discoveries essential for the advancement of physiological chemistry were also made in other fields. Berzelius had recognized that traces of foreign elements and compounds often speed up chemical reactions. He also realized that something similar was occurring during reactions such as fermentation, and investigations of the role of the living yeast cell in fermentation led to the recognition of 'enzymes', a term which was introduced by Wilhelm Kühne in 1878. The first enzyme was discovered in 1833 by Payen and Persoz who found a constituent in malt, which converted starch into sugar. This was named diastase and its discovery was soon followed by other similar cases. For example, pepsin was extracted from gastric juice by Theodore Schwann in 1836. Recognition of the action of these natural compounds preceded the introduction of the term 'catalysis' by Berzelius in 1837. Linked to the early work on enzymes was the realization that many reactions in which they take part occur among living cells. Cagniard de Latour showed in 1838 that fermentation was due to living organisms and this was reinforced by Pasteur's work. Liebig proposed a mechanical explanation of fermentation, which depended on the putrefaction of matter rather than life-processes and this was adopted by many chemists who resisted the introduction of vital action to the explanation of chemical change. Until 1860, there was no thought that the enzymes might operate *inside* living cells, but in that year Berthelot extracted the enzyme from macerated yeast cells and showed that it hydrolysed cane sugar, indicating that the enzyme was contained within the cells. In 1897, Buchner obtained a cell-free juice from yeast, which showed all the fermenting properties of living yeast cells and finally confirmed that enzyme action was wholly chemical and did not, as Pasteur had thought, require the presence of living matter.[163]

Many nineteenth-century organic chemists worked on natural substances, concentrating on the chemistry of compounds such as carbohydrates and proteins. The leader in this field was, undoubtedly, Emil Fischer, who had studied chemistry under Kekulé and von Baeyer. Fischer devised methods for determining the structures of carbohydrates, beginning with the constitution of glucose.[164] He later worked on the proteins, their formation and degradation in animal tissues and organs due to the action of enzymes, themselves found to be proteins. Fischer suggested that the specificity of each enzyme for a particular chemical change could be compared to the mechanism of a lock and its key.[165] This analogy has since been found to be very powerful as an aid to understanding the mechanisms on which the behaviour of enzymes depends. In the twentieth century, it has been shown that most of the known enzymes occur *within* the cells and cell-biochemistry has led developments in understanding life processes. Together with molecular biology, cell biochemistry has made considerable contributions to medical diagnosis and treatment.[166] At the same time, these studies have caused some of the most difficult and poignant human and ethical problems, the full effects of which are yet to be seen.

Malcolm Dixon, who had worked with Gowland Hopkins at Cambridge, briefly reviewed the development of enzyme chemistry in 1970.[167] Each enzyme is specific to a particular biochemical change and some hundreds of enzymes are now known, each with a specific biochemical function.[168] So ubiquitous have enzymes been found to be that they have been considered the very foundation of biochemistry.[169] Enzymes control biochemical reactions in living organisims for while they are chiefly concerned with promoting chemical changes, it is also known that they are sometimes involved in inhibiting such changes as well as promoting them. Many enzyme reactions have been minutely studied using isotopes.[170] The history of enzymology in relation to fermentation chemistry has been reviewed in a recent study in which the importance of chemistry for the development of biochemistry around 1900 was investigated, especially considering the work of E. Fischer, A von Baeyer and E. Buchner in Berlin.[171]

In the later nineteenth century Claude Bernard, discoverer of the glycogenic function of the liver, was perhaps the most influential among physiological chemists.[172] Bernard was a pupil of the French physiologist Françoise Magendie, from whom he learned vivisectional methods and techniques of experimental pharmacology, but Bernard also spent time studying chemistry in Pelouze's laboratory. Consequently, he was better equipped than most of his contemporaries to deal with the integration of chemical and physiological methods. Recognizing that physiological chemistry rested on the foundations laid by the animal chemists from the late eighteenth century, Bernard began his studies with a critical analysis of their work. F. L. Holmes has examined these influences on Bernard's work.[173] Bernard went much further, elaborating the liver's glycogenic function between 1848 and 1853,[174] and showing the importance of the fluids in which the cells of the body are bathed. He recognized that there were continual osmotic exchanges through the cell walls between the contents of the cells and their milieu; from this, he developed his theory of an internal environment controlling the balance between these fluids and the contents of the cells themselves. Joseph Barcroft, a member of Hopkins' research team at Cambridge later revived this concept.[175] Bernard pioneered an experimental approach to medicine that inspired others to use chemical methods in physiological research.[176] Frederick William Pavy, a physician at Guy's Hospital who studied carbohydrate metabolism and the symptoms of diabetes, also followed Bernard's lead. Pavy observed that Bernard's 'glycogen', the saccharine matter found in the liver, was not present in life. Bernard had himself observed this on numerous occasions, but had chosen to explain away the problem. Pavy argued that sugar developed in the liver rapidly after death and tried to account for this as a consequence of the circulation continuously removing it or preventing its formation during life. Bernard's recognition that the liver secretes glucose, which is then released into the blood stream, began the search for other organs with a similar function. Charles Edward Brown-Séquard, an American working in Paris, followed Bernard's work on the internal secretions and in 1878 succeeded Bernard as professor of experimental medicine at the Collège de France. Brown-Séquard is best known as one of the founders of endocrinology. He also introduced some of his methods to American medicine.[177]

In Germany, physiological chemistry was advanced by the work of former students of the physiologist Johannes Müller. This group included Theodor Schwann, Emil du Bois-Reymond, Ernst Wilhelm von Brücke, Herman Helmholtz and Rudolf Virchow. In the 1840s, Carl Ludwig, a friend of several of these men, recognized the importance of chemistry in biological studies. The value of chemical methods in physiological research was gradually accepted by physiologists, but the limitations of contemporary chemistry were also emphasized. Liebig's summary equations were patently inadequate for the detailed investigation of biochemical change. The chemical constitution and reactions of albuminoid substances, considered the main constituents of protoplasm, remained undefined. The view that the chemical processes by which cellular products like glycogen are linked to life processes, and the recognition, following Pasteur's work, that some reactions, like fermentations, are caused by the life-processes of microbial organisms, ensured that the chemical study of such biological problems did not figure prominently in the work of Liebig's chemical contemporaries and followers. The emphasis in physiology had shifted from chemistry towards biology.

However, a very important outcome of the controversies that raged over Liebig's work was the stimulation given to biochemical research. This led to the establishment of chemical research schools across Europe, beginning with Liebig's own school in Giessen. Thus, in the 1870s, physiological chemistry was again becoming accepted. The nineteenth-century research schools were characterized not only by their different subject matter but even more by the personalities and styles of the leaders. Joseph Fruton, himself a leading biochemist, turned historian, has explored this in detail. In his study of the different styles adopted in research schools from Liebig to Gowland Hopkins, Fruton shows why some were more successful not only in producing experimental results, but also in turning out able research students who would go on to head their own schools and produce valuable work. He lists a large number of these with some details of their work.[178–180]

Many advances in biochemistry and medical chemistry came not from chemical departments but from physiological and pathological studies in medical institutes. The leaders in this field in the nineteenth century were Felix Hoppe-Seyler, Wilhelm Kühne, Emil Fischer, and Franz Hofmeister. The contributions made by each of these have been explored in some detail by Fruton; together they laid the foundations on which twentieth-century biochemistry would be built.[181] The applications of chemistry to medicine increased during the nineteenth century[182] and by the 1880s physiological chemistry was recognized as an independent academic discipline, which would become a branch of biochemistry before the end of the century.[96] Despite these developments, there has always been in biochemistry a desire to see the phenomena of life in anthropomorphic and even theological terms. The question of reductionism in biochemistry has also been hotly debated. This is fostered by the increasing extent to which chemical and even purely physical techniques are necessary to explore the structures and complex behaviour of biochemical molecules. Fruton has argued against reductionism and in favour of the pursuit of empirical results as the only way to develop biochemistry further.[183]

8.6 Proteins and Peptides

A recent autobiography by B. Merrifield, who won a Nobel Prize in 1984, describes his work on peptides, chiefly in association with the Rockefeller Institute and University.[184] A broader review of the history of peptide chemistry was published in 1991.[185] The discovery of heparin and its connection with the peptones, together with the late nineteenth-century work on peptone-shock, has been recalled in a recent paper.[186] In another connection, the discovery of glyceropeptides is described.[187] The long series of researches by M. A. Ondetti at the Squibb Institute for Medical Research led from peptides to peptidases and ultimately to a new drug, the 2-methyl analogue of 3-mercaptopropanoyl-L-proline, or Captopril, now used in the treatment of heart failure.[188] Neuropeptides have also recently come to the forefront of research in this field.[189,190] The neuropeptides have emerged as a class of highly specific physiologically active compounds possessing unique properties. Both biochemistry and neurochemistry were initiated within the ambience of animal chemistry in the nineteenth century, and together they have made many valuable contributions to progress in this field.[191]

D. E. Koshland and J. T. Edsall have recorded their own experiences in early protein research. Koshland stresses the value of his induced fit theory,[192] while Edsall describes his long career at Harvard after two years with Hopkins at Cambridge.[193] Fruton has also described the work of T. B. Osborne (1859–1928), noted for his analyses of amino-acids from seed proteins.[194] The idea that enzymes might be proteins was a matter of heated debate among chemists from about 1915, including arguments between Willstätter and James Sumner, who in 1926 isolated the enzyme urease and showed it to be a protein.[195] However, while it is most important to emphasize protein chemistry, the contributions made to protein science by physics and physical chemistry must not be overlooked,[196,197] nor should the development of studies of the nature and behaviour of macromolecules.[198,199]

The role of certain enzymes in cell control and human disease has also received attention, especially in regard to the specificity of enzymes and to protein phosphatases.[200] There has also been intense interest in biocatalysis as an application of enzyme chemistry in organic synthesis,[201] in the uses of proteolytic enzymes as shown by the work of Perutz,[202] and in nucleotide enzymology.[203] New methods of investigating the complexities of enzyme action have been described,[204,205] as have new methods of determining enzyme structures using studies of bacterial nutrition.[206] Other studies have investigated the relations between allosteric proteins and enzymes,[207] and the mechanisms by which they function.

8.7 Biochemistry Since 1900

From the 1880s, there were moves within the universities to establish chairs and departments of biochemistry.[208] This began in Germany and America, but after 1900 the development of biochemistry in British universities also happened rapidly, although progress was not always steady, especially in the years between 1910 and 1930.[209] In the rest of this chapter we consider some of the wide range of new work that has appeared in this vast field in recent years. At the beginning of the twentieth century, studies of the internal secretions led to the discovery of hormones, a term coined in 1905 by E. H. Starling, who introduced it in his Croonian Lectures to the Royal College of Physicians. The idea arose out of his work on the pancreatic juice with W. M. Bayliss, published in 1902, in which they named the pancreatic hormone 'secretin'. Starling thought of the hormones as chemical messengers carried by the blood. T. B. Aldrich and J. J. Abel had already identified adrenalin[210] in 1901, thyroxine was discovered by E. C. Kendall in 1914, and insulin was isolated in 1922 by F. G. Banting[211,212] and C. H. Best.[213] Insulin treatments now about seventy-five years old have recently attracted renewed attention by medical historians.[214, 215] It is now known that hormones are chemicals of various kinds, including proteins, peptides and steroids. They control biochemical functions, promote growth, and are necessary to maintain a healthy balance in the body.[216] Their effects explain the rapid rise of comparative endocrinology in the twentieth century.

A brief, but still valuable, review of the discovery of vitamins was given by Leslie J. Harris in 1970.[217,218] Early researches by C. Eijkman on the cause of beri-beri have also been discussed.[219] The complex molecule vitamin B_{12} has been studied for forty years; its structure as a cobalt-containing corrin was established by Dorothy Hodgkin in 1952, whose life and work were reviewed following her death in 1994.[220] Since then work by Battersby and many others has led ultimately to an understanding of all the stages of its biosynthesis, 'nature's route to vitamin B_{12} has finally been mapped out'.[221] 1955 was the 200th anniversary of an Admiralty decision to issue every sailor with a daily ration of lime-juice as a preventive against scurvy, an important anniversary for vitamin C. The same year saw the 25th anniversary of Pauling's controversial advocacy of huge doses of vitamin C to ward off the common cold and other illnesses.[222] The history of the discovery of the nutritional properties of vitamin C has also been reviewed.[223] A new account of the early uses of it in the prevention of scurvy at sea was given at the Royal Society of Medicine in January 1997,[224] provoking some lively discussion in the ensuing months. A comparison of the British and Russian experiences of the incidence of scurvy in the eighteenth century has also appeared.[225] The results of vitamin deficiencies continue to arouse interest and attract nutritional and metabolic studies.[226] Reviews of historical work on vitamin D deficiency related to rickets have appeared,[227] together with new assessments of earlier work,[228] the use of vitamin-rich cod-liver oil,[229] and projections on the uses of vitamin D in nutrition and health for the next century.[230] However, if rickets was a common social disease in nineteenth-century industrial societies, the general decline in population was even more problematical. This has been ascribed to vitamin E deficiency in the diet.[231]

From the beginning of the nineteenth century it had been realized that the energy required to maintain life-processes must come from oxidation. In 1840, following his discovery of ozone, C. F. Schönbein suggested that the conversion of oxygen in to ozone was an essential step in these processes and this idea came to be widely accepted. Later it was thought that the active agents were peroxides but from 1920, with the realization that biological oxidations are dependent on enzymes, two other theories were proposed. One due to O. Warburg postulated a respiratory enzyme that was identified with cytochrome in 1925; the other led to the discovery of the dehydrogenases, enzymes that bring about dehdrogenation, rather than oxidation. These two theories, though opposed to each other, were linked by several intermediate steps brought about by the cytochromes and discovered by D. Keilin using spectroscopic techniques.[232] Researches on intermediate metabolism have figured prominently in the twentieth century.[233] Metabolic pathways have been studied using radioactive isotopes as labelled atoms whose fate can be traced in the body. Proteins, lipids, carbohydrates and other dietary constituents have separate metabolic pathways, which are crosslinked at many points. The metabolism of the carbohydrates led Sir Hans Krebs in 1937 to his work on the tricarboxylic acid, or Krebs, cycle, a complex series of chemical changes by which lactic acid derived from glucose in the muscles is oxidized, releasing large amounts of energy. Krebs' work and the origins of the Krebs cycle have been studied in detail by F. L. Holmes.[234]

The revolution in molecular biology of the late 1940s was based on a new understanding of the concept of biochemical specificity. It was seen to be essential to determine the precise molecular structures of the nucleic acids,[235] but Frederick Sanger and Erwin Chargaff also showed that the *sequence* of amino acid subunits was crucially important.[236] At the same time, it was gradually becoming clear that genes were made, not of protein, but of DNA and that it was in fact genes that determined the structure of proteins. This completely changed the science of genetics.[237] The history of ideas and experiments to show that DNA is the genetic material has been reviewed in a recent paper.[238] DNA is the best-known and most important discovery in molecular biology; a complete volume of the *Annals* of the New York Academy of Science is dedicated to it, beginning with facsimile reprints of three original papers and including many portraits of early workers in the field.[239] The difficulties in accepting these ideas are vividly conveyed by Max Perutz[240] and a short historical survey of the work done on DNA in the ten years *before* recognition of the double helix is also given.[241] Other workers involved in the search before Crick and Watson's discovery include Gulland and Jordan at Nottingham.[242] Brief accounts of the discovery of the double helix have been given,[243,244] together with two autobiographical notes by Francis Crick himself in which he gives credit to the scientists whose early work prepared the ground for discovery of the double helix.[245,246] Robert Olby's classic account of the researches leading to the double helix, first published in 1974, has been reprinted,[247] and subsequent reflections on the genetic role of DNA by others have also included a measure of historical analysis.[248,249] James D. Watson has always been his own premier publicist and has continued by lectures, writing, and television to present the story of DNA.[250,251] In an autobiographical account by S. Nishimura work on nucleotide modification and base conversion in RNA since the 1950s has been reviewed.[252,253] Studies in this field have been pursued at many levels and in diverse university departments,[254] with far-reaching consequences.[255] The full implications of genetic studies in medical, ethical, and legal terms are yet to be revealed. Michel Morange has recorded the history of molecular biology to the late 1990s, but it is by no means complete.[256]

Another field of intense recent activity has been that of steroid chemistry. Although cholesterol, the earliest known steroid was discovered in the eighteenth century,[257] little research was done on these compounds until their medical implications became clear in the 1940s. A review by Sir Ewart Jones tells the story of early steroid research in England.[258] The discovery of vitamin D in fish-liver oils and of oestrogens in pregnancy urine seem to have triggered an explosion of research activity in this field. Notable research was carried out by Ian Heilbron at Liverpool, Otto Rosenheim at the National Institute of Medical Research in London and Jones himself at Imperial College, Manchester, and Oxford. A different kind of stimulus came to A. J. Birch at Oxford, who was encouraged in general by Sir Robert Robinson and in particular by a rumour that Luftwaffe pilots were dosed on cortical hormones to promote resistance to shock.[259] During this work, Birch devised a technique for the partial hydrogenation of aromatic systems, using sodium in liquid ammonia with ethanol as a proton donor. One application of the Birch reduction led to the development of oral contraceptives, a field in which Djerassi has been prominent among others.[260,261] His partial syntheses of steroids, and later studies in marine steroids, paved the way to his celebrated work in optical rotatory dispersion and circular dichroism.

There has been renewed interest recently in early work leading to the discovery of the contraceptive pill,[262] particularly on the pharmacological effects of the oestrogens[263] and progestins.[264] Many firms working in this field have featured in personal reminiscences in the journal *Steroids*. They include BDH until its take-over by Glaxo in 1968;[265] Searle, with its early research leading to the contraceptive pill itself;[266] Ciba in Basle where the search was on for oestrogens, then aldosterone and finally progestins.[267] It was at Syntex in Mexico, whose founders were R. R. Marker, E. Somlo and F. A. Lehmann,[268] that Djerassi carried out his pioneering work on the contraceptive pill.[269] Ruzika's early work on terpenes was connected with the production of corticoids and oestrogens from sterols of the Mexican yam.[270] There are also accounts of steroid research at Upjohn in Michigan, where the search for the synthesis of rare corticosteroids was prompted by the discovery in 1949 of the beneficial effects on rheumatoid arthritis of cortisone at the Mayo Clinic.[271] The Schering Corporation of America supported work on anti-inflammatory steroids,[272] while cortisone was investigated at Merck.[273] Early work on comparative endocrinology has also been described: at Organon in the Netherlands, the first

European Company to produce insulin and important suppliers of oestrone,[274] Lederle,[275] and Roussel Uclaf and Schering AG in Europe.[276] Another development arising from this early research on steroids was Barton's pioneering studies in conformational analysis. The reminiscences include those by Fried on the discovery of the fluorosteroids at the Squibb Institute for Medical Research in New Jersey,[277] and of studies of the biochemistry of steroid hormones linked to genetics and the history of cancer.[278] Clearly, the problems of unravelling the complex mechanisms of steroid action have aroused considerable interest.[279]

References

1. R. P. Multhauf, *The Origins of Chemistry*, Gordon & Breach, Williston, VT, 1993.
2. W. Dymock, C. J. H. Warden and D. Hooper, *Pharmacographica Indica, History of the Principal Drugs of Vegetable Origin Met with in British India*, 3 vols, Kegan Paul, Trench Tübner, London, 1890–93; Facsimilie Reprint, Low Price Publications, Delhi, 1995.
3. C. Balaban, J. Erlen and R. Siderits, eds, *The Skilful Physician*, Harwood Academic, New York, 1998.
4. B. K. Holland, ed., *Prospecting for Drugs in Ancient and Medieval European Texts*, Harwood Academic, New York, 1996.
5. R. I. McCallum, *Antimony in Medical History*, Pentland Press, Durham, 1999.
6. J. M. Stillman, *Paracelsus, his Personality and Influence as a Physician, Chemist and Reformer*, new edn., Kessinger Publications, Kila, MT, 1997.
7. T. U. Hoogenraad and K. D. Rainsford, eds, 'History of zinc therapy', in *Copper, Zinc Inflammatory Degenerative Diseases*, Kluwer, Dordrecht, 1998, pp. 1–5.
8. B. T. Moran, 'The alchemical world of the German Court, occult philosophy and chemical medicine in the circle of Moritz of Hessen (1572–1632)', *Südhoffs Arch.*, Beiheft, 1991, **29**, pp. 7–193.
9. B. T. Moran, 'Court authority and chemical medicine, Moritz of Hessen, Johannes Hartmann, and the origins of academic chemiatria', *Bull. Hist. Med.*, 1989, **63**, 225–246.
10. A. G. Debus, *The Chemical Philosophy, Paracelsian Science and Medicine in the Sixteenth and Seventeenth Centuries*, 2 vols, Science History Publications, New York, 1977.
11. P. Rattansi, 'Recovering the Paracelsian milieu', in *Revolutions in Science, their Meaning and Relevance*, ed. W. R. Shea, Science History Publications, Canton, MA, 1988, pp. 1–26.
12. W. Pagel, *Paracelsus, an Introduction to Philosophical Medicine in the Era of the Renaissance*, 2nd edn., S. Karger, Basle, 1982.
13. W. Pagel, *The Smiling Spleen, Paracelsianism in Storm and Stress*, S. Karger, Basle, 1984.
14. J. G. O'Shea, 'Two minutes with venus, two years with mercury', Mercury as an antisyphilitic chemotherapeutic agent', *J. Roy. Soc. Med. London*, 1990, **83**, 392–395.
15. W. Pagel, *J. B. van Helmont, Reformer of Science and Medicine*, Cambridge University Press, Cambridge, 1982.

16. W. Pagel, *From Paracelsus to van Helmont, Studies in Renaissance Medicine and Science*, ed. M. Winder, Variorum Reprints, London, 1986.

17. A. G. Debus, 'The Paracelsians and the chemists, the chemical dilemma in Renaissance medicine', *Clio Med.*, 1972, **7**, 185–199.

18. B. T. Moran, 'A survey of chemical medicine in the 17th century, spanning court, classroom and cultures', *Pharm. Hist.*, 1996, **38**, 121–133.

19. O. P. Grell, ed., *Paracelsus, the Man and his Reputation, his Ideas and their Transformation*, Brill, Leiden, 1998.

20. A. G. Debus, *The English Paracelsians*, Oldbourne, London, 1965, ch. IV, pp. 137–174; with a bibliography of 16th and 17th-century medical titles.

21. A. G. Debus, *The French Paracelsians, the Chemical Challenge to Medical and Scientific Tradition in Early Modern Europe*, Cambridge University Press, Cambridge, 1992.

22. J. Shackelford, *Paracelsianism in Denmark and Norway in the 16th and 17th Centuries* (On Petrus Severinus 1540/2-1602), *Diss. Abstr. Int.*, 1990, **50**, 3342–A.

23. J. Shackelford, 'Paracelsianism and patronage in early modern Denmark', in *Patronage and Institutions, Science, Technology and Medicine at the European Court 1500–1750*, ed. B. T. Moran, Boydell Press, Rochester, NY, 1991.

24. R. Multauf, 'Medical chemistry and "The Paracelsians"', *Bull. Hist. Med.*, 1954, **28**, 101–126.

25. B. T. Moran, *Chemical pharmacy enters the university, Johannes Hartmann and the didactic care of chymiatria in the early 17th century*, Publications of the American Institute of History of Pharmacy. No. 14, Madison, WI, 1991.

26. A. G. Debus, 'Alchemy and iatrochemistry, persistent traditions in the 17th and 18th centuries', *Quim. Nova*, 1992, **15**, 262–268.

27. A. B. Davis, 'The circulation of the blood and chemical anatomy', in *Science, Medicine and Society in the Renaissance*, ed. A. G. Debus, 2 vols, Science History Publications, New York, 1972, vol. 2, pp. 25–37.

28. N. G. Coley, 'Cures without care', 'Chymical physicians' and mineral waters in seventeenth-century English medicine', *Med. Hist.*, 1979, **23**, 191–214.

29. M. Crillo *et al.*, 'A history of salt', *Am. J. Nephrol*, 1994, **14**, 426–431.

30. P. Astrup, P. Bie and H. C. Engell, *Salt and Water in Culture and Medicine*, trans. by K. Skovgjerg and A. L. Cameron-Mills, Munksgaard, Copenhagen, 1993.

31. S. A. M. Adshead, *Salt and Civilisation*, Macmillan, Basingstoke, 1992.

32. D. L. Cowen and W. H. Helfand, *Pharmacy, an Illustrated History*, Abrams, New York, 1990.

33. C. Friedrich, 'The influence of pharmacists on the development of pharmacy and chemistry as academic disciplines', *Pharmazie*, 1992, **47**, 541–546 [in German].

34. J. P. Sousa Dias, 'Secret medicines and pharmaceutical chemistry in Portugal in the 17th and 18th centuries', in *Revolutions in Science, their Meaning and Relevance*, ed. W. R. Shea, Science History Publications, Canton, MA, 1988, pp. 221–238.

35. J. W. Estes, *Dictionary of Protopharmacology, Therapeutic Practices, 1700–1850*, Science History Publications, Canton, MA, 1990.

36. J. Duffin and P. René, '"Antimoine; Anti-biotique": The public fortunes of the secret properties of antimony potassium tartrate (tartar emetic)', *J. Hist. Med.*, 1991, **46**, 440–456.

37. E. V. Tyler, 'A history of pharmacy, future opportunities', *Pharm. Hist.*, 1993, **35**, 163–168.

38. J. H. Gregory and E. C. Stroud, *The History of Pharmacy, an Annotated Bibliography*, Garland, New York, 1995.

39. H. M. Dingwall, 'Making up the medicine, apothecaries in 16th and 17th-century Edinburgh', *Caduceus*, 1994, **10**, 121–130.

40. J. Simon, 'The chemical revolution and pharmacy, a disciplinary perspective', *Ambix*, 1998, **45**, 1–13.

41. A. G. Debus, 'Quantification and medical motivation, factors in the interpretation of early modern chemistry', *Pharm. Hist.*, 1989, **31**, 3–11.

42. P. Turner, 'Pharmaceutical medicine, a new specialty comes of age', *J. Pharm. Med*, 1993, **3**, 77–83.

43. J. Liebenau, G. J. Higby and E. C. Stroud, eds, *Pill Peddlers, Essays on the History of the Pharmaceutical Industry*, American Institute of the History of Pharmacy, Madison, WI, 1990.

44. J. P. Swann, 'The evolution of the American pharmaceutical industry', *Pharm. Hist.*, 1995, **37**, 76–86.

45. G. J. Higby, 'Chemistry and the 19th-century American pharmacist', *Bull. Hist. Chem.*, 2003, **28**, 9–17.

46. M. T. Robson, 'The pharmaceutical industry in Britain and France 1919–1939', PhD (London) thesis, LSE, 1993.

47. R. Porter and D. Porter, 'The rise of the English drug industry, the role of Thomas Corbyn', *Med. Hist.*, 1989, **33**, 277–295.

48. G. Tweedale, *At the Sign of the Plough, 275 Years of Allen & Hanburys and the British Pharmaceutical Industry, 1715–1990*, Murray, London, 1990.

49. R. P. T. Davenport-Hines, *Glaxo, a History to 1962*, Cambridge University Press, Cambridge, 1992.

50. C. Weir, *Jesse Boot of Nottingham*, Boots, Nottingham, 1994.

51. P. J. Egenolf, ed., *Pharmaziehistorische Bibliographie*, Govi-Verlag, Eschborn, Germany (formerly *Pharmaziegeschichtliche Rundschau*), 1993.

52. E. Costa, A. G. Karczmar and E. S. Vesell, 'Bernard B. Brodie and the rise of chemical pharmacology', *Annu. Rev. Pharmacol.*, 1989, **29**, 1–21.

53. J. Liebenau, 'Paul Ehrlich as a commercial scientist and research administrator', *Med. Hist.,* 1990, **34**, 65–78.

54. J. Liebenau, 'Ethical Business, the formation of the pharmaceutical industry in Britain, Germany and the United States before 1914', *Bus. Hist.*, 1988, **30**, 116–129.

55. M. Ramsey, 'Academic medicine and medical industrialism, the regulation of secret remedies in nineteenth-century France', *Clio Med.*, 1994, **25**, 25–78.

56. A. I. Marcus, 'From Ehrlich to Waksman, and the seamed web of the past', in *Beyond History of Science, Essays in Honor of Robert E. Schofield*, ed. Elizabeth Garber, Lehigh University Press, Bethlehem, PA, 1990, pp. 266–283.

57. J. Parascandola, 'Form and function, early efforts to relate chemical structure and pharmacological activity', *Can. Bull. Med. Hist.*, 1988, **5**, 61–72.

58. J. Parascandola, 'To bond or not to bond: chemical versus physical theories of drug action', *Bull. Hist. Chem.*, 2003, **28**, 1–8.

59. V. Rein, *Drugs Looking for Diseases, Innovative Drug Research and the Development of the Beta Blockers and the Calcium Antagonists*, Kluwer Academic, Dordrecht, 1991.

60. C. T. Collery, 'Medicine and the pharmacological revolution', *J. Roy. Coll. Physici., Lond.*, 1994, **28**, 59–69.

61. M. Weatherall, *In Search of a Cure, a History of Pharmaceutical Discovery*, Oxford University Press, Oxford, 1990.

62. H. Schadewaldt, 'Historical aspects of pharmacological research at Bayer, 1890–1990', *Stroke*, 1990, **21**, (no. 12 Suppl.), IV5–IV8.

63. G. S. Marks, 'The history of pharmacology in Canada', *Trends Pharmacol. Sci.*, 1994, **15**, 205–210.

64. X. Gao, 'Lectures on Chinese pharmacology, genuine and high-quality drugs', *J. Tradit. Chin. Med*, 1994, **14**, 147–151.

65. J. R. Vane and R. M. Botting, 'The History of Aspirin', in *Aspirin and Other Salicylates*, Proceedings of Conference, Chapman & Hall, London, 1992, pp. 3–16.

66. S. Jourdier, 'A miracle drug', *Chem. Br.*, 1999, **35**, 33–35.

67. T. Hedner and B. Everts, 'The early clinical history of salycilates in rheumatology and pain', *Clin. Rheumatol.*, 1998, **17**, 17–25.

68. R. L. Mueller and S. Scheidt, 'History of drugs for thrombotic disease, discovery, development, and directions for the future', *Circulation*, 1994, **89**, 432–449.

69. K. D. Rainsford, *Aspirin and the Salicylates*, Butterworths, London, 1984.

70. R. Alstaedter, ed., *Aspirin, the Medicine of the Century*, Bayer AG, Germany, 1984.

71. M. Viktorin, 'The remedy from willow-bark: 100 years of aspirin', *LaborPraxis*, 1999 (July-August), 82–85 [in German].

72. M. D. Mashkovsky, 'Acetylsalicylic acid among currently available drugs', *Khim-Farm Zh.*, 1994, **28**, 4–8.

73. D. Martinetz, 'Acetylsalicylic acid: drug with history and future', *Wiss. Fortschr.*, 1992, **42**, 339–341.

74. M. Sanjurjo, 'Aspirin, a legacy of traditional medicine', *Educ. Quim.*, 1996, **7**, 13–15.

75. H. D. Schwarz, 'Fiftieth anniversary of the death of Felix Hoffmann', *Dtsch. Apoth. Ztg.*, 1996, **136**, 673–674.

76. E. Verg, P. Gottfried and H. Scultheis, *Milestones, the Bayer Story 1863–1988*, Bayer AG, Leverkusen, 1988.

77. Q. Shi, 'Aspirin, an old medicine and a new phytohormone', *Guangxi Shifan Daxue Xuebao*, 1993, **11**, 69–74 [in Chinese].

78. J. C. Henquin, 'Fifty years of hypoglycemic sulphonamides', *Rev. Fr. Endocrinol. Clin. Nutr. Metab.*, 1993, **34**, 255–259 [in French].

79. C. R. Ramsden, 'Medicinal chemistry, a multidisciplinary science', *Sch. Sci. Rev.*, 1994, **75**, 49–58.

80. J. Büttner, 'Interrelationships between clinical medicine and clinical chemistry, illustrated by the German-speaking countries in the late 19th-century', *J. Clin. Chem. Clin. Biochem*, 1982, **20**, 465–471.

81. J. S. King, 'Clinical chemistry, a fragmentary history (1969–77)', *Clin. Chem.*, 1994, **40**, 2106–2110.

82. E. Kaiser, 'Clinical chemistry in Austria, past-present-future, reflections on the occasion of the 25th anniversary of the Austrian Society for Clinical Chemistry', *Eur. J. Clin. Chem. Clin. Biochem.*, 1994, **32**, 579–582.

83. J. Büttner, 'Clinical chemistry as scientific discipline, historical perspectives', *Clin. Chim. Acta*, 1994, **232**, 1–9.

84. J. Büttner, 'Evolution of clinical enzymology', *J. Clin. Chem. Clin. Biochem.*, 1981, **19**, 529–538.

85. A. Cornish-Brown, 'Two centuries of catalysis', *J. Biosci. (Bangalore, India)*, 1998, **23** (2), 87–92.

86. J. Drews, 'Intentions and coincidences in drug research. The influence of biotechnology', *Pharm. Ind.*, 1995, **57**, 970–976 [in German].

87. J. Drews, 'Drug discovery: a historical perspective', *Science*, 2000, **287**, 1960–1964.

88. D. M. Perrine, *Chemistry of Mind-Altering Drugs, History, Pharmacology and Cultural Context*, American Chemical Society, Washington DC, 1997.

89. J. F. Spillane, *Modern Drug, Modern Menace, the Legal Use and Distribution of Cocaine in the United States 1880–1920*, Carnegie Mellon University, Pittsburgh, PA, 1994.

90. M. Weatherall, *In Search of a Cure, a History of Pharmaceutical Discovery*, Oxford University Press, Oxford, 1990.

91. J. W. Nicholson, 'The story of strychnine', *Educ. Chem.*, 1993, **30**, 46–47.

92. G. Fodor, 'New chemical achievements in the struggle against cocaine', *Magyar Kém. Lapja*, 1995, **3**, 93–101.

93. F. Eden, 'A historical trip: quinine and other quinoline alkaloids. Part 2. elucidation of the stereochemistry of quinoline alkaloids', *Pharm. Zt.*, 1999, **28**, 11–20 [in German].

94. A. R. Battersby, 'The marvels of biosynthesis, tracking down nature's pathways', *Biorg. Med. Chem.*, 1996, **4**, 937–964.

95. S. Bindra and D. Lednicer, eds, *Chronicles of Drug Discovery*, No. 2, Wiley, New York, 1983. The following papers are of particular interest here: H. Kawaguchi and T. Naiti, 'Amikacin' (an aminoglycide antibiotic), pp. 207–234; B. Ehström and G. Sjäbey, ' Bacampicillin' (penicillin-related antibiotic), 109–132; M. A. Ontelli, D. W. Chapman and B. Rubin, 'Captopril' (inhibitor of Angiotensin-Converting Enzyme (ACE), 1–32; M. Gorman, S. Kukolja and R. R. Chauvette, 'Cefaclor' (a cephalosporin antibiotic), 49–85.

96. C. L. Moberg, *Launching the Antibiotic Era, Personal Accounts of the Discovery and Use of the First Antibiotics*, Rockefeller University Press, New York, 1990.

97. N. Puig, 'Networks of innovation or networks of opportunity? The making of the Spanish antibiotics industry', *Ambix*, 2004, **51**, 167–185.

98. J. Nielsen, 'History of penicillin', *Dan. Kemi.*, 1995, **75**, 24–28 [in Danish].

99. C. H. Evans, 'Medical uses of the rare earths', *Digestion*, 1996, **57**, 205–228.

100. P. Meyer, 'Discovering new drugs, the legacy of the past, present approaches and hopes for the future', *Pract. Med. Chem.*, 1996, 11–24.
101. C. M. Cimarusti, 'Aztreonam' [an antimicrobial agent], in *Chronicles of Drug Discovery*, Conference Proceedings, American Chemical Society, Washington DC, 1993, pp. 239–297.
102. H. Inoue and T. Nagao, 'Diltazem' [used in many cardiac conditions], in *Chronicles of Drug Discovery*, Conference Proceedings, American Chemical Society, Washington DC, 1993, pp. 207–238.
103. L. B. Slater, 'Malaria chemotherapy and the 'kaleidoscopic' organisation of biomedical research during World War II', *Ambix*, 2004, **51**, 107–134.
104. J. S. Ceccatti, 'Biology in the chemical industry: scientific approaches to the problem of insecticide resistance, 1920s–1960s', *Ambix*, 2004, **51**, 135–147.
105. M. Florkin, *A History of Biochemistry Pt I Proto-biochemistry; Pt 2 Proto-biochemistry to Biochemistry*, vols 30–33 of *Comprehensive Biochemistry*, ed. M. Florkin and E. H. Stoltz, 34 vols, Elsevier, Amsterdam, London, New York, 1979.
106. A. G. Debus, 'Fire analysis and the elements in the sixteenth and the seventeenth centuries', *Ann. Sci.*, 1967, **23**, 127–147.
107. F. L. Holmes, 'Analysis by fire and solvent extractions, the metamorphosis of a tradition', *Isis*, **62**, 1971, 129–148.
108. F. L. Holmes 'Elementary analysis and the origins of physiological chemistry', *Isis*, 1963, **54**, 50–81.
109. M. C. Usselman, 'Liebig's alkaloid analyses: the uncertain route from elemental content to molecular formulae', *Ambix*, 2003, **50**, 71–89.
110. E. Glas, *Chemistry and Physiology in their Historical and Philosophical Relations*, Delft University Press, Delft, 1979.
111. I. Sencar-Cupovic, 'The rise and breakdown of vitalism in nineteenth-century South Slavic chemical literature', *Hist. Phil. Life Sci.*, 1984, **6**, 183–198.
112. G. K. Hunter, *Vital Forces; the Discovery of the Molecular Basis of Life*, Academic Press, San Diego, CA and London, 2000.
113. N. G. Coley, 'Studies in the history of animal chemistry and its relation to physiology', *Ambix*, 1996, **43**, 164–187.
114. A. Cunningham and R. French, eds, *The Medical Enlightenment of the 18th Century*, Cambridge University Press, Cambridge, 1990.
115. A. L. Blum, 'Solitary views of the stomach', *Digestion*, 1996, **57**, 287–298.
116. W. A. Smeaton, 'Bergman', in *Dictionary of Scientific Biography*, ed. C. C. Gillispie, Scribners, New York, vol. 2, 1970, pp. 4–8.
117. W. A. Smeaton, 'Torbern Olof Bergman, from natural history to quantitative chemistry', *Endeavour*, 1984, NS **8**, 71–74.
118. F. Kohl, 'Carl W. Scheele - discoverer of oxygen', *Pharm. Ztg.*, 1992, **137**, 48–50.
119. W. A. Smeaton, 'Carl Wilhelm Scheele (1742–1786), provincial Swedish pharmacist and world-famous chemist', *Endeavour*, 1992, NS **16**, 128–131.
120. U. Boklund, 'C. W. Scheele', in *Dictionary of Scientific Biography*, ed. C. C. Gillispie, Scribners, New York, vol. 12, pp. 143–150.
121. J. S. Fruton, 'The emergence of biochemistry', *Science*, 1976, **192**, 327–334.

122. M. Teich and D. M. Needham, *A Documentary History of Biochemistry 1770–1940*, University of Leicester Press, Leicester & London, 1991; Fairleigh Dickinson University Press, Madison, NJ, 1992.

123. J. S. Fruton, *Selected Bibliography of Biographical Data for the History of Biochemistry since 1800*, American Philosophical Society, Philadelphia, PA, 1974.

124. J. S. Fruton, *A Supplement to a Bio-bibliography for the History of the Biochemical Sciences since 1800*, American Philosophical Society, Philadelphia, PA, 1985.

125. L. T. Morton and R. J. Moore, *A Bibliography of Medical and Biomedical Biography*, Scholar Press, Aldershot, 1989.

126. M. G. Ord and L. A. Stocken, 'Biochemistry then and now', in *Foundations of Modern Biochemistry*, ed. M. G. Ord and L. A. Stocken, 4 vols, Greenwich, CT; JAI Press, London, England, 1995–1998, vol. 4, 1998, pp. 267–280.

127. F. L. Holmes, *Lavoisier and the Chemistry of Life; An Exploration of Scientific Creativity*, University of Wisconsin Press, Madison, WI, 1985.

128. M. Beretta, *Imaging a Career in Science: The Iconography of Antoine Laurent Lavoisier*, Science History Publications, Canton, MA, 2001, pp. 43–52.

129. A. J. Rocke, 'From physiology to organic chemistry (1805–1814)', in *Enlightenment Science in the Romantic Era, the Chemistry of Berzelius and its Cultural Setting*, ed. E. M. Melhado and T. Frängsmyr, Cambridge University Press, Cambridge, 1992, pp. 107–131.

130. W. F. Bynum, *Science and the Practice of Medicine in the Nineteenth Century*, Cambridge University Press, Cambridge, 1994.

131. N. G. Coley, 'Early blood chemistry in Britain and France', *Clin. Chem.*, 2001, **47**, 2166–2178.

132. N. G. Coley, 'Animal chemists and the urinary stone', *Ambix*, **28**, 1971, 69–93.

133. H. A. Ellis, *A History of Bladder Stone*, Blackwell Scientific Publications, Oxford, 1969.

134. K. Lonsdale, 'Human stones', *Science*, 1969, **159**, 1199–1207.

135. D. Berry, *Richard Bright, 1789–1858, Physician in an Age of Revolution and Reform*, Royal Society of Medicine, London, 1992.

136. L. S. King, 'Richard Bright', in *Dictionary of Scientific Biography*, ed. C. C. Gillispie, Scribners, New York, vol. 2, 1970, pp. 463–465.

137. P. Bright, *Dr Richard Bright (1789–1858)*, Bodley Head, Oxford, 1983.

138. S. J. Peitzman, 'Bright's disease and Bright's generation, toward exact medicine at Guy's Hospital', *Bull. Hist. Med.*, 1981, **55**, 307–321.

139. A. B. Costa, *Michel Eugene Chevreul, 1786–1889: Pioneer of Organic Chemistry*, University of Wisconsin Press, Madison, WI, 1962.

140. D. B. Tower, *Brain Chemistry and the French Connection (1791–1841)*, Raven Press, New York, 1994.

141. D. L. Drabkin, *Thudichum, Chemist of the Brain*, University of Pennsylvania Press, Philadelphia, PA, 1958.

142. T. L. Sourkes, *The Life and Work of J. L. W. Thudichum 1829–1901*, McGill University, Osler Library Studies in the History of Medicine, Montreal, 2003.

143. W. H. Brock, 'The life and work of William Prout', *Med. Hist*, 1965, **IX**, 101–126.

144. R. B. Nelson, *Beaumont, America's First Physiologist*, Grant House Press, Geneva, IL, 1990.

145. J. J. Bylebyl, 'William Beaumont, Robley Dunglisson, and the "Philadelphia physiologists"', *J. Hist. Med.*, 1970, **25**, 3–21.

146. W. Beaumont, *Experiments and Observations on the Gastric Juice and the Physiology of Digestion* (1833), Dover Reprint, Plattsburg, NY, 1959; Oxford Historical Books, Abingdon, 1989.

147. A. L. Blum, 'Solitary views of the stomach', *Digestion*, 1996, **57**, 287–298.

148. H. W. Davenport, *A History of Gastric Secretion and Digestion, Experimental Studies to 1975*, Oxford University Press, Oxford, 1992.

149. W. H. Brock, 'William Prout', in *Dictionary of Scientific Biography,* ed. C. C. Gillispie, Scribners, New York, vol. 11, 1975, pp. 172–174.

150. W. H. Brock, *From Protyle to Proton; William Prout and the Nature of Matter 1785–1985*, Adam Hilger, Bristol and Boston, 1985, 12–59, 109–142.

151. E. Glas, 'The Liebig-Mulder controversy. On the methodology of physiological chemistry', *Janus*, 1976, **63**, 27–46.

152. F. L. Holmes, 'Introduction' to J. Liebig, *Animal Chemistry or Organic Chemistry in its Application to Physiology and Pathology*, trans. W. Gregory, Cambridge, 1842; Facsimile Ed., Johnson Reprint Corporation, New York and London, 1964.

153. F. W. J. McCosh, *Boussingault. Chemist and Agriculturalist*, Reidel, Dordrecht & Boston, 1984.

154. F. L. Holmes, 'The intake-output method of quantification in physiology', *Hist. Stud. Phys. Biol. Sci.*, 1987, **17**, 236–270.

155. L. Rosenfeld, 'Henry Bence Jones (1813–1873), the best 'chemical doctor' in London', *Clin. Chem.*, 1987, **33**, 1687–1692.

156. F. W. Putnam, 'Henry Bence Jones, the best chemical doctor in London', *Perspect. Biol. Med.*, 1993, **36**, 565–579.

157. L. G. Fine, 'Henry Bence Jones (1813–1873): on the influence of diet on urine composition', *Kidney Int.*, 1990, **37**, 1019–1025.

158. N. G. Coley, 'Henry Bence Jones MD, FRS, (1812–1873)', *Notes Rec. Roy. Soc. Lond.*, 1973, **28**, 31–56.

159. N. G. Coley, 'George Owen Rees, M. D., FRS. (1813–89), pioneer of medical chemistry', *Med. Hist.*, 1986, **30**, 173–190.

160. W. Coleman and F. L. Holmes, eds, *The Investigative Enterprise, Experimental Physiology in 19th-century Medicine*, University of California Press, Berkeley, CA, 1988.

161. D. M. Needham, *The Biochemistry of Muscle*, Methuen, London, 1932.

162. D. M. Needham, *Machina Carnis; the Biochemistry of Muscular Contraction in its Historical Development*, Cambridge University Press, Cambridge, 1971.

163. K. Manchester, 'Eduard Buchner and the origins of modern biochemistry', *S. Afr. J. Sci.*, 1998, **94**, 100–102.

164. M. Engel, 'On the 100th anniversary of the determination of the constitution of glucose by Emil Fischer, comments on a change in paradigms', *Mitt.-Ges. Dtsch .Chem., Fachgruppe Gesch. Chem.*, 1991, **6**, 44–55.

165. R. U. Lemieux and U. Spohr, 'How Emil Fischer was led to the lock and key concept for enzyme specificity', *Adv. Carbohydr. Chem. Biochem.*, 1994, **50**, 1–20.

166. M. Perutz, *Protein Structure, New Approaches to Disease and Therapy*, Freeman, New York, 1992.

167. M. Dixon 'The history of enzymes', in *The Chemistry of Life, Eight Lectures on the History of Biochemistry*, ed. Joseph Needham, Cambridge University Press, Cambridge, 1970, pp. 15–37.

168. H. Beinert, 'Looking at enzymes in action in the 1950s', *Protein Sci.*, 1994, **3**, 1605–1612.

169. R. E. Kohler, Jr., 'The enzyme theory and the origin of biochemistry', *Isis*, 1973, **64**, 181–196.

170. I. A. Rose, 'Isotopic strategies for the study of enzymes (1900–1995)', *Protein Sci.*, 1995, **4**, 1430–1433.

171. M. Engel, 'Enzymology and fermentation chemistry. Alfred Wohl's and Carl Neuberg's reaction schemes of alcoholic fermentation. The "chemical point of view" as characteristic of the Berlin school', *Mitt.-Ges. Deutsch. Chem., Fachgruppe Gesch. Chem.*, 1996, **12**, 3–29 [in German].

172. M. D. Grmek, 'First steps in Claude Bernard's discovery of the glycogenic function of the liver', *J. Hist. Biol.*, 1968, **1**, 141–154.

173. F. L. Holmes, *Claude Bernard and Animal Chemistry*, Harvard University Press, Cambridge, MA, 1974.

174. P. Lefebvre, N. Pacquot and A. Scheen, 'Role of the liver in sugar homeostasis, a recollection of the work of Claude Bernard', *J. Annu. Diabetol., Hotel-Dieu*, 1996, 1–9 [in French].

175. F. L. Holmes, 'Joseph Barcroft and the fixity of the internal environment', *J. Hist. Biol.*, 1969, **2**, 89–122.

176. M. D. Grmek, *Raisonnement expérimental et recherches toxicologiques chez Claude Bernard*, Droz, Paris, 1973.

177. M. Borell, 'Brown-Séquard's organo-therapy and its appearance in America at the end of the nineteenth century', *Bull. Hist. Med*, 1976, **50**, 309–320.

178. J. S. Fruton, *Molecules and Life: Historical Essays on the Interplay of Chemistry and Biology*, Wiley, New York, 1972.

179. J. S. Fruton, 'The emergence of biochemistry', *Science*, 1976, **192**, 327–334.

180. J. S. Fruton, *Contrasts in Scientific Style; Research Groups in the Chemical and Biochemical Sciences*, American Philosophical Society, Philadelphia, PA, 1990.

181. M. Weatherall, *Dynamic Science, Biochemistry in Cambridge, 1898–1949*, Wellcome Unit for History of Medicine, Cambridge, 1992.

182. J. Büttner, 'Wechselbeziehungen zwischen Chemie und Medizin im 19 Jahrhundert', in *Jahrbuch des Instituts für Geschichte der Medizin der Robert Bosch Stiftung*, ed. Renate Wittern, Bd. 2, Hippokrates Verlag, Stuttgart, 1983, pp. 7–24.

183. J. Fruton, *A Skeptical Biochemist*, Harvard University Press, Cambridge, MA, 1992.

184. B. Merrifield, *Life during a Golden Age of Peptide Chemistry, the Concept and Development of Solid-Phase Peptide Syntheses* (in the series 'Profiles,

Pathways and Dreams: Autobiographies of Eminent Chemists'), American Chemical Society, Washington DC, 1993.

185. T. Wieland and M. Bodanszky, *The World of Peptides, a Brief History of Peptide Chemistry*, Springer-Verlag, Berlin, New York, 1991.

186. J. A. Marcum, 'The discovery of heparin revisited, the peptone connection', *Perspect. Biol. Med.*, 1996, **39**, 610–625.

187. R. C. Yeo and L. W. Crandall, 'Glyceropeptides, classification, occurrence and discovery', *Drugs. Pharm. Sci.*, 1994, **34**, 1–16.

188. M. A. Ondetti, 'From peptides to peptidases, a chronicle of drug discovery', *Annu. Rev. Pharmacol. Toxicol.*, 1994, **63**, 11–27.

189. M. M. Klavdieva, 'The history of neuropeptides. 1', *Frontiers in Neuroendocrinol.*, 1995, **16**, 293–321.

190. M. M. Klavdieva, 'The history of neuropeptides. 2', *Frontiers Neuroendocrinol.*, 1996, **17**, 155–179; 247–280.

191. H. McIlwain, 'Biochemistry and neurochemistry in the 1800s, their origins in comparative animal chemistry', *Essays Biochem.*, 1990, **25**, 197–224.

192. D. E. Koshland, 'The joys and vicissitudes of protein science', *Protein Sci.*, 1993, **2**, 1364–1368.

193. J. T. Edsall, 'Memories of early days in protein science, 1926–1940', *Protein Sci.*, 1992, **1**, 1526–1530.

194. J. S. Fruton, 'Thomas Burr Osborne and chemistry', *Bull. Hist. Chem.*, 1995, **17/18**, 1–8.

195. A. B. Costa, 'James Sumner and the urease controversy', *Chem. Br.*, 1989, **25**, 788–790.

196. H. A. Scheraga, 'Contribution of physical chemistry to an understanding of protein structure and function', *Protein Sci.*, 1992, **1**, 691–693.

197. W. Kauzmann, 'Reminiscences from a life in protein physical chemistry', *Protein Sci.*, 1993, **2**, 671–691.

198. C. Tanford, 'Macromolecules', *Protein Sci.*, 1994, **3**, 857–861.

199. B. Ranby, 'Svedberg - discoverer of protein macromolecules', *Macromol. Symp.* no. 98, 1995, 1227–1245.

200. P. Cohen, 'The discovery of protein phosphatases, from chaos and confusion to an understanding of their role in cell regulation and human disease', *BioEssays*, 1994, **16**, 583–588.

201. M. K. Turner, 'Biocatalysis in organic chemistry, Part 1, Past and present (1800–1995)', *Trends Biotechnol.*, 1995, **13**, 173–177.

202. H. Neurath, 'Proteolytic enzymes past and present, the second golden era, recollections, special section in honor of Max Perutz', *Protein Sci.*, 1994, **3**, 1734–1739.

203. J. M. Buchanan, 'Aspects of neucleotide enzymology and biology', *Protein Sci.*, 1994, **3**, 2151–2157.

204. M. Cohn, 'Atomic and nuclear probes of enzyme systems', *Annu. Rev. Biophys. Biomol. Struct.*, 1992, **21**, 1–24.

205. I. A. Rose, 'Isotopic strategies for the study of enzymes (1900–1995)', *Protein Sci.*, 1995, **4**, 1430–1433.

206. E. E. Snell, 'From bacterial nutrition to enzyme structure, a personal odyssey', *Annu. Rev. Biochem*, 1993, **62**, 1–27.

207. J-P. Changeux, 'Allosteric proteins, from regulatory enzymes to receptors, personal recollections', *BioEssays*, 1993, **15**, 625–634.

208. R. E. Kohler, *From Medical Chemistry to Biochemistry*, Cambridge University Press, Cambridge, 1982.

209. N. Morgan, 'The strategy of biological research programmes, reassessing the "dark age" of biochemistry 1910–1930', *Ann. Sci*, 1990, **47**, 139–150.

210. M. E. Bowden, 'The rush to adrenaline', *Chem. Heritage*, 2003, **21** (1), 12–13; 38–40.

211. M. Bliss, *The Discovery of Insulin*, Allen and Unwin, London, 1996.

212. M. Bliss, *Banting, a Biography*, University of Toronto Press, Toronto, 1993.

213. H. A. Bern, 'The development of the role of hormones in development. A double remembrance', *Endocrinology*, 1992, **131**, 2037–2038.

214. C. Ionescu-Tirgoviste, 'Insulin, the molecule of the century', *Arch. Physiol. Biochem*, 1996, **104**, 807–813.

215. M. Bliss, 'Discovering the insulin documents, an archival adventure', *Watermark*, 1998, **21**, 516–528.

216. V. C. Medvei, *History of Clinical Endocrinology, A Comprehensive Account of Endocrinology from Earliest Times to the present Day*, Parthenon, Pearl River, NY, 1993.

217. L. J. Harris, 'The discovery of vitamins', in *The Chemistry of Life, Lectures on the History of Biochemistry*, ed. Joseph Needham, Cambridge University Press, Cambridge, 1970, pp. 156–170.

218. L. Rosenfeld, 'Vitamine - vitamin, the early years of discovery', *Clin. Chem.*, 1997, **43**, 680–685.

219. K. Carpenter and A. Teunis, 'The birth of vitamin studies, Eijkman in Java (1920–1929)', *Chem. Heritage*, 1996, **13**, 32.

220. J. P. Glusker, 'Dorothy Crowfoot Hodgkin (1910–1994)', *Protein Sci.*, 1994, **3**, 2465–2469.

221. L. R. Milgrom, 'Vitamin B_{12}, the view from the summit', *Chem. Br.*, 1994, **30**, 923–927.

222. J. Emsley, 'A life on the high Cs', *Chem. Br.*, 1995, **31**, 946–948.

223. R. A. Jacob, 'Three eras of vitamin C discovery', *Sub-cell. Biochem.*, 1996, **25**, 1–16.

224. D. P. Thomas, 'Sailors, scurvy and science', *J. Roy. Soc. Med*, 1997, **90**, 50–54.

225. R. Bartlett, 'Britain, Russia and scurvy in the eighteenth century', *Oxford Slavon. Pap.*, 1996, **29**, 23–42.

226. K. J. Carpenter, Beriberi, *White Rice, and Vitamin B: A Disease. A Cause, and a Cure*, University of California Press, Berkeley, CA, 2000.

227. P. Hernigou, 'Historical overview of rickets, osteomalacia and vitamin D (1600–1995)', *Rev. Rheum. Engl. Ed.*, 1995, **62**, 261–270.

228. A. Hardy, 'Rickets and the rest, child-care, dirt and the infectious children's diseases, 1850–1914', *Soc. Hist. Med.*, 1992, **5**, 389–412.

229. P. Wilton, 'Cod-liver oil, vitamin D and the fight against rickets', *Can. Med. Assoc. J.*, 1995, **152**, 1516–1517.

230. M. F. Holick, 'Vitamin D, new horizons for the 21st century', *Am. J. Clin. Nutr.*, 1994, **60**, 619–630.

231. D. Verrinder, 'Why did England and Wales suffer accelerated population decline in the late nineteenth century?', *Nutr. Health*, 1992, **8**, 223–226.

232. D. Keilin, *The History of Cell Respiration and Cytochrome*, ed. J. Keilin, Cambridge University Press, Cambridge, 1966.

233. F. L. Holmes, *Between Biology and Medicine: The Formation of Intermediary Metabolism*, Office for the History of Science and Technology, University of California at Berkeley, Berkeley, CA, 1992.

234. F. L. Holmes, *Hans Krebs*, 2 vols, Oxford University Press, Oxford, 1994.

235. M. H. Wilkins, A. R. Stokes and H. R. Wilson, 'Molecular structure of nucleic acids, molecular structure of deoxypentose nucleic acids [1900–1953], *Ann. N.Y. Acad. Sci.*, 1995, **758**, 13–16.

236. H. F. Judson, 'Frederick Sanger and Erwin Chargaff, and the metamorphosis of specificity', *Gene*, 1993, **135**, 19–23.

237. F. Jacob, 'Genetics and the twentieth century', *Gene*, 1993, **135**, 1–2; 5–7.

238. F. Kohl, 'Concepts and experiments to demonstrate DNA as the genetic material', *Deutsch. Med. Wochenschr.*, 1996, **121**, 1066–1069 [in German].

239. J. D. Watson and F. H. C. Crick, 'Molecular structure of nucleic acids, a structure for deoxyribose nucleic acid [1900–1953]', *Ann. N.Y. Acad. Sci.*, 1995, **758**, 13–14.

240. M. F. Perutz, 'Before the double helix', *Ann. N.Y. Acad. Sci.*, 1995, **758**, 9–13.

241. D. Hotchkiss, 'DNA in the decade before the double helix', *Ann. N.Y. Acad. Sci.*, 1995, **758**, 55–73.

242. H. Booth and M. J. Hay, 'Before Watson and Crick - the pioneering studies of J. M. Gulland and D. O. Jordan at Nottingham', *J. Chem. Educ.*, 1996, **73**, 928–931.

243. T. Travis, 'The twisted molecule', *Educ. Chem.*, 1993, **30**, 152–155.

244. K. L. Manchester, 'British Rail and the discovery of the double helical structure of DNA', *S. Afr. J. Sci.*, 1993, **89**, 525–527.

245. F. Crick, 'Looking backwards, a birthday card for the double helix', *Gene*, 1993, **135**, 15–18.

246. F. Crick, 'DNA, a co-operative discovery', *Ann. N.Y. Acad. Sci.*, 1995, **758**, 198–199.

247. R. Olby, *The Path to the Double Helix, the Discovery of DNA*, Dover, New York, 1994 (reprint of the 1974 edn. with additions).

248. G. S. Stent, 'The aperiodic crystal of heredity', *Ann. N.Y. Acad. Sci.*, 1995, **758**, 25–31.

249. M. McCarty, 'A fifty-year perspective on the genetic role of DNA', *Ann. N.Y. Acad. Sci.*, 1995, **758**, 48–54.

250. J. D. Watson, *A Passion for DNA: Genes, Genomes and Society*, Cold Spring Harbor Laboratory Press, New York, 2000.

251. J. D. Watson, *DNA, the Secret of Life*, Arrow Books, London, 2003.

252. D. A. Chambers, 'Forty years of DNA', *Ann. N.Y. Acad. Sci.*, 1995, **758**, 1–11.

253. S. Nishimura, 'Studies of modified nucleotides in RNA, past and future reflections on my work for the last three decades', *Biochemie*, 1994, **76**, 1105–1108.

254. M. H. F. Wilkins, 'DNA at King's College, London', *Ann. N.Y. Acad. Sci.*, 1995, **758**, 200–204.

255. J. Lederburg, 'What the double helix has meant for basic medical research, a personal commentary', *Ann. N.Y. Acad. Sci.*, 1995, **758**, 182–193.

256. M. Morange, *A History of Molecular Biology*, Harvard University Press, Cambridge, MA, 1997.

257. B. M. Ganzer, 'Cholesterol - the history of a controversial steroid', *Pharm. Ztg.*, 1993, **138**, 9–14; 16–17 [in German].

258. E. R. H. Jones, 'Early English steroid history', *Steroids*, 1992, **57**, 357–362.

259. A. J. Birch, 'Steroid hormones and the Luftwaffe, a venture into fundamental strategic research and some of its consequences, the Birch reduction becomes a birth reduction', *Steroids*, 1992, **57**, 363–377.

260. C. Djerassi, *Steroids Made It Possible*, (in the series 'Profiles, Pathways and Dreams: Autobiographies of Eminent Chemists'), American Chemical Society, Washington DC, 1990.

261. C. Djerassi, *The Pill, Pigmy Chimps and Degas' Horse*, HarperCollins, New York, 1992.

262. B. Asbell, *The Pill, a Biography of the Drug that Changed the World*, Random House, New York, 1995.

263. J. W. Goldzieher, 'The oestrogens', in *Pharmacology of the Contraceptive Steroids*, ed. J. W. Goldzieher and K. Fotherby, Raven, New York, 1994, pp. 21–25.

264. N. Perone, 'The progestins', in *Pharmacology of the Contraceptive Steroids*, ed. J. W. Goldzieher and K. Fotherby, Raven, New York, 1994, pp. 5–19.

265. V. Petrow and F. Hartley, 'The rise and fall of British Drug Houses', *Steroids*, 1996, **61**, 476–482.

266. F. B. Colton, 'Steroids and "the pill", early steroid research at Searle', *Steroids*, 1992, **57**, 624–630.

267. K. Heusler and J. Kalvoda, 'Between basic and applied research, Ciba's involvement in steroids in the 1950s and 1960s', *Steroids*, 1992, **57**, 492–503.

268. P. A. Lehmann, 'Early history of steroid chemistry in Mexico, the story of three remarkable men', *Steroids*, 1992, **57**, 403–408.

269. C. Djerassi, 'Steroid research at Syntex, "the pill" and cortisone', *Steroids*, 1992, **57**, 631–641.

270. G. Rosenkranz, 'From Ruzika's terpenes in Zurich to Mexican steroids via Cuba', *Steroids*, 1992, **57**, 409–418.

271. J. A. Hogg, 'Steroids, the steroid community and Upjohn in perspective, a profile of innovation', *Steroids*, 1992, **57**, 593–616.

272. H. Herzog and E. P. Oliveto, 'A history of significant steroid discoveries and developments originating at the Schering Corporation (USA) since 1948', *Steroids*, 1992, **57**, 617–623.

273. R. Hirschmann, 'The cortisone era: aspects of its impact: some contributions of the Merck laboratory', *Steroids*, 1992, **57**, 579–592.

274. S. A. Szpilfogel and F. J. Zeelan, 'Steroid research at Organon in the golden 1950s and the following years', *Steroids*, 1992, **57**, 483–491.

275. S. Bernstein, 'Historic reflection on steroids, Lederle and personal aspects', *Steroids*, 1992, **57**, 392–402.

276. V. Petrow, 'A history of steroid chemistry, some contributions from European industry', *Steroids*, 1992, **57**, 473–475.

277. J. Fried, 'Hunt for an economical synthesis of cortisol, discovery of the fluorosteroids at Squibb (a personal account)', *Steroids*, 1992, **57**, 384–391.
278. E. B. Thompson, 'Steroids, gene expression, and apoptosis, recollections of contributions and controversies', *Steroids*, 1998, **67**, 368–374.
279. S. Wright, 'Steroids and their mechanisms of action', *Proc. Roy. Coll. Physici. Edin.*, 1995, **25**, 34–39.

CHAPTER 9

Instruments and Apparatus

R. G. W. ANDERSON

Honey Hill House, Honey Hill, Cambridge CB3 0BG

9.1 Introduction

Since the publication of *Recent Developments in the History of Chemistry*,[1] there has been no clearly definable change in the way in which instruments and apparatus are regarded by historians of chemistry. There are still few scholars who take a significant interest in this field compared with, say, those who write on the history of mathematical or astronomical instruments. Often chemical instruments are merely mentioned in publications on the practice of chemistry, the main interests lying elsewhere. Furthermore, the matter is complicated by the poor survival rate of chemical apparatus. Laboratories and experiments are more often considered by those whose interests lie in the sociology of knowledge than by those working in the field of material culture. This must clearly be the conclusion of those who consult a useful editorial in *Isis* written by one whose own work has touched on both approaches.[2] The disparate directions from which instrument studies are being approached is exemplified by the published papers of a conference held in London in 1991, whose stated purpose was to examine the relationship between scientific instruments and the political and economic context of science.[3] The instruments themselves tended to lie outside the main foci of interest.

However, an encyclopaedia focusing on instruments, which includes material on chemical instruments and apparatus, was published in 1998. It incorporates 327 entries on a diverse range of historical items.[4] Its strengths are its fairly comprehensive coverage and inclusion of more recent and complex instruments. Descriptions of the historical development of each item are necessarily short. Though it does not concentrate particularly on chemistry, there is some useful material to be found in a German volume of 37 essays, including a chapter on a subject rarely treated, industrial reaction vessels.[5]

Sometimes papers that consider specific instruments are found not in well-circulated, established periodicals, but in ephemeral publications that can be frustratingly difficult to locate. There has been some useful progress, however. The *Bulletin for the History of Chemistry*, published by the Division of the History of Chemistry of the American Chemical Society, has, since its inception in 1988, included regular short articles on artefacts and laboratories. There are now also two specialized periodicals dealing with the history of scientific instruments: the *Bulletin of the Scientific Instrument Society*, published in Great Britain, originated in 1983, while *Rittenhouse Journal of the American Scientific Instrument Enterprise* first appeared in 1986. There continues to be an outpouring of inventories, directories and guides, which will be of significant value to those who will wish to pursue instrumental research topics in the future. Some important papers on archaeological discoveries which relate to early chemical practices have appeared. There have been a number of conferences on instruments in general and three conferences have been held that concentrated substantially on chemical instrumentation.

The XXth International Congress of the History of Science was held in Liège, Belgium, in July 1997. In the introduction to a volume arising from contributions to this meeting, mustered under the title *Scientific Instruments and Museums*, the editor remarks:

> ... instruments from the first decades of the 20th century are sometimes scientifically exciting instruments, related to very important discoveries in science of that period. If we do not write down the history of these instruments now, all related data will be lost. Preserving 20th century instruments – together with the scientific notes – should receive the full attention of the 'modern' science historian.[6]

Several essays are concerned with the recording of historical instruments.

A conference held at the Dibner Institute, Cambridge, Massachussets led to the production of what is probably the most substantial treatment of the history of chemical instruments in recent years.[7] Fourteen essays cover practices extending from medieval alchemy to the X-ray diffraction work of Michael Polanyi in the 20th century. Unlike the encyclopaedia mentioned above, contributions are discursive and diverse in approach (though there is a useful introductory essay and link passages). Included are pieces on evidence from archaeology, assaying, gas collection, eudiometry, hydrometry, thermometry, Lavoisier's apparatus and organic analysis.

Two helpful bibliographic works have appeared. The first, issued by the Beckman Center for the History of Chemistry in Philadelphia, is a short pamphlet for those who wish to start to work in the field.[8] The other is a survey of all books, pamphlets, catalogues and articles that have appeared on the history of scientific instruments between 1983 and 1995.[9] This bibliography has been published by the Scientific Instrument Commission of the International Union of the History and Philosophy of Science. It is an on-going project and an on-line cumulative bibliography is now available.[10]

The arrangement of the topics that follow continues that developed for *Recent Developments in the History of Chemistry*: General Works, Instruments and Apparatus of Particular Chemists and Laboratories; Instruments and Apparatus by Type and Function; Museum and Laboratory Collections; and Instrument Makers.

9.2 General Works

Information on instruments preserved in collections, on ephemeral printed material connected with instruments and on the makers of instruments is not easily available. Before substantial studies on instruments can be undertaken, basic curatorial information needs to be available. Over the past few years there has been some progress in compiling material in all of these categories.

Instrument makers catalogues can be extremely helpful though they need to be handled with care. An instrument appearing in a catalogue may not have been available from stock, may have been left on the list long after it went out of production or may have been constructed by a maker other than the one whose name appears on the title page (or, indeed, engraved on the instrument itself). Instrument makers catalogues have rarely been added to the stock of libraries in a systematic manner. Sometimes they survive, but in an uncatalogued state. One exercise attempted to list as many instrument makers catalogues, issued up to 1914, as could be discovered.[11] The published listing includes locations, and it is indicative of the destruction of this evidence that, within the 107 libraries surveyed, 75% of catalogues were found only as single examples. The production of the list has been well received and a new edition, incorporating recent discoveries, is being contemplated.

One of the roles of the Scientific Instrument Commission of the International Union of the History and Philosophy of Science has been to encourage publication of national inventories of preserved historical scientific instruments in public collections. *Recent Developments in the History of Chemistry* listed six of these in 1985. Now there are five more. It must be pointed out that only a small proportion of their content is of a chemical nature. The long-awaited inventory for the British Isles has at last appeared.[12] This lists approximately 2800 instruments in 200 collections, though the six largest institutions have been dealt with differently because of the very large number of objects which they each contain and because information about them should be available from expert curators on their staffs. A much fuller listing of instruments preserved in the Republic of Ireland and Northern Ireland appeared in its final version in 1995.[13] A very large proportion of entries is for instruments produced for teaching and demonstration purposes. Lastly there is the Danish national inventory with, perhaps, unexpected riches.[14] Much of Europe is now covered by these inventories, which, it has to be said, are very variable in quality and coverage. Although a general work on German scientific instruments has been published,[15] the great lacuna as regards inventories is Germany, a nation which should be particularly rich in chemical artefacts.

A major, systematic piece of research has resulted in a directory that lists British scientific instrument makers who were active over the period 1550 to 1851.[16] This has been compiled mainly from the records of the London companies and guilds. What is particularly fascinating is that it provides an insight into the 'family' relationships of makers by revealing who was apprenticed to whom.

Finally, in this section, a guide has appeared that is very pertinent to those who need to seek out historical apparatus and instruments.[17] This is a list of museums in Europe that possess collections which are of, or have a bearing on, the history of chemistry. It is a useful addition to the literature (earlier versions appeared in 1981 and 1982, the better 1982 publication being in Dutch) but it remains a rather personal response to what has been encountered on the author's travels.

9.3   Instruments and Apparatus of Particular Chemists and Laboratories

Relatively little has been published that is concerned particularly with the instruments and apparatus of particular chemists (as opposed to more general treatments of their work). The most significant work of this type concerns Antoine Laurent Lavoisier, whose demonstration apparatus, much of which is illustrated in his *Traité*, survives in fine condition at the Conservatoire National des Arts et Mètiers in Paris. But what, exactly, is its status? Trevore Levere and Jan Golinski confront this, and other issues, in separate papers.[18,19]

In most ways there are fewer problems with the apparatus of the improbable and isolated figure of the Rev. Nicholas Callan, a natural philosopher who taught at the Catholic seminary of St Patrick's College, Maynooth in Ireland and who did fundamental research in electromagnetism at the beginning of the 19th century. Callan is by no means unknown, but the fine collection of his scientific relics has been unexamined until now. Charles Mollan and John Upton have disinterred it and though most items are physical rather than chemical there is a section on "Chemistry and Analysis" in their catalogue.[20]

There are two minor papers that deal with the practice of chemistry at the University of Edinburgh in the latter half of the 18th century, one concerning the availability and manufacture of chemical glassware (an under-studied topic)[21] and the second with the supply of heat sources in laboratories (also under-studied) and, in particular, with the new form of chemical furnace designed by Joseph Black (1728–1799).[22] Though once very widely used, no original form of the furnace appears to have survived. This is a typical example of the poor survival rate, which may well be a factor in the paucity of papers on chemical apparatus.

There are a few publications on laboratories, two in the nature of guides to Liebig's laboratory at the University of Giessen.[23,24] Liebig's combustion apparatus has been discussed in a paper concerning his analyses of alkaloids.[25] There is also a substantial publication on the Laboratory of the Government Chemist in London.[26] This was established in 1842 as The Excise Laboratory to determine the strength of spirits to protect the revenue. Later its main focus was the detection of adulterants in food. A symposium held at the Royal Institution in London led to a volume on the development of laboratories of various types.[27] Four of the thirteen papers deal with chemistry laboratories in particular: Humphry Davy and his Royal Institution laboratory, the laboratories at the University of Glasgow (1747–1818), the laboratories of Finsbury Technical College (1878–1926), and portable chemical chests, developed for amusement, education and mineral analysis.

9.4   Instruments and Apparatus by Type and Function

There have been significant additions to the literature of chemistry and archaeology. Two new and major sites have been identified. One is in the centre of Paris, associated with the Louvre Palace, although at a slight distance from it.[28] The laboratory, which is identified as such by ceramic and glass alembics, cucurbits, distillation bases and a pelican, all of which were cast down a well in about 1350, seems to have been part of the College of St Nicholas. The actual purpose of the laboratory is not entirely clear, though as the site is adjacent to a medieval garden, its output may have been pharmaceutical products.

The other site of significance is at the Castle of Oberstockstall, near Vienna.[29–34] Under one of the floors was found what may be a major portion of a metallurgical laboratory, consisting of 800 objects dating from *c.* 1550–1600. These include alembics, aludels, cucurbits, matulae, distillation columns, crucibles and cupels. It appears to have been associated with Christoph von Trenbach, Vicar of Kirchberg and Canon of Passau, who was the likely patron.

Also from the German-speaking world, the alchemical laboratory of Graf Wolfgang II at Schloss Weikersheim has been described in considerable detail, from architectural studies and inventories. The operations of this laboratory date from 1587 to 1610.[35] Remarkably, the earliest physical evidence of a mineralogical laboratory in the New World is contemporary with this. Fragments of glass and ceramic vessels have been discovered at the Fort Raleigh site on Roanoke Island, North Carolina, where it is known from literary sources that chemical activities were conducted from 1585. This is briefly announced in a popular publication.[36] Archaeological work at Martin's Hundred, Williamsburg, Virginia had earlier produced a fine ceramic alembic probably dating from 1623 to 1635.[37] Proper archaeological reports have still to be published for both these sites. Considering early apparatus from the point of view of texts, a German translation of "De Sceuastica Artis" has appeared, a rare tract which Andreas Libavius affixed to the second edition of his *Alchymia* of 1606.[38] This includes illustrations of a wide variety of furnaces, distillation and sublimation apparatus. More generally, alchemical instrumentation has been considered,[39] though we await a comprehensive survey of surviving apparatus which could be thus categorized. The significance of the laboratory design of Andreas Libavius has been compared with that of Tycho Brahe.[40] A source of evidence of early chemical practices, but one that needs to be handled with care, is 17th- and 18th-century genre paintings of alchemists. The Chemical Heritage Foundation in Philadelphia, a lively body which was founded by the American Chemical Society and the Institute of Chemical Engineers, and which traces its origins to 1982, has built up a large collection of these enticing representations of early chemists surrounded by their paraphernalia and has published a selection of them with an interpretative text.[41]

There has been no systematic assessment of instruments or apparatus since Ernest Child's survey of 1940.[42] To some extent, what gets published is that which is visually attractive, intriguing or collectable. In the first category, there has been a work on equal-arm balances, though only a small proportion of those illustrated could be considered to be associated with chemical practices.[43–45] Somewhat unattractive, though unquestionably useful, were the chainomatic balances, the origin of which has been elucidated.[46] Other papers have dealt with specific categories of instrument, some modest though widely used and of great utility. Thus the test-tube clamp has been treated,[47] as have the urinometer (a hydrometer adapted for urine analysis),[48] and the chemical slide rule,[49] developed by W. H. Wollaston for the determination of equivalent weights of chemical substances.

More attention is being paid to the 20th century than hitherto. An overview of the place of instruments in research and their relationship to the chemical profession concludes: "Instruments and their production profoundly affect the very identity of chemical research and remain crucial in the ongoing stratification of scientific nations."[50] A rather more specific review regards the rise in importance of scientific instrumentation in developments in analytical chemistry between 1920 and 1950, causing what is referred to as "a major revolution".[51] The role played by new instrumental techniques in the development of analytical chemistry is discussed in Chapter 7. Specific aspects of such change are considered in papers dealing with the Melloni thermopile (used in some early infrared spectroscopes),[52] the grating spectrograph,[53] and process control instruments generally over the period 1900 to 1940, over which "the process industries moved from the spasmodic use of automatic controllers to accepting them as the norm".[54] Infrared spectroscopy, occasionally used for investigating organic compounds from as early as the 1880s, was not adopted to any extent by chemists until the late 1940s. After that, as instruments became simple, reliable and relatively inexpensive, the technique flourished.[55] Tiselius apparatus for the elecrophoretic analysis of colloidal mixtures was developed in the 1930s and 1940s. The joint effort between The Svedberg's group at the University of Uppsala and scientists at the Rockefeller Institute in New York to develop the technique has been examined.[56] A good review considers the introduction of nuclear magnetic resonance techniques to Australia.[57] The first experiments were conducted by physicists in 1952; in 1958, chemists were using experimental instruments and in 1962 the first standard commercial instrument was supplied.

Three meetings have brought together scholars who have a mutual interest in chemical instruments. All resulted in the publication of proceedings. The first was an American Chemical Society meeting in Chicago in September 1985, whose History of Chemistry Division held a symposium on the subject "The History and Preservation of Chemical Instrumentation".[58] Contributions were from historians, collectors, and scientists who take an interest in the historical background to their own field. Though the approach varied considerably and the quality of contribution was, unsurprisingly, uneven, there is much of interest, including a paper on the process of preservation itself.[59,60] Particularly interesting papers from this collection include one on the choice of raw materials in the manufacture of scientific instruments over the period 1880 to 1920[61] and a survey of the development of blowpipe analysis.[62]

The second meeting was also held under the auspices of the American Chemical Society, in Toronto in June 1988.[63] The subject was the general one of electrochemistry; only a proportion dealt with historical aspects of instrumentation. One paper provided an overview of electrochemical instrumentation[64] and others considered the specific topics of the pH meter,[65] the glass electrode,[66] and the dropping mercury polarograph of Jaroslav Heyrovsky.[67]

The third meeting was the most significant and resulted in an important and scholarly work. The publication stems from a conference held in London in 2000, which was titled "From the Test Tube to the Autoanalyzer: The Development of Chemical Instrumentation in the Twentieth Century". The approaches by the authors of the chapters vary: some deal with specific techniques (spectrophotometry, mass spectroscopy, chromatography), others consider the impact of instrumentation in the development of particular areas of chemical advance, while some consider commercial aspects. The first three chapters are key papers that have been reprinted from earlier works: Yakov Rabkin on infrared spectroscopy, Davis Baird on the instrumental revolution of 1920–1950, and Peter Morris (the overall editor) and Anthony Travis on the role of instrumentation in structural organic chemistry.[68]

Finally, in this section, publications concerning specific laboratories are considered. A monograph has been devoted to the Mineral Chemistry Laboratory of the Polytechnic School of Lisbon over the decade 1884 to 1894; the heavily sociological approach adopted is somewhat eccentric.[69] More straightforward are short papers that deal with American chemistry laboratories at the University of New Hampshire,[70] Columbia University [for whom the industrial chemist Charles Frederick Chandler (1836–1925) purchased the equipment–and is connected with relics and samples which survive],[71] and Transylvania University at Lexington Kentucky, one of the earliest of the medical schools in the New World.[72] Another laboratory, that at the University of Kazan in Russia, connected with the work of Aleksandr Milhailovich Butlerov (1828–1886) has been described.[73] Additionally, there is a paper on the fascinating and unregarded history of the lecture demonstration. A published lecture surveys this crucial aspect of chemistry teaching from the 16th century to the present.[74,75]

9.5   Museum and Laboratory Collections

The only substantial surviving British cabinet of scientific instruments is that usually associated with King George III (though its sources are, in fact, various). Originally housed in the Kew Observatory, it was removed to King's College, London, in the mid-19th century and in 1927 it transferred to the Science Museum. The collection has recently been published in the form of a lavish catalogue with valuable introductory essays.[76] Though there are relatively few chemical instruments, those representing the subject are of some interest, for example Lorenz's Instantaneous Light Medicine, Wedgwood's Pyrometer and Nooth's apparatus. A brief guide to the science in the 18th century, illustrated with items from the King George III Collection, has also been issued.[77]

Italy continues to publish details of the vast number of instrument collections that have been preserved or which have simply survived. Those from Piedmont, housed in Turin institutions, have been described.[78] Likewise, those surviving in the University of Siena have been published.[79] A useful volume on early scientific glassware there has appeared.[80] A particularly fine collection of instruments is to be found in the Institute of Chemistry of the University of Genoa.[81] It is largely of 19th-century origin and some of it can be associated with Stanislao Cannizzaro (1826–1910), who held the chemistry chair from 1855 to 1861. When he arrived in Genoa, the laboratory was in the most rudimentary of states. He developed an innovative teaching course and purchased apparatus (though chemistry was treated meanly compared with physics). The full study, which has now been published, brings together evidence from various sources and is a significant contribution.[82]

Three works elucidate chemical instrument collections in Spain (where one suspects that there is still much to discover). One deals with Andalusian scientific heritage and includes sections on pharmacy and medicine, and on laboratory instruments.[83] Another is a book based on the Museo de la Farmacia Hispana at the Complutense University of Madrid.[84] This well-illustrated work has parallel texts in English, French and Spanish. The material culture of pharmacy is divided up in sections, relevant chapters for historians of apparatus and instruments being "Glass and Crystal", "Mortars and Scales", and "Scientific Instruments". Last there is a catalogue of an exhibition entitled "La Casa de la Quimica. Ciencia, Artilleria e Ilustracion".[85] This was held in Segovia in 1992 to celebrate the bicentenary of the Royal Chemistry Laboratory of the Royal College of Artillery where Louis Joseph Proust (1754–1826) taught.[86] In the 19th century the University of Tartu had close links with universities in Western Europe, especially Jena; some 400 individual objects and several collections of chemical samples have been preserved from this time.[87]

Teyler's Museum at Haarlem, The Netherlands, retains one of the most complete collections of scientific instruments amassed by an institution for the purpose of teaching and research. An earlier volume, published in 1973, elucidates the collection acquired by Martinus van Marum between 1784 and 1837. The current volume, compiled by the same author, deals with the period *c.* 1840 to 1915.[88] A somewhat smaller proportion of the 454 entries deal with chemical apparatus compared with the earlier period, but what is to be found is not without interest, including an extensive set of equipment for testing the quality of town gas.

The technology of science teaching is rarely addressed, but an entertaining paper recently considered the role of the blackboard in classrooms in the United States.[89] An exhibition of blackboards, held in the Museum of the History of Science, Oxford, in 2005 was stimulated by their holding of an example inscribed by Albert Einstein.

A major new collection of artefacts has arisen from an initiative by the Chemical Heritage Foundation (CHF) in Philadelphia. As part of its operation, significant instruments have been preserved, together with personal and commercial papers, ephemeral publications and photographs. This undertaking is supported by the establishment of a "Hall of Fame" which honours contemporary scientific instrument makers and designers. Suitable premises to allow for the public exhibition of key objects is being created. The CHF has an active programme of publication and several books deal with instrumental techniques and their development, for example a history of the Pittsburgh Conference (a commercial exhibition of analytical instrumentation),[90] Arnold O. Beckman and his firm,[91] and the history of mass spectroscopy.[92]

9.6   Instrument Makers

No general surveys of chemical instrument makers have been published though a review of 19th-century scientific instrument-making in France has appeared.[93] Perhaps more useful to those interested in instruments themselves is a series of some thirteen papers that considers firms of French scientific instrument makers, singly or in small groups.[94,95] These papers draw attention to the vicissitudes of what was often a fragile industry.

The Irish chemical instrument-making trade has been surveyed.[96] Here the market was small and a good deal of what was sold was imported, even if the instrument bore the name of an Irish practitioner. The author concludes: "There is no real evidence for a successful specialist chemical instrument trade in Ireland". The rise of the American instrument-making profession is particularly interesting because of the shift from imports to home-made goods in the 19th century. One of the early makers, as opposed to suppliers, was Alva Mason who traded as a philosophical instrument maker in Philadelphia between 1817 and 1859. An 1837 broadsheet has been discovered which indicates that he offered blowpipes, balances and sulphuretted hydrogen machines as well as physics demonstration apparatus. A valuable paper contrasts Mason's business to that of George Washington Carpenter, a contemporary who was merely a supplier of laboratory chemicals and apparatus.[97]

Studying the sources of supply of instruments to particular institutions can be revealing about the state of the trade. A paper considers the provision of Lavoisier-type pneumatic apparatus to The Netherlands.[98] It shows that, at the end of the 18th century and early in the 19th century, various specialist suppliers were constructing these complex pieces, including George Adams of London, J. H. Tiedemann of Stuttgart, and J. H. Onderdewyngaart Canzius of Brussels and Delft. Instrument makers are not usually well served in general biographical dictionaries, but the 2004 edition of the British *Dictionary of National Biography* has considered this category of professional and a few makers and dealers of chemical instruments and apparatus are included: new entries[99,100] include Friedrich Christian Accum, John Joseph Griffin, and Ludwig Oertling.

Two small firms of instrument suppliers whose businesses struggled and eventually collapsed have been investigated. The more successful was the firm of Kemp & Co., which traded between 1829 and 1899 in Edinburgh.[101] The firm largely depended on the market offered by Edinburgh University and the flourishing extramural teaching profession. It diversified into the supply of photographic apparatus (from 1864) but it did not expand sufficiently to take advantage of the rapidly developing market in the UK as other firms did. A very brief episode in instrument making in an undeveloped market is explained in a paper that describes the Australian firm established by W. Russell Grimwade in May 1920. By November 1922 the firm had been wound up, a single example of the Grimwade Milligram Chemical Balance surviving today to illuminate the story of a failure.[102] Another study charts the development of the Australian industry.[103]

In contrast is the hugely successful American firm of Beckman Instruments, which constructed and marketed pH meters from 1935 and the DU Spectrophotometer from 1941. Papers of a biographical nature based on interviews with Arnold Beckman have been published.[104,105] The development of nuclear magnetic resonance spectrometers by the firm of Varian is considered in another paper, with emphasis on the introduction of the Varian A-60, the first commercial instrument intended for the broadly trained chemist as opposed to the custom-built tools for the research specialist.[106]

The role of instruments has never been generally recognized as a core pursuit in history of science studies. Issues which explain this have been tackled in a book, *Thing Knowledge,* by Davis Baird. It is significant that Baird, of the Philosophy Department of the University of South Carolina, is the son of Walter S. Baird, who founded the firm of Baird Associates in 1936, its first important product being grating spectrographs for quantitative chemical analysis. He writes:

> Part of the reason instruments have largely escaped the notice of scholars and others interested in our modern techno-scientific culture is language, or rather its lack. Instruments are developed and used in a context where mathematical, scientific and ordinary language is neither the exclusive vehicle of communication nor, in many cases, the primary vehicle of communication. Instruments are crafted artefacts, and visual and tactile thinking and communication are central to their development and use. Herein lies a big problem and a primary reason why instruments have been ignored by those who write about science and technology. Writers, reasonably enough, understand language to be the primary vehicle of communication. Other modes of communication are not recognized or, if they are, are not well understood.[107]

References

1. R. G. W. Anderson, 'Instruments and apparatus', in *Recent Developments in the History of Chemistry*, ed. C. A. Russell, Royal Society of Chemistry, London, 1985, pp. 217–237.
2. J. L. Sturchio, 'Artifact and experiment', *Isis*, 1988, **79**, 369–372.
3. R. F. Bud and S. E. Cozzens, eds, *Invisible Connections. Instruments, Institutions and Science*, Spie Optical Engineering Press, Bellingham, Washington, 1992.
4. R. F. Bud and D. J. Warner, eds, *Instruments of Science, an Historical Encyclopedia*, Garland Publishing, New York, 1998.
5. C. Meinel, ed., *Instrument–Experiment Historische Studien*, Verlag für Geschichte der Naturwissenschaften und Technik, Berlin, 2000.
6. M. Dorikens, ed., *Scientific Instruments and Museums*, Proceedings of the XXth International Congress of History of Science, Liège, 1997, Vol. XVI, Brepols, Turnhout, Belgium, 2002.
7. F. L. Holmes and T. H. Levere, eds, *Instruments and Experimentation in the History of Chemistry*, MIT Press, Cambridge, MA, 2000.
8. J. L. Sturchio and B. V. Lewenstein, 'The history of chemical instrumentation: an introduction', *Information Pamphlet No. 8*, Beckman Center for the History of Chemistry, Philadelphia, PA, 1987.
9. G. L'E. Turner and D. J. Bryden, *International Union of the History and Philosophy of Science. Scientific Instrument Commission. A Classified Bibliography on the History of Scientific Instruments*, Published by the Commission and distributed by the Museum of the History of Science, Oxford OX1 3AZ, 1997.
10. IUHPS, Scientific Instrument Commission, 'Cumulative Bibliography', http://www.sic.iuhps.org/in_bibrm.htm [accessed 14 oct 2005].
11. R. G. W. Anderson, J. Burnett and B. Gee, *Handlist of Scientific Instrument-Makers' Trade Catalogues 1600–1914*, National Museums of Scotland, Edinburgh, 1990.

12. M. Holbrook, with additions and revisions by R. G. W. Anderson and D. J. Bryden, *Science Preserved: A Directory of Scientific Instruments in Collections of the United Kingdom and Eire*, HMSO, London, 1992.

13. C. Mollan, *Irish National Inventory of Historic Scientific Instruments*, Samton Ltd, Blackrock, Co. Dublin, 1995.

14. H. Andersen, *Historical Scientific Instruments in Denmark*, Det Kongelige Danska Videnskabernes Selskab, Copenhagen, 1995.

15. C. Meinel, ed., *Instrument–Experiment Historische Studien*, Verlag für Geschichte der Naturwissenschaften und Technik, Berlin, 2000.

16. G. Clifton, *Directory of British Scientific Instrument Makers 1550–1851*, Zwemmer, London, 1995.

17. J. W. van Spronsen, *Guide of European Museums with Collections on History of Chemistry*, Federation of European Chemical Societies, Antwerp, 1996.

18. T. H. Levere, 'Lavoisier: Language, instruments and the chemical revolution', in *Nature, Experiment and the Sciences: Essays on Galileo and the History of Science*, ed. T. H. Leavere and W. R. Shea, Kluwer Academic, Amsterdam, 1990, pp. 207–233.

19. J. Golinski, 'Precision instruments and the demonstrative order of proof in Lavoisier's Chemistry', *Osiris*, 1994, **9**, 30–47.

20. C. Mollan and J. Upton, *St Patrick's College, Maynooth: The Scientific Apparatus of Nicholas Callan*, St Patrick's College, Maynooth and Samton Limited, Blackrock, Co. Dublin, 1994; "Chemistry and analysis", pp. 18–35; 'Descriptions of chemical balances', 214–216.

21. R. G. W. Anderson, 'A Source for 18th-century chemical glass', in *Proceedings of the Eleventh International Scientific Instrument Symposium*, ed. G. Dragoni, A. McConnell and G. L'E. Turner, Grafis Edizioni, Bologna, 1994, pp. 47–52.

22. R. G. W. Anderson, 'Joseph Black and his chemical furnace', in *Making Instruments Count: Essays on Historical Scientific Instruments presented to Gerard L'Estrange Turner*, ed. R. G. W. Anderson, J. A. Bennett and W. F. Ryan, Variorum, Aldershot, 1993, pp. 118–126.

23. S. Heilenz, *Das Liebig-Museum in Giessen*, Giessen, 1991.

24. G. K. Judd, *Die Geschichte des Liebig-Museums in Giessen*, Liebig Museum, Giessen, 1996.

25. M. C. Usselman, 'Liebig's alkaloid analyses: the uncertain route from elemental content to molecular formulae', *Ambix*, 2003, **50**, 71–39.

26. P. W. Hammond and H. Egan, *Weighed in the Balance. A History of the Laboratory of the Government Chemist*, HMSO, London, 1992.

27. F. A. J. L. James, ed., *The Development of the Laboratory: Essays on the Place of Experiment in Industrial Civilization*, Macmillan Press, London, 1989.

28. I. Rouaze, 'Un Atelier de Distillation du Moyen Age', *Bull Archéol. Comité Travaux Hist. Sci.*, 1989, new series, **22**, 159–271.

29. S. von Osten, 'Das Alchemistenlaboratorium Oberstockstall', unpublished dissertation, Institut für Ur-und Frühgeschichte, Vienna University, 1992.

30. R. W. Soukup and S. von Osten, 'Das Alchemistenlaboratorium von Oberstockstall. Ein Vorbericht zum Stand des Forschungsprojekts', *Mitt. - Ges. Dtsch. Chem., Fachgruppe Gesch. Chem.*, 1992, **7**, 11–19.

31. R. W. Soukup, S. von Osten and H. Mayer, 'Alembics, cucurbits, phials, crucibles: a 16th-century docimastic laboratory excavated in Austria', *Ambix*, 1993, **40**, 39.

32. S. von Osten and R. W. Soukup, 'Alchemistenlaboratorium Oberstockstall. Vorbericht über einen Fundcomplex des 16. Jahrhunderts aus Niederösterreich, Archäologie Österreichs', *Mitt. Öster. Ges. Ur- und Frühgeschichte*, 1992, **3**, 61–66.

33. S. von Osten, R. W. Soukup and H. Mayer, 'Das Laboratoriums-Inventar von Oberstockstall aus dem 16. Jahrhundert-ein Zwischenbericht', *Anschnitt*, 1996, **16**, 100–102.

34. R. W. Soukup, 'Probierkunst and Chemiatrische Laboratoriumspraxis 1560–1600 in Niederösterreich', *Georgius Agricola Gedenkveranstaltung. Res Monanarum. Zeit. Des Montanhistorischen Vereins für Österreich*, 1994, **9**, 10–11, 34–37 (catalogue entries 62–71).

35. J. Weyer, *Graf Wolfgang 11. von Hohenlohe und die Alchemie: Alchemischer Studien in Schloss Weikersheim 1587–1610*, Jan Thorbecke Verlag, Sigmorigen, 1992; see particularly "Das alchemistische Laboratorium in Schloss Weikersheim", pp. 64–120, and "Chemisch-alchemistische Geräte and Apparate", 121–165.

36. I. N. Hume, 'Roanoke Island: America's first science center', nd; originally published in *Colonial Williamsburg: J. Colonial Williamsburg Foundation*, Spring 1994.

37. I. N. Hume, *Martin's Hundred*, Alfred A. Knopf, New York, 1982, pp. 101–105, 194.

38. B. Meitzner, *Die Gerätschaft der Chymischen Kunst. Der Traktat 'De Sceuastica Artis' des Andreas Libavius von 1606*, Franz Steiner Verlag, Stuttgart, 1995.

39. M. Ron, 'The instruments of the alchemists', in *Proceedings of the Eleventh International Scientific Instrument Symposium*, ed. G. Dragoni, A. McConnell and G. L'E. Turner, Grafis Edizioni, Bologna, 1994, pp. 35–37.

40. O. Hannaway, 'Laboratory design and the aim of science', *Isis*, 1986, **77**, 585–610.

41. L. M. Principe and L. De Witt, *Transmutations: Alchemy in Art*, Chemical Heritage Foundation, Philadelphia, PA, 2002.

42. E. Child, *Tools of the Chemist*, Reinhold Publishing Corporation, New York, 1940.

43. G. Luppi, ed., *Bilance a bracci uguali*, Museo della Bilancia, Modena, 1993.

44. L. Grosai, 'La Collezione di Bilance de Laboratorio', in *Bilance a Bracci Uguali*, ed. Giulia Luppi, Museo della Bilancia, Modena, 1993, pp. 27–38.

45. L. Grosai, 'Bilance in Vetrina', in *Bilance a Bracci Uguali*, ed. Giulia Luppi, Museo della Bilancia, Modena, pp. 177–194.

46. J. T. Stock, 'Victor Serrin and the origins of the chainomatic balance', *Bull. Hist. Chem.*, 1990, **8**, 12–15.

47. W. D. Williams, 'Brief history of the test tube clamp', *Bull. Hist. Chem.*, 1991, **9**, 37–39.

48. J. Burnett, 'William Prout and the urinometer: some interpretations' in *Making Instruments Count: Essays on Historical Scientific Instruments presented to*

Gerard L'Estrange Turner, ed. R. G. W. Anderson, J. A. Bennett and W. F. Ryan, Variorum, Aldershot, 1993, pp. 118–126.

49. W. D. Williams, 'Some early chemical slide rules', *Bull. Hist. Chem.*, 1992, **12**, 24–29.

50. Y. M. Rabkin, 'Uses and images of instruments in chemistry', in *Chemical Sciences in the Modern World*, ed. S. H. Mauskopf, University of Pennsylvania Press, Philadelphia, PA, 1993, pp. 25–42.

51. D. Baird, 'Analytical chemistry and the "big" scientific instrumentation revolution', *Ann. Sci.*, 1993, **50**, 267–290.

52. E. Schettino, 'A new instrument for infrared radiation measurements: the thermopile of Macedonio Melloni', *Ann. Sci.*, 1989, **46**, 511–517.

53. D. Baird, 'Baird Associates's commercial three-meter grating spectrograph and the transformation of analytical chemistry', *Rittenhouse*, 1991, **5**, 65–80.

54. S. Bennett, 'The development of process control instruments', *Trans. Newcomen Soc.,* 1991–92, **63**, 113–164.

55. Y. M. Rabkin, 'Technological innovation in science: the adoption of infrared spectroscopy by chemists', *Isis*, 1987, **78**, 31–54.

56. L. E. Kay, 'Laboratory technology and biological knowledge: the Tiselius electrophoresis apparatus 1930–1945', *Hist. Phil. Life Sci.*, 1988, **10**, 51–72.

57. K. Marsden and I. D. Rae, 'Nuclear magnetic resonance in Australia 1952–1986', *Hist. Recs. Aust. Sci.,* 1990, **8**, 119–150.

58. J. T. Stock and M. V. Orna, eds, *The History and Preservation of Chemical Instrumentation*, D. Reidel, Dordrecht, 1986.

59. J. T. Stock, 'Historic instruments: survival or disappearance', in ref. 58, pp. 239–248.

60. J. T. Stock, 'Historical chemical instrumentation: from the cellar upwards', *Bull. Hist. Chem.*, 1994, **15/16**, 1–8.

61. J. Burnett, 'The use of new materials in the manufacture of scientific instruments, c. 1880-c. 1920', in ref. 58, pp. 217–238.

62. W. B. Jensen, 'The development of blowpipe analysis', in ref. 58, pp. 123–149.

63. J. T. Stock and M. V. Orna, eds, *Electrochemistry, Past and Present*, American Chemical Society Symposium Series 390, American Chemical Society, Washington DC, 1989.

64. H. Gunasingham, 'Development of Electrochemical Instrumentation', in ref. 63, pp. 236–253.

65. B. Jaselskis, C. E. Moore, and A. von Smolinski, 'Development of the pH Meter', in ref. 63, pp. 254–271.

66. C. E. Moore, B. Jaselskis and A. von Smolinski, 'Development of the glass electrode', in ref. 63, pp. 272–285.

67. P. Zuman, 'With the Drop of Mercury to the Nobel Prize', in ref. 63, pp. 339–369.

68. P. J. T. Morris, ed., *From Classical to Modern Chemistry: The Instrumental Revolution*, Royal Society of Chemistry, Cambridge, 2002.

69. A. L. Janeira et al., *Demonstrar ou Manipular? O Laboratório de Química Mineral da Escola Politécnica de Lisboa na sua Época (1884–1894)*, Livraria Escolar Editoria, Lisboa, 1996.

70. P. Jones, 'Chemical artifacts', *Bull. Hist. Chem.*, 1988, **1**, 8–10.

71. L. Fine, 'The Chandler chemical museum', *Bull. Hist. Chem.*, 1988, **1**, 19–21.

72. G. M. Bodner, 'The Apparatus Museum at Transylvania University', *Bull. Hist. Chem.*, 1988, **1**, 22–27.

73. J. H. Wotiz, 'The Butlerov museum at the University of Kazan', *Bull. Hist. Chem.*, 1988, **1**, 24–26.

74. W. B. Jensen, 'To demonstrate the Truths of "Chymistry": an historical and pictorial celebration of the art of the lecture demonstration in honor of Dr Hubert Alyea', *Bull. Hist. Chem.*, 1991, **10**, 3–15.

75. H. Toftlund, 'History of the lecture demonstration', *Educ. Chem.*, 1988, **25**, 109–111.

76. A. Q. Morton and J. A. Wess, *Public & Private Science: The King George III Collection*, Oxford University Press, Oxford, 1993.

77. A. Q. Morton, *Science in the 18th Century: The King George III Collection*, Science Museum, London, 1993.

78. G. di Modica, "Chimica", in *Strumenti Ritrovati. Materiali della ricerca Scientifica in Piemonte tra Settecento e Ottocento*, Archivo di Stato di Torino, Turin, 1991, pp. 37–41.

79. F. Vannozzi, *Inventario del Patrimonio dell'ateneo Senese gli Strumenti Scientifici*, Università degli Studi di Siena, Siena, 1992, which describes items from the Dipartimento di Chimica, pp. 18–25; and items from the Dipartimento Farmaco-Chimico-Technologico, 47–55.

80. N. Nicolini and G. Terenna, 'La Collezione di Vetreria Scientifica', in *Patrimonio Storico-Scientifico dell'Università degli Studi di Siena*, Nuova Immagine Editrice, Siena, 1999.

81. G. Rambaldi Morchio, 'Inventory of the instruments belonging to Cannizzaro's laboratory', in *Scientific Instruments and Museums*, ed. M. Dorikens, Proceedings of the XXth International Congress of History of Science, Liège, 1997, Brepols, Turnhout, Belgium, 2002, vol. XVI, pp. 237–251.

82. G. Rambaldi, *Istrumenti di Chimica. Un Laboratorio del XIX Secolo*, Pirella Editore, Genoa, 1996.

83. *El Legado Científico Andelusí*, Museo Arqueológico Nacional, Madrid, 1992.

84. *El Museo de la Farmacia Hispana*, Consejo Social de la Universidad Complutense de Madrid, 1993.

85. *La Casa de la Quimica. Ciencia, Artillería e Ilustracion*, Ministerio de Defensa, [Madrid], 1992.

86. R. G. Bohorquez, 'Louis Proust y el laboratorio del Real Colegio de Artillería de Segovia', in *La Casa de la Quimica. Ciencia, Artillería e Ilustracion*, Ministerio de Defensa, [Madrid], 1992, pp. 73–84.

87. L. Kriis-Ilves, 'Chemical instruments and collections from the 19th century in the History Museum of Tartu', in *Scientific Instruments and Museums*, ed. M. Dorikens, Proceedings of the XXth International Congress of History of Science, Liège, 1997, Brepols, Turnhout, Belgium, 2002, vol. XVI, pp. 261–270.

88. G. L'E. Turner, *The Practice of Science in the Nineteenth Century: Teaching and Research Apparatus in the Teyler Museum*, Teyler Museum, Haarlem, 1996.

89. P. A. Kidwell, 'An erasable surface as instrument and product: the blackboard enters the American classroom, 1800 – 1915', *Rittenhouse*, 2003, **17**, 85–98.

90. J. Wright, *Vision, Venture, and Volunteers: Fifty Years of the Pittsburgh Conference on Analytical Chemistry and Applied Spectroscopy*, Chemical Heritage Foundation, Philadelphia, PA, 1999.

91. A. Thackray and M. Myers, Jnr., *Arnold O. Beckman: One Hundred Years of Excellence*, Chemical Heritage Foundation, Philadelphia, PA, 2000.

92. M. A. Grayson, ed., *Measuring Mass: From Positive Rays to Proteins*, Chemical Heritage Press, Philadelphia, PA, 2002.

93. J. Payen, 'Les constructeurs d'instruments scientifique en France au XLXe siecle', *Arch. Int. d'Hist. Sci.*, 1986, **36**, 84–161.

94. P. Brenni, '19th century French scientific instrument makers I: H. P. Gambey', *Bull. Sci. Instrum. Soc.*, 1993, **38**, 11–13.

95. P. Brenni, 'Soleil, Duboscq and their successors,' *Bull. Sci. Instrum. Soc.*, 1996, **51**, 7–13.

96. A. D. Morrison-Low, 'Irish instrument makers and chemical instrumentation', *Bull. Sci. Instrum. Soc.*, 1990, **26**, 11–15.

97. D. J. Bryden, 'Alva Mason, The Franklin Institute, and the origins of philosophical and chemical instrument manufacture in the United States, *Proc. Am. Phil. Soc.*, 1988, **132**, 400–419.

98. T. H. Levere, 'Pneumatic apparatus and the spread of the chemical revolution', Bull. Sci. Instrum. Soc., 1996, **49**, 14–16.

99. A. McConnell, 'New Instrument Makers in the *New Oxford Dictionary of National Biography*', *Bull. Sci. Instrum. Soc.*, 2003, **77**, 25.

100. A. McConnell, 'Instrument Makers in the *Oxford Dictionary of National Biography*', *Bull. Sci. Instrum. Soc.*, 2005, **84**, 3–8.

101. A. D. Morrison-Low, 'Kemp & Co., Laboratory Suppliers', in *The History and Preservation of Chemical Instrumentation*, ed. J. T. Stock and M. V. Orna, D. Reidel, Dordrecht, 1986, pp. 163–186.

102. H. C. Bolton, J. Holland and N. H. Williams, 'The Grimwade Milligram Chemical Balance: an early Australian attempt to establish a scientific instrument industry', *Hist. Recs. Aust. Sci.*, 1992, **9**, 107–117.

103. J. B. Willis, 'Three little companies–the birth of a major Australian scientific instrument industry', *Hist. Recs. Aust. Sci.*, 2003, **14**, 403–429.

104. J. L. Sturchio and A. Thackray, 'Arnold Beckman and the revolution in instrumentation', *Today's Chem.*, 1988, **1**, 8–10 and 16, 18, 31.

105. A. Thackray and J. L. Sturchio, 'The education of an entrepreneur: the early career of Arnold Beckman', in *The Beckman Symposium on Biomedical Instrumentation*, ed. C. M. Moberg, The Rockefeller University, New York, 1986, pp. 3–17.

106. T. Lenoir and C. Lecuyer, 'Instrument makers and discipline builders: the case of nuclear magnetic resonance', *Perspectives Sci.*, 1995, **3**, 276–345.

107 D. Baird, *Thing Knowledge: A Philosophy of Scientific Instruments*, University of California Press, Berkeley, CA, 2004, xv.

Subject Index

Name Index